Implementing Safety Management Systems in Aviation

EDITED BY

ALAN J. STOLZER
Embry-Riddle Aeronautical University, USA

CARL D. HALFORD
The MITRE Corporation, USA

JOHN J. GOGLIA
Independent Safety Consultant

ASHGATE

Published by
Ashgate Publishing Limited
Wey Court East
Union Road
Farnham, Surrey
GU9 7PT England

Ashgate Publishing Company
110 Cherry Street
Suite 3-1
Burlington, VT 05401-3818
USA

www.ashgate.com

British Library Cataloguing in Publication Data
Implementing safety management systems in aviation. --
   (Ashgate studies in human factors for flight operations)
   1. Aeronautics--Safety measures. 2. Aeronautics--Safety
   regulations. 3. System safety--Management.
   I. Series II. Stolzer, Alan J., 1960- III. Halford, Carl
   D., 1950- IV. Goglia, John Joseph, 1944-
   363.1'247-dc22

   ISBN: 978-1-4094-0165-0 (hbk)
   ISBN: 978-1-4724-1279-9 (pbk)
   ISBN: 978-1-4094-0166-7 (ebk – PDF)

Library of Congress Cataloging-in-Publication Data
Stolzer, Alan J., 1960-
   Implementing safety management systems in aviation / by Alan J. Stolzer, Carl D. Halford, and John J. Goglia.
       p. cm. -- (Ashgate studies in human factors for flight operations)
   Includes bibliographical references and index.
   ISBN 978-1-4094-0165-0 (hardback) -- ISBN 978-1-4094-0166-7
   (ebook) 1. Aeronautics--Safety measures. 2. Aeronautics--Safety regulations. 3. System
   safety. 4. Safety education. I. Halford, Carl D.,
   1950- II. Goglia, John Joseph, 1944- III. Title.
   TL553.5.S7429 2011
   387.7068'3--dc22

2011013273

Printed and bound in Great Britain
by MPG PRINTGROUP

Contents

List of Figures

List of Tables

List of Abbreviations

AAP	Accident Analysis and Prevention (FAA program)
AAW	Anti-air Warfare
ACSF	Air Charter Safety Foundation
ADDIE	analysis, design, development, implementation and evaluation
ADM	aeronautical decision making
ADS-B	automatic dependent surveillance – broadcast
AEB	ASIAS Executive Board
AEP	airport emergency plan
ALARP	as low as reasonably practicable
ALoS	acceptable level of safety
ALPA	Airline Pilots Association
AO	audit organization
AOA	airport operations area
Aoc	Air Operators Certificate
AOM	Aircraft Operating Manual
APE	accident prevention effort
AQP	Advanced Qualification Program
ARC	Aviation Rulemaking Committee
ARFF	Aircraft rescue fire fighters
ASAP	Aviation Safety Action Program
ASIAS	Aviation Safety Information Analysis and Sharing System
ASO	Aviation Safety Officers
ASRS	aviation safety reporting system
ASUW	Anti-surface Warfare
ASW	Anti-submarine Warfare
ATA	Air Transport Association
ATC	air traffic control
ATM	Air Traffic Management
ATO	Air Traffic Organization
ATOS	air transport oversight system
ATSAP	Air Traffic Safety Action Program
BPET	business process engineering team
CAA	Civil Aviation Authority
CAASD	MITRE Corporation's Center for Advanced Aviation and System Development
CAP	corrective action plan
Car	corrective action record/request
CASA	Civil Aviation Safety Authority

CASS	continuing analysis and surveillance system
CAST	Commercial Air Safety Team
CAVE	consequences, alternatives, reality, external factors
CDT	central daylight time
CEO	Chief Executive Officer
CFR	Code of Federal Regulations
CI	continuous improvement
CICTT	CAST/ICAO Common Taxonomy Team
CMO	Certificate Management Office
COO	chief operating officer
COQ	cost of quality
COSO	Committee of Sponsoring Organizations (of the Treadway Commission)
CRM	crew resource management
CRM	community resource management
CRO	corporate risk officer
DCTs	data collection tools
DECIDE	detect, estimate, choose, identify, do, evaluate
DECIIDE	detect, estimate, choose, identify, input, do, evaluate
DFDR	digital flight data recorder
DiMS	dispatch management system
DMS	documentation management system
DNAA	Distributed National ASAP Archive
EASA	European Aviation Safety Agency
EDT	eastern daylight time
EFB	electronic flight bag
EGPWS	enhanced ground proximity warning systems
ELT	emergency locator transmitters
EMASS	Engineered Materials Arrestor System
EMS	environmental management system
EMT	emergency medical technique
EO	enabling objective
EPA	Environmental Protection Agency
EPI	element performance inspection
ERC	event review committee
ERM	enterprise risk management
ERP	emergency response plan
FAA	Federal Aviation Administration
FAR	Federal Aviation Regulation
FBO	fixed base operator
FDA	flight data analysis
FDM	flight data monitoring
FFRDC	federally funded research and development center
FITS	FAA Industry Training Standard

FLIR	forward-looking infra red (camera)
FMEA	failure mode effects analysis
FMS	flight management system
FMS	financial management system
FO	flying officer
FOD	foreign object debris
FOHE	fuel-oil heat exchanger
FOIA	Freedom of Information Act
FOM	flight operations manual
FOQA	flight operations quality assurance
GA	general aviation
GAIN	Global Aviation Information Network
GOM	general operations manual
GPWS	ground proximity warning systems
HF	human factors
HFACS	Human Factors Analysis and Classification System
HRO	high-reliability organization
IAMS	integrated airline management system
IAPs	Incident Action Plans
IATA	International Air Transport Association
IBAC	International Business Aircraft Council
ICAO	International Civil Aviation Organization
ICOM	input, control, output, mechanism
ICS	Incident Command System
ID	instructional design
IDEFØ	integration definition for function modelling
IEP	internal evaluation program
IFR	instrument flight rules
IHST	International Helicopter Safety Team
IMS	integrated management system
IS-BAO	International Standard – Business Aircraft Operations
ISARPS	IATA Standards and Recommended Practices
ISD	instructional system design
IOSA	IATA Operational Safety Audit
ISM	*IOSA Standards Manual*
ISO	International Standardization Organization
ISS	information-sharing system
JPDO	Joint Planning and Development Office
JSA	job safety analyses
JQR	joint quality review
LABSKI	Lufthansa, Air France, BA, Swiss, KLM and Iberia
LEO	law enforcement officer
LOIs	letters of intent
LOSA	line operations safety audit

LTI	lost time injury
LTIFR	lost time injury frequency rate
MAAs	mutual aid agreements
MALSR	medium intensity approach light system
MDT	mobile data terminal
MEC	Master Executive Council
MEL	minimum equipment list
MMS	marketing management system
MOU	memorandum of understanding
NAS	national airspace system
NASA ARS	National Aeronautics and Space Administration Aviation Safety Reporting System
NATCA	National Air Traffic Controllers Association
NIMS	National Incident Management System (USA)
NextGen	Next Generation Air Transportation System
NOTAM	Notice to Airmen
NOTOC	Notification to Captain
NTSB	National Transportation Safety Board
OHSMS	occupational health and safety management systems
OODA	observation, orientation, decision and action
ORM	operational risk management
OSHA	Occupational Health and Safety Act
OSHA	Occupational Safety and Health Administration
OTP	on time performance
PAPI	precision approach path indicator
PAVE	pilot, aircraft, environment, external factors
PDCA	plan-do-check-act
PETN	pentaerythritol tetranitrate
PPE	personal protection equipment
PRA	probabilistic risk assessment
PSRC	Pilot Safety Resource Committee
QA	quality assurance
QAP	quality assurance program
QAR	quick access recorder
QMS	quality management system
QSM	quantum safety metrics
QSR	quarterly safety review
QSRs	quality and safety requirements
RAA	Regional Airline Association
RATS	recommended actions tracking system
RJ	regional jet
RM	risk mitigation
ROE	return on equity
SA	safety assurance

SAI	safety attribute inspection
SARP	Standards and Recommended Practice
SCBAs	self-contained breathing apparatus
SeMS	security management system
SIC	Standard Industrial Classification
SIDA	security identification display area
SMC	safety management committee
SME	subject matter expert
SMS	safety management system
SMS QRG	SMS quick reference guide
SOPs	standard operating procedures
SPINs	special instructions
SPO	supporting proficiency objective
SRM	safety risk management
SRMD	safety risk management document
SRMDM	safety risk management decision memo
SRMP	safety risk management panel
SSP	State Safety Program
SuMS	supplier management system
TATP	triacetone triperoxide
TAWS	terrain awareness and warning system
TC	Transport Canada
TCAS	traffic collision avoidance systems
TEAM	transfer, eliminate, accept, mitigate
TEM	threat and error management
TIC	thermal imaging camera
TSA	Transportation Security Administration
TSB	Transport Safety Board
USOAP	Universal Safety Oversight Audit Program
VOC	Voice of the Customer
WBS	work breakdown structure

About the Editors

Alan J. Stolzer, Ph.D. is Chair and Professor of Doctoral Studies at Embry-Riddle Aeronautical University, Daytona Beach, Florida, USA. He holds a Ph.D. in Quality Systems from Indiana State University, and several professional certifications: Quality Engineer, Quality Manager, and Quality Auditor from the American Society for Quality; Project Management Professional from the Project Management Institute; Airline Transport Pilot, Flight Instructor with Instrument and Multi-Engine Ratings, and Airframe and Powerplant Mechanic from the Federal Aviation Administration. Dr. Stolzer has several thousand hours in more than 40 makes and models of fixed-wing aircraft. His research interests include safety, quality, flight operations quality assurance, safety management systems, and emergency preparedness. He is a prolific author and has been awarded and managed numerous grants and contracted training programs. Dr. Stolzer is involved in academic accreditation activities and has served as an expert witness and legal consultant on aviation accidents.

Carl D. Halford has wide experience in many aspects of airline operations, including safety and quality, airline flight and simulator instruction, maintenance, management and union representation. Mr. Halford holds an Airline Transport Pilot certificate, with an assortment of type ratings, and has over 9000 hours of flight time in a variety of aircraft. Mr. Halford has held a variety of airline positions, including Manager of Voluntary Aviation Safety Programs, Manager of Flight Operations Quality Assurance (FOQA), Manager of Aviation Safety Action Partnership (ASAP), and Manager of Part 121 Training. He has also been an Airline Pilots Association Master Executive Committee Chairman and holds several professional certifications. Mr. Halford has completed a Masters degree from DePaul University, with a specialization in Safety Management Systems (SMS). He has assisted in the construction of the Distributed National Archive for airline safety information, and is presently engaged in research with MITRE Corporation in this endeavor.

John J. Goglia is an active consultant, author, and educator in the field of transportation safety. Mr. Goglia served as a Member of the National Transportation Safety Board from August 1995 to June 2004 and was the first Board Member to hold an FAA aircraft mechanic's certificate. As a Board Member, Mr. Goglia distinguished himself in numerous areas of transportation safety. In particular, he was instrumental in raising awareness of airport safety issues, including the importance of airport crash fire and rescue operations and the dangers of wildlife at airports, and played a key role in focusing international attention on the increasing significance of aircraft maintenance in aviation accidents. Mr. Goglia has been recognized many times for his contribution to aviation safety; awarding bodies include the National Air Disaster Alliance, Aviation Week and Space Technology, The Society of Automotive Engineers and the Air Transport Association (ATA). He currently serves on a number of Boards, including the Aviation Technical Training College in Singapore, and advises a number of organizations, including the Professional Aviation Maintenance Association.

About the Contributors

Anthony Adamski, Ph.D. is a Professor in the College of Technology and Program Coordinator for the Aviation Flight Technology program at Eastern Michigan University. He teaches courses in aviation management, system safety, crew resource management, and human factors. He has served as a certification consultant to a number of air carrier applicants that were required to meet ATOS guidelines. He also consults with the aviation industry in areas of training design and organizational issues. Dr. Adamski is a past military and civilian professional pilot and holds an ATP rating.

Dr. Don Arendt, Ph.D. specializes in aviation risk management, human factors and systems engineering. He is currently Manger, FAA Flight Standards SMS Program Office. In his current work, he is involved in development of policy and guidance, training requirements, and oversight and implementation processes for Safety Management Systems for aviation service providers. Earlier, he conducted and directed test, evaluation, and system analysis activities for reliability, maintainability, maintenance logistical support and human factors for a variety of aviation, air defense, and command and control systems for the U.S. Army. He holds degrees in Industrial/Organizational Psychology (Doctorate), Operations Research (Masters), and industrial Technology (Bachelor's). He also holds FAA Airline Transport Pilot and Gold Seal Flight Instructor certificates. The information and opinions expressed in his contributions to this book are his and not the official position of the U.S. Government.

Dr. Mark Friend, Ed.D., CSP is a Certified Safety Professional (CSP) and a professional member of the American Society of Safety Engineers (ASSE). Friend has over thirty-five years of experience in higher education, having taught at six institutions of higher learning. He serves as Dean of the Worldwide Campus of Embry-Riddle University (ERAU) in San Antonio, Texas. Prior to Embry-Riddle, he was Professor of Safety and Director of the Center for Applied Technology (CAT) at East Carolina University (ECU). He also served as Chair of the Department of Occupational Safety and Health at Murray State University after teaching at West Virginia University, Fairmont State College, and Waynesburg College. His text, *Fundamentals of Occupational Safety and Health*, a top-selling book in the field, is in its fifth edition. Friend serves on the editorial board of Occupational Hazards and has published several articles in safety and safety management. Friend received the Academics Practice Specialty Safety Professional of the Year 2006–2007 award from ASSE. He was given the Charles V. Culbertson Outstanding Service Award by ASSE at the 2005 Professional Development Conference. Friend received the 2003 PPG Safety Educator of the Year award from ASSE, The West Virginia University/National Safety Management Society Achievement Award, and has been listed in Who's Who in America, and Who's Who in American Education.

Don Gunther is the staff vice president of safety for Continental Airlines Inc. Previously, as the Manager of Human Factors, his team developed the Threat & Error Management (TEM) training programs. Don graduated from the U.S. Naval Academy in 1971, flew the Navy's P-3 Orion for 20 years and retired from the Navy Reserves in 1993. He was hired by Continental Airlines in 1977 and has qualified on the Boeing 727, Boeing 737, DC-10, Boeing 757, Boeing 767 and Boeing 777 aircraft. Don has completed terms as Chairman of the Air Transport Association (ATA) Human Factors Committee, Chairman of the International Air Transport Association (IATA) Human Factors Working Group and Chairman of the ATA Safety Council. He is currently a member of the Flight Safety Foundation Board of Governors, Industry Co-chair for the Aviation Safety Information Analysis and Sharing (ASIAS) program and Industry Co-chair for the Commercial Aviation Safety Team (CAST).

Michelle L. Harper, Ph.D. is a Lead Multi-Discipline Systems Engineer with the MITRE Corporation where she supports the development and expansion of the Aviation Safety Information Sharing and Analysis program (ASIAS). ASIAS is a collaborative program developed to support the Federal Aviation Administration (FAA) and the commercial aviation community to proactively analyze a broad range of public and proprietary data sources towards the discovery of vulnerabilities in the national air transportation system. Michelle has helped lead the development of a range of safety programs and original research supported by the FAA, NASA, the Department of Defense, and the National Society for Behavioral Research. Her research and associated projects are used to support safety initiatives led by the commercial aviation industry, including commercial aviation airlines, air traffic management organizations, airline manufacturers and labor unions towards the establishment, analysis and utilization of data stemming from voluntary aviation safety programs. Michelle's current work is focused on establishing processes for the fusion and exploration of safety reporting data, digital flight data, and census-based data sources. Michelle holds a Masters in Experimental Psychology from the University of Colorado and is scheduled to receive her doctoral degree from the University of Texas in April 2011.

Captain James Hobart is a veteran major airline, corporate, and general aviation pilot, with over 16,000 hours of flight time and over 9,000 hours in command of jet airliners. In addition to his extensive involvement in global air safety, he has flown as a check airman and instructor pilot. Hobart holds a Master of Aeronautical Science from Embry-Riddle Aeronautical University. He is a graduate of the University of Southern California Air Safety Program Human Factors and Safety Program Management Courses, and is an International Air Transport Association

Accredited Airline Auditor (IOSA) and an International Business Aviation Council Accredited Auditor (IS-BAO). Captain Hobart is the Vice President of Training Programs for JDA Aviation Technology Solutions of Bethesda, MD.

Richard Komarniski is President and CEO of Grey Owl Aviation Consultants Inc., a leading provider of human factors and safety management systems training. Richard draws on his 36 years of experience as a certified maintenance technician, Transport Canada Airworthiness Inspector, Quality Assurance Manager and Aviation Consultant. To date, Richard has developed seven human factors workshops focusing on error prevention strategies and several workshops for SMS familiarization and implementation.

Richard was rewarded in 2008 for his ongoing efforts in error prevention by receiving the NAASCO Aircraft Maintenance Technician of the Year Award. Richard enjoys sharing his ideas with peers and has strongly supported industry organizations such as NBAA, PAMA, HAI, and CAMEA. He has been a featured speaker at every Bombardier Annual Safety Stand down since its inclusion of maintenance. Richard's articles on human factors and SMS can be read in trade publications like AMT, Maintenance UpDate and DOM magazine.

Jack Kreckie is a 31 year veteran of the fire service, serving his last 28 years at Boston Logan International Airport where he retired as Deputy Fire Chief in 2007. He was the co-founder of the Logan Airport Safety Alliance and was recognized by the Flight Safety Foundation for that work with the Airport Safety Award at their 2006 conference in Moscow. Jack is a past Chairman of the Aircraft Rescue and Fire Fighting (ARFF) Working Group and has served as a Director or Officer in the organization for the past 15 years, currently serving as the Regulatory Affairs Officer. He is one of 4 people in the world holding the distinction and title, "ARFF Legend", a lifetime achievement award by the ARFF Working Group. His consulting firm, ARFF Professional Services LLC (www.APSSafety.net) provides consultation services relative to ARFF, Emergency Planning, Emergency Response and Airport Safety. Jack is an OSHA Outreach Instructor and has served as an ARFF Subject Matter Expert for the Science Applications International Corporation and SRA International supporting FAA ARFF Research and Training projects. His firm has provided ARFF Consulting services to a number of Airports in the United States and abroad. Jack serves as an Overseas Guest Lecturer for the Singapore Aviation Academy.

Jack is the author of the Aircraft Rescue and Fire Fighting Chapter in the *NFPA Fire Protection Handbook 20ᵗʰ Edition*. He is also a contributing author to *Safety Management Systems in Education* published by Ashgate Publishing.

Curt Lewis served with American Airlines/AMR Corporation (AA) for seventeen years as the Director System Safety and previously as the Corporate Manager of Flight Safety and Flight Operational Quality Assurance (FOQA) for American Airlines. Prior to American, he was a Lead System Safety Engineering

Manager for LTV Aerospace Company, a Product Safety Manager for Texas Instruments Inc., a Flight Training Instructor for Boeing Commercial Airplane Group, and served as Chief Corporate Pilot and Safety Director for various industrial corporations. Additionally, he is currently an Assistant Professor with Embry-Riddle Aeronautical University instructing aviation, system, human factors, and industrial safety courses (Outstanding Faculty Award – 2003).

He has in excess of thirty years of safety experience as a professional pilot, safety engineer/director, and air safety investigator. Mr. Lewis holds an Airline Transport Pilot License (ATPL), Certified Flight Instructor Certificate (Airplane Single, Multiengine & Instrument); with over 10,000 hours of flight experience. In addition, he has earned technical Bachelors degrees in Aeronautical Engineering and Physics, and a Masters degree in Aviation & System Safety. He is currently pursuing a Ph.D. in Aeronautical Science Management (completed 21 graduate hours in Management). He has earned the designation of Certified Safety Professional (CSP) from the Board of Certified Safety Professionals (BCSP) and served as a Board Director. He was elected the BCSP Vice President for 2004 after serving as the Secretary-Treasurer (Managed $2.8 mil budget). Mr. Lewis is a Licensed Professional Engineer (P.E.) in the discipline of safety engineering (Mass. # 40394), a Texas Professional Safety Source (TWCC # 1941), and an ISO-9001:2000 Certified Quality Auditor (PA).

Kent Lewis is an Air Transport Pilot currently flying with Delta Air Lines in Atlanta, Georgia. He has international flying experience on the Boeing 777, and currently flies the MD-88. Kent graduated with Honors from the U.S. Navy Postgraduate School's Aviation Safety Officer course, was the Director of Safety and Standardization at MCAS Yuma—the largest aviation training facility in the United States, and was previously the Safety Department Head at VT-27—the Navy's busiest aviation training squadron. His safety programs have been recognized as the best in the Navy/Marine Corps and have a zero mishap rate. He currently volunteers as the Director of Safety at the Vintage Flying Museum in Ft Worth, TX and owns the safety website www.signalcharlie.net.

Kent also volunteers as a FAA Safety Team Lead Representative for the Ft. Worth FSDO, and he was selected as the 2009 National FAASTeam Representative of the Year. He has attended the Air Line Pilots Association Basic Safety School, Safety Two School, Accident Investigation 2 (AI2), Advanced Accident Investigation course (AI3), FAA SMS Standardization course and FAA Root Cause Analysis Training. Kent's focus is Human Factors and System Safety. Current committee work includes ALPA SMS and Human Factors, and FAA Runway Safety Root Cause Analysis Team, as well as teaching the Human Factors module in ALPA's AI2 course. As the ALPA Atlanta Council 44 Safety Chairman, he represents over 3,500 of his fellow pilots on aviation safety matters.

Kent is a graduate of the University of Texas at Austin, with a Bachelor of Arts degree in History. He was awarded a Masters in Library Science at Texas Woman's University in December of 2010.

Captain Larry McCarroll is a Professional Aviator and Aviation Safety Advocate. His career encompasses General Aviation and Air Transport. Since 1978 Larry has continued to fly for a legacy U.S. carrier. At the airline he worked 20 years in various training and safety capacities; as a check airman, FAA Air Program Designee, International Standards Chairman, ATA International Operations representative and taught instructor facilitation skills. As a member of his pilot's association he worked as the safety co-chair facilitating the understanding of SMS and its utility to fellow pilots and promoted the active role all employees play in assuring an effective safety culture.

Larry remains involved in general aviation as a FAASTeam representative emphasizing Practical Risk Management. He is a Co-Founder of CAP Aviation Consulting Group LLC (CAPACG), specializing in Flight Data Management and Safety Management solutions.

Daniel M. McCune is currently the Associate Vice President for safety for Embry-Riddle Aeronautical University. Mr. McCune retired from the United States Army after serving 27 years. Mr. McCune's last flying position in the Army was a Citation Jet Captain at Dobbins Air Force Base. He has flown over 6500 accident free hours. Mr. McCune has been involved in Safety Management System (SMS) development since 2004. Mr. McCune has attended the FAA calibration session which discussed the first concept of SMS. Mr. McCune currently is a member of the advisory committee for the Aviation Rulemaking Committee (ARC) for SMS.

Jay Pardee is Chief Scientific and Technical Advisor for Vulnerability Discovery and Safety Measurement Programs. Mr. Pardee has been with the FAA for over 30 years and has held numerous positions within the Aviation Safety (AVS) Line of Business. He began his career as a Branch Manager in the Aircraft Certification Service (AIR) Engine Certification Office (ECO), where he went on to become the ECO Manager and eventually the Assistant Manager for the Engine and Propeller Directorate. Mr. Pardee became the Directorate Manager in 1994. In 2006, Mr. Pardee accepted the position as the Safety Integrated Product Team Director, in support of the Joint Planning and Development Office (JPDO). Then in 2007, he agreed to support AVS by establishing the Aviation Safety Analytical Services (ASA) organization, which eventually became the Office of Accident Investigation and Prevention (AVP), where he most recently served as Director. In his current position with the FAA, Mr. Pardee advises government and industry on collaborative data-driven safety enhancement development and implementation. He is also an expert in the field of safety information systems using voluntarily supplied and publicly available safety information systems capable of vulnerability discovery and system safety indicator monitoring.

Mr. Pardee has won several industry and government awards including: Collier Trophy, as part of the "Commercial Aviation Safety Team" (CAST), for achieving an unprecedented safety level in U.S. commercial airline operations by reducing risk of a fatal airline accident by 83 percent, resulting in two consecutive years of no commercial scheduled airline fatalities" (2008); Flight Safety Foundation President's Citation for Outstanding Achievement in Safety Leadership (2001); Award for contributions and leadership of the Certification Process Improvement Team (2001); FAA Administrator's Certificate of Accomplishment for Leadership of the Safer Skies Initiative for Commercial Aviation (1999). Mr. Pardee holds a B.S. in Aerospace Maintenance Engineering from Parks College of Aeronautical Technology.

Captain Nicholas Seemel began flying in 1980 and with over 15,000 logged hours is currently flying the line as a line training captain at a Canadian airline. He began his career in various northern Canadian bush pilot charter companies involved in a wide variety of operations including medivac, fire support, exploration and sport fishing. His involvement with the 50,000 member Airline Pilots Association (ALPA) began in 1980, first as an elected official representing the membership pilots at his local airline in various portfolios including contact negotiations. Subsequently he was appointed the Chair of the local ALPA Air Safety Committee. He has attended all of the safety and accident investigation schools offered by ALPA. Nick has assisted in accident investigations and was awarded an IFALPA Presidential Citation for his contribution on the IFALPA team assisting on the Swissair 111 investigation. His passion for improving safety management began approximately 12 years ago during the development of SMS both nationally in Canada and within his airline. His efforts continue today as he manages the ALPA safety team in partnership with his airline's safety investigation team. His role includes risk-assessment, teaching, investigating, promotion and whatever else is needed for the ongoing success of SMS.

D Smith served nearly twenty-nine years in the US Army earning the designation of "Master Army Aviator". He has over 26 years experience designing, developing, implementing, and managing organizational aviation safety programs, including extensive experience in Risk Management and Human Factors. He has held several key positions in the field of aviation safety including, Director of Aviation Safety Officer Qualification Training for the US Army and Director of Safety for Operation Iraqi Freedom and its thirty four coalition countries. He is currently employed as a manager and senior instructor by the US Department of Transportation providing Safety Management Systems (SMS) instruction to private industry, US federal agency, and many international government personnel. Known for his innovative and sometimes unconventional approach to safety and accident prevention program management he is an often sought speaker. Described by many as passionate, he is considered a leading subject matter expert and forward thinker in the field of Aviation Safety and related topics.

Lieutenant Commander (LCDR) Roberto H. Torres is presently serving as Command Safety Officer of the US Coast Guard's Aviation Training Center in Mobile, Alabama. A 19-year veteran of the Coast Guard, he is a career military officer, an aviation safety professional, and both a rotary- & fixed-wing pilot for the life-saving service. A former instructor pilot in the HH-65C "Dolphin" helicopter, LCDR Torres recently qualified in both the MH-60J "Jayhawk" helicopter and the HC-144A "Ocean Sentry" airplane. As an aviator, he has been stationed at Coast Guard Air Stations in Borinquen, Puerto Rico and New Orleans, Louisiana. Under the sponsorship of the Coast Guard's Aviation Safety Division, LCDR Torres earned a Master of Science in Aeronautics (with Distinction) from Embry-Riddle Aeronautical University in Daytona Beach, Florida. He also completed the Aviation Safety & Security Certificate Program at the University of Southern California and is a graduate of the Naval Postgraduate School's Aviation Safety Officer Course. Married to Jennifer L. Monie de Torres, they have three children: Ariana E. (12), R. Mateo (10), and David A. (9).

Bill Yantiss began his aviation career as a U.S. Air Force T-38 Instructor Pilot. During his 20-year military career, he had an opportunity to train pilots in the F-4D/E, F-5E, and F-15A/B. He has served in many leadership capacities that include Check Pilot, Operational Test & Evaluation Test Pilot, F-15 Program Manager, Academic/Simulator Training Squadron Commander, and Fighter Squadron Commander. He accepted an assignment as F-15 Program Manager, responsible for managing student training worldwide. In this role, Bill prepared and implemented the game plan for the transition of 6 F-106 squadrons to the F-15. Bill concluded his military career as the Assistant Deputy Commander for Operations, responsible for the daily supervision of three fighter squadrons, an academic training squadron, and an air weapons controller training squadron. He was a "Distinguished Graduate" from Air Force pilot training, "Instructor of the Year" and received an Air Training Command "Well Done" safety award. He retired from the Air Force in 1989 as a Lieutenant Colonel.

Bill joined United Airlines as a Pilot Instructor in 1989 and was quickly promoted to Department Manager. Shortly thereafter, he assumed staff responsibilities for both Flight Standards and Flight Safety. These short assignments have led to a 16-year career in developing Safety, Security, Quality, and Environmental programs and management systems. As Manager—Flight Safety, Bill played a key leadership role in the implementation of United's industry-leading Flight Operations Quality Assurance (FOQA) data analysis program. He also developed and implemented three of United's Aviation Safety Action Programs (ASAP).

Bill was promoted to Managing Director—Quality Assurance, responsible for developing United's Quality Assurance program for Airline Operations which served as a model for US air carriers. He developed the Star Alliance safety audit program, a concept that was to serve as the model of a safety auditing system that would eventually become the global standard. As an IOSA team leader, Bill has conducted over 65 audits of international carriers, recommending them for IOSA registration.

Bill was selected as Vice President – Corporate Safety, Security, Quality, & Environment in December, 2006. He partnered with the Federal Aviation

Administration to publish the first Safety Management System (SMS) Advisory Circular, a guide to developing and implementing an SMS in any organization.

Bill has nearly 20-years commercial aviation experience as both a Captain and First Officer in the B737, B757, and B767 aircraft. He holds a Bachelor of Science degree in Zoology/Botany and a Masters of Business Administration. He has completed the University of Southern California Aviation Safety Program and achieved registration as an American Society for Quality (ASQ) Certified Quality Auditor (CQA), a Certified Quality Manager for Organizational Excellence (CQMgr), IOSA Auditor, BARS Auditor, and IS-BAO Auditor. He has also achieved certification as a Lean Six Sigma Black Belt. He has been invited to speak at a number of international aviation safety forums, focusing on safety and quality management systems. Bill has accepted an assignment as Executive Vice President, PRISM Solutions, LLC, a wholly-owned subsidiary of ARGUS International, Inc. Bill resides in Denver, Colorado with his wife Glennda.

Preface

In theory there is no difference between theory and practice. But, in practice, there is.

Jan L. A. van de Snepscheut/Yogi Berra

We wrote our first book on the subjective of Safety Management Systems (SMS) in an *authored* format. In this book our focus is on implementation, so it seemed natural to turn to experts in the field and ask them to relate to the reader their experience in actually implementing SMSs. Thus, this book is written in a multiple author/editor(s) format.

Editor might be too generous of a term for our role, however. From the outset we decided to allow our various authors to speak their minds on a particular topic without our attempting to align their words with other points of view. Thus, if you see in this book some differing views on a subject, we think that simply reflects the fact that there are differing views of the subject – and that's a good thing because none of us has perfect knowledge about this phenomenon called SMS.

Fact is not enough, opinion is too much.

Todor Simeonov

We also wanted to get a practical viewpoint of SMS. Our first book has been very well received and for that we are immensely grateful. We did our best to present the theory and underpinnings of SMS so that the reader was well-grounded in the subject. While the editors are involved in industry, consulting, and academia, we thought that turning to the day-to-day practitioners of SMS provides yet another perspective for the student of SMS. Our authors are among the most experienced practitioners in the industry today.

Like the first book, this one should by no means be considered a complete guide to implementation. Rather, it is selective, but does cover topics from all four elements of SMS and many of the processes. We are confident that the range of material covered will offer the reader insight into some of the thornier topics related to implementation.

The trouble with organizing a thing is that pretty soon folks get to paying more attention to the organization than to what they're organized for.

Laura Ingalls Wilder

The book is at best *loosely* organized around the International Civil Aviation Organization (ICAO) implementation model. There is no real beginning and end to the topics; rather they flow from one to another with (intentional) gaps in certain areas. Thus, the reader is not discouraged from finding the author or the topic they are presently interested in, and reading that section. There is no *out of order* in regards to this book – experience it how it suits you best.

> To me a story is true not because it happened, but because it happens to make sense.
>
> Unknown

In our first book, we included a fictional story as an instructional device, intended to provide something of a case study, thought-experiment on SMS. We hoped the story would be an entertaining departure from the (albeit important) textbook presentation of SMS. The feedback we received from the story was extremely positive, so we decided to continue that theme in this book with another episode from Quest Airlines. Hopefully, the reader will be both enlightened and entertained with this one as well.

Acknowledgements

We are also indebted to our friends and colleagues who contributed chapters for this book. We are confident that their experiences in safety management systems will benefit the reader, and we are grateful for the investment of time and energy they expended to write their respective chapters.

We would also like to thank two Embry-Riddle Aeronautical University students, Theresa Brown and Keri Younger, for their excellent work in editing and generating figures.

Finally, much thanks to Guy Loft and the entire staff at Ashgate for their professionalism, support, and patience as we failed to meet a single deadline.

Prologue – Quest Airlines

Jim Price, pilot flying for this short hop from Washington to Cleveland, carefully watched the automation as the glideslope captured and the DME ticked down 7.9, 7.8, then 7.7. "OK, there's TUBOE, gear down, flaps 30, before landing checklist."

"Squanto 853, visibility now 3/8 mile, mid RVR 1800."

"Roger, Approach, we're inside TUBOE, continuing the approach."

"Squanto 853, contact tower 120.9, have a good night."

"120.9, thanks," responded Harry Bledsoe, co-captain flying right seat and accomplishing pilot monitoring duties for this flight. Price quickly reviewed the most important element of the approach briefing accomplished earlier just after top of descent, saying to Bledsoe, "Viz has been going up and down all evening. Remember we're planning for a go around at 200 feet, but maybe we'll be pleasantly surprised."

"Good plan." Bledsoe flipped frequencies in VHF1 and transmitted, "Cleveland Tower, Squanto 853 is with you TUBOE inbound."

"Squanto 853, Cleveland Tower, cleared to land runway 6R, wind 010 at 10 gusting 15. Last aircraft to land was a Southwest 737, reported braking action good. Visibility right now ¾ mile, blowing snow, RVR has been varying quite a bit."

"Squanto 853 is cleared to land," responded Bledsoe, who then began the litany of standard approach callouts and crosschecks that had become ingrained after ten thousand hours of flying. "Missed approach altitude is 3000, set, first step of missed is fly heading 072, radios are set for missed."

Price and Bledsoe were near the end of a long day. Dispatched at 7 a.m. from Nashville, the home base of HighJet Charter, the two pilots had accomplished a routine flight to DCA where they were to meet Bob Holmes, one of their occasional customers, to transport him to CLE. To call Holmes a VIP was an understatement. Senator Holmes had served Ohio for 26 years in Congress and had risen to become the Chair of the Senate Ways and Means Committee. He was flying to his hometown of Cleveland in order to be present the next morning for the arrival of 15 of Ohio's sons and daughters, all members of an Army Guard unit, all killed in an ambush a few days before. Holmes had a car ready as backup for the flight – being there was the least he could do; he was not going to miss the solemn occasion. But if at all possible, he needed the efficiency that HighJet's service offered. An important vote was scheduled for tomorrow evening, and that too was something he could not afford to miss. It was pricey, but he could afford to pay for it – from his own money.

Price and Bledsoe had met up with the senator at the entrance to Signature Aviation at Reagan National at 10 a.m., and proceeded to wait – and wait – for the weather to clear up at CLE. The city was being pounded by the trailing edge of a winter storm, but that was only part of the reason for the delay. Virtually the entire fleet of the airport's snow plows were down, incapacitated by contaminated fuel. The airport's emergency preparedness plan had contingencies for such an occurrence: surrounding communities dispatched equipment from their own fleets to assist in snow removal from CLE's two main runways. The additional resources were a great help, but still the airport was down to one runway operation for the entire day as the snow removal effort struggled to keep up.

Price and Bledsoe had almost taken off with the senator on board once, around sundown. They were number two for departure when tower informed them that CLE had just been closed again. One of the borrowed plows had just taken out some runway edge lights, and the authorities had closed the airport while they were sorting out the situation.

Luckily the delay was not too protracted. Though not required, HighJet had opted to run its pilot scheduling in accordance with FAR Part 121 rules, which meant that Price and Bledsoe were going to turn into pumpkins around 8 p.m. Central Standard Time, 14 hours after their report time. Price had modified their preflight risk assessment twice already, and skirting the edge of duty time caused another reassessment. He called the HighJet dispatcher after the reassessment was complete, and both had agreed that though the risk index number was high, it was still reasonable to proceed.

Finally, a little after 6 p.m., their Citation X was wheels up out of DCA. The flight had proceeded routinely according to plan. A little ice in the descent, but not bad. If the visibility cooperated they would be on the ground in five minutes, and they and the senator at their respective hotels in an hour.

"500 above, crosscheck no flags. On speed, on localizer, on glideslope," chanted Bledsoe. A few seconds later, tower transmitted "Squanto 853, visibility now ½ mile, runway 6 Right touchdown and mid RVR 2400, rollout 1800." Bledsoe responded "Roger, Squanto 853."

"100 above, ground contact, on speed on loc, on glideslope," Bledsoe informed Price. Five seconds later, he said "Minimums, runway in sight."

Price responded, "Visual, landing. I'm outside, you're in."

Price maneuvered the jet to a firm landing in the touchdown zone, and carefully began braking, alert to any sensations of control problems. Braking seemed fine. But ahead he could tell that the visibility was worsening. Clouds of fine snow were blowing across the runway from the left, created by the clearing activities presently in progress on 6L.

Suddenly he saw a red light emerging from the obscurity, straight ahead. He quickly saw that the light was a position light, attached to a B757, sitting right in the middle of the runway.

Bledsoe saw it too, loudly announcing "Aircraft on runway!" Price was already reacting, pushing hard on the left rudder in an attempt to zoom past the nose of the 757 intruding on his runway.

If it weren't for the patches of snow and ice on the runway, they might have made it. But almost does not count in the aviation business. The nose of the Citation plowed into the Boeing just in front of the left wing, the Citation's right wing slicing into the Boeing's left fuel tank.

A Few Minutes Earlier ...

Glenn Seyfat settled back into seat 16A on Quest Airlines Flight 242, from CLE to STL. He had just completed a day's visit with the new General Manager of Quest's Cleveland operation. The new GM had asked Glenn to visit to help him understand his safety responsibilities in the company's Risk Management Plan, and Glenn was enthusiastically happy to comply with the request. As the airline's Vice President of Safety, he was always happy to help an operational manager stay on top of the Safety Plan. Having a new GM reach out of his own accord and ask for help, especially in the first few days in the position, was a very good sign, and bode well for the Cleveland operation. Jeri Tippets, the ASAP manager, had accompanied Glenn to talk to the Cleveland ground crew about organizing a ramp ASAP program, and was staying over another day to finalize the plan. It had been a good visit after a long week, and Glenn had decided to take tomorrow off.

The flight had been delayed over two hours because of the snow and CLE's operational problems, and at the gate the natives had been getting restless. Seyfat had watched with admiration as the Quest gate agents deftly handled one irate customer after the other. Halfway through the second hour he had fetched a couple of lattes for the harried gate agents, and had spent ten minutes or so talking to the flight crew. Glenn knew most all of the pilots at Quest, but he was especially fond of both of these guys. Captain Hargrove was a check airman, a writer for several aviation publications, and a frequent contributor of articles to the company's safety website. First Officer Arnut was a new hire, an Inuit from Alaska who proceeded to tell Glenn the Inuit words for the various types of snow that Cleveland had been experiencing throughout the day. By the time the discussion of snow was complete, Captain Hargrove had received an update to the expect departure clearance time, and the gate agents began the cattle drive to board the airplane.

Despite the thousands of hours Glenn had spent in the left seat of an airliner, he always picked a window seat. First Class was full tonight, so he chose a window seat in an exit row. From seat 16A he watched as the snow whirled past in great clouds as the aircraft slowly rolled along one of the parallel taxiways. At times the visibility was bad enough to obscure the runway just a few hundred feet from the taxiway. A heads-up taxi for the guys in front tonight. He settled back in his seat. Someone else's problem.

After stopping for a short delay, he felt the left yaw indicating that a 90 degree turn was in progress. *Must be crossing the left and departing on 6R*, he thought to himself. His eyes once again aimlessly wandered to the view out the window.

As the aircraft rolled to a stop and a transient cloud of blowing snow continued east beyond the plane, his eyes suddenly picked out something that brought him to full alert. Out of the window he saw a thick white line fading into the distance – *Must be the runway centerline. Why are we stopping here?* He began to get a bit nervous despite himself; being on a runway in low visibility conditions brings any pilot's hackles up, and Glenn was no exception. *Just get across, so I can go back to reading my book*, he said to himself.

When he saw the bright white lights in the distance the first strange observation he made was that it was amazing how quickly it was – instantaneously it seemed –that he understood what the lights were, what they meant, and what was about to happen. He let out an involuntary grunt, "Ughh", as he considered then just as quickly dismissed the idea of calling the attendant and asking that a message be relayed to the flight deck. There was no time for that, no time for anything, really. He decided to make his last thoughts about his family. Not in the nuclear sense – his wife had died in a car accident two years before, and they had no kids. His family, he realized, were his co-workers at Quest. Seyfat regretted that he had not figured out how to tell the travelling public how lucky they were to have such a dedicated professional bunch of people providing service for them, the irony of this thought not being lost on him in his last seconds. And as he considered his brothers and sisters in his airline family, he started speaking to them one word, over and over – "learn … Learn … LEARN" – loud enough that the passengers close by had time to look at him and wonder what was going on, until their curiosity was ended in a mercifully violent slam of aluminum into aluminum.

Event Plus Six Hours, St. Louis

Ken Tiller strode toward the temporary podium situated along one wall of the Quest Airline dispatch center. The dispatch manager on duty had just asked for everyone's attention in the busy room, and as the MOD introduced Tiller as the president and CEO of Quest Airlines, the quiet, intense buzz that characterizes an airline dispatch center was replaced by hushed whispers. Everyone knew this would be about Flight 242. Everyone was anxious to know more, and rumors were not good.

"Ladies and gentlemen," Tiller began, "I won't take up much of your time. I know you have a job to do. But I know you are as vitally interested in learning about the crash tonight as everyone else is. I'll tell you what I know, and do a quick review of our plan moving forward."

"The news is not good. As you know, Flight 242, nose number …" Ken looked at his notes "… 123 was hit by a Citation X business jet as it taxied out for takeoff. The wreckage is on runway 6R at CLE. The Citation was apparently landing on 6R. Weather was definitely a factor, with the major snowstorm that is presently tracking across the Midwest and approaching the East Coast. We don't know any further details concerning clearances, causes, etc. There were 120 souls on board, including 2 flight crew, 4 flight attendants, one deadhead crew of six from a cancellation earlier in the day, and 5 Quest non-revs." Ken paused, "There were only three survivors on our aircraft. They were all paying passengers. We lost all of our fellow employees." Ken stopped for a moment; the silence in the room was palpable. "There were six people aboard the Citation, two flight crew, one flight attendant and three customers, one of them being US Senator Robert Holmes, of Ohio. He was traveling with two of his aides. Rest assured that this accident will be closely watched, as will we."

"I've just told you just about all we know. We're setting up a page on the employee-only website that will be updated every six hours for the next few days with the latest information we have, so please refer to that site for the latest. As you guys know, our go team aircraft was dispatched to CLE about 90 minutes after the crash, and the go team has been onsite now for about two hours."

"Now for the main reason I'm talking to you. As all of you also know, our VP of Safety, Glenn Seyfat, was on this flight, and like all of our brothers and sisters, was killed. But among many, many other things, Glenn has a legacy that is with us, right here, in this room." Tiller held up a thick loose leaf binder. "More than any other person, Glenn was responsible for this Emergency Response Plan. We've had an ERP for years, of course, but in the last four years, since Glenn was in charge of it, he made it his first priority. Years ago he kept telling me," Ken's voice cracked a bit, then quickly regained its authority, "Tiller, I'm just sayin', when you need this thing, you really need it, and it has to be right."

"Thanks to Glenn, we have a plan, and it's a very good one. Unless we see a compelling reason to deviate, we will all follow it. And the most important part of the plan is the Continuing Operations section. All of you in this room now have the most important role spelled out in the plan. You are responsible for maintaining the safety of our operation. It's you, not the Event Command Center, who has first priority on company resources. If you are not getting what you need, I want you to personally call me and tell me. The emergency staffing plan is in effect, so your hour-to-hour workload should be decreased. There will be a short-term decrease in the number of days off, but I ask for all of you to have patience with this, and we'll return to normal as quickly as we safely can. Finally, we should all expect of one another the highest level of professionalism, with the focus on the task at hand."

"Thanks for your attention. Let's get back to work." Tiller left the podium and exited the dispatch center into the adjoining event command center room, joining a small, intense cadre of coworkers. "Okay guys, the plan says it's time for the six-hour review. Ops, you first."

Don Doppermeyer, VP of Operations, began running down his six-hour review checklist. "Passenger manifest was reviewed, verified, secured and forwarded to the onsite commander from the CLE Fire Department and to the NTSB at 9:15 p.m., event plus two hours, as scheduled in the plan. Go team departed at 9:30 p.m., 45 minutes before planned time. Joe Quick is our Investigator in Charge." Formerly the company ASAP manager, Joe had decided to return to line flying but remained a fixture in the safety department. "Joe was able to herd about half of the reps he wants for the NTSB functional teams. The rest will be heading to CLE tomorrow. We'll decide whether we need a number two go bird then, but I don't think we'll need it."

Doppermeyer continued, "Joe's first report as company IIC came in about an hour ago. He dispatched Martin Goldstein to the fire department to be the Quest liaison. Goldstein reports that the fire department's first responders are extraordinarily well organized. The three survivors are all identified, at local hospitals, all critical. Thanks to Glenn, we already had links established to the CLE ERP so we'll be able to follow victim ID and survivor status in real time. I'll leave the CARE Team brief to Mark," nodding to Mark Andrews, managing Director of Emergency Response Planning. Mark had been appointed by Glenn Seyfat four years previously, when the push to create Quest's SMS had first begun.

"The flight plan and training records of all personnel associated with the flight have been locked," Don concluded. "Checklist is complete, plan on track."

A similar litany was joined by the others around the table, from Todd Jacks, VP of Maintenance, from John Simmons, Managing Director of Dispatch … from all aspects of Quest's operation. All told they had spent hundreds upon hundreds of hours perfecting the Quest ERP, and some had to admit that they had balked at the man-hour expense that went into the very detailed plan. For some, like Todd Jacks, Glenn had had to ask Ken Tiller to intercede to get Jacks to send maintenance representatives to the ERP working group meetings. In the end, everyone had fallen in line and done the work, but truthfully only because Tiller told them to, and even more truthfully, the CEO had told them to only because Seyfat had nagged him until he did. Tiller reflected that "obtaining top management support" for Quest's Safety Management System didn't just happen; it took team work and the willingness of members of the executive level to push each other.

"Okay, we've saved the most important piece for last," Tiller said. "How's the CARE Team going, Mark?"

Quest Airlines" Customer Assistance Relief Effort, or CARE Team, was a part of the Quest ERP required by the Aviation Disaster Family Assistance Act of 1996. Patterned similar to most airline's teams, it was composed of employee volunteers who, once having been appointed to the team, received considerable training (during normal working hours) to prepare them to provide any assistance that might be needed by the families of victims of an aviation disaster involving a Quest aircraft.

"It's going remarkably well. It's proving the value of our drills. We're six hours into this event and over 90 percent of our CARE volunteers have reported in, far more than we need right now. Every one of them is ready to go if we need them. We've done a tentative match of team members with the victim list; that match will probably change, but as we've been doing the family notification it's helped to be able to give a name and phone number of a team member to everyone we've been able to reach, which as of right now is …" Mark leaned over and toggled his laptop to display on the wall LCD, "57 of the 103 paying passengers on board."

Don Doppermeyer and Jan Firth, managing Director of In-flight Services, had had the terrible duty of notifying the flight crew and flight attendant families.

"The CARE Onsite Command Center has been established by the CARE Go Team, at the Airport Radisson, and we had coordinated long ago with the NTSB to co-locate with their Family Assistance Center, so they are there too. The IICs and Investigation Command Center is at Airport Sheraton, about 5 miles away from the CARE team and families."

"Okay," said Tiller, "barring anything that requires a change of plans, I'll be staying with the CARE Team for the next week, so Don, you're in charge of the Event Team from now on. Let's all remember that we're running a flight operation, and we'll do nothing to overtax them."

"Finally, though most of you know him already, just to be sure, let me introduce Greg Murphy. I hired Greg two months ago to fill the newly created position of VP of Quality. I stole Greg from the car company I used to run. He's a pretty unique guy, with a Ph.D. in economics and a specialization in quality management. In addition to running the new quality area, he is going to help us implement our integrated risk management strategy."

"I've asked Greg to be an observer of what we're doing here because we need to be sure that it's working, and so that any integrated risk management strategy that we might come up with in the future will incorporate the realities of nights such as tonight. I hope this doesn't sound callous to anyone, using this as an opportunity." Tiller paused. "But if it does, too bad. That's my job. And we owe it to the friends we lost. Following the actions of this team is Greg's only assignment for the next few weeks. I expect everyone to help him understand how all this works."

As they started to get up from the table, Doppermeyer interjected, "I know it's premature, and it's in the Go Team's court to investigate, but I might as well say it. There's a good chance we were at fault. The news is saying that pilots of other aircraft on freq definitely heard the landing clearance for the Citation."

"I know," Tiller said, pausing, and again, "I know. Let's take care of what we can do right now. Don, make sure we stay close to the Captain's and FO's families. It's going to be rough."

After the Event Plus Six Hour Review meeting, everyone went their separate ways, referring to their piece of the plan often, using the outline generated in calmer more clear-thinking times as a foundation on which to walk through the days ahead, knowing they were all doing the best they could.

Event Plus Six Hours, Nashville

Mike Kramer sat at his desk in his office at the HighJet hangar, overlooking the spotless hangar floor illuminated by the bright Klieg lights overhead. An equally spotless Challenger 604 was parked just below, with HighJet maintenance technicians meticulously finishing up an inspection. He glanced down at his ERP checklist. All items that needed to be completed by this time were done, and he had a few minutes to think.

Mike had joined HighJet as their Director of Safety three years ago. Two years ago, he had won the battle with the Board of Directors to establish an SMS program for HighJet, and the Emergency Response Plan sitting in front of him was one product of that victory. He knew that he would very likely be going off the deep end right now had they not made the commitment – and provided the resources needed – to do SMS right. He had friends at other bizjet operations that were doing his job on a shoestring. HighJet had made a good decision. Thank God that at least he had that going for him, with three friends dead, and the company facing its biggest crisis by far in its ten-year history.

Within an hour of the crash, every one of the ten pilots that remained in Nashville, not out on a trip, had appeared at the hangar. Only two of them had go team duties, but all the rest were there to help any way they could. About three hours into their ordeal, reading a bit ahead in the ERP, Mike met with the pilots to find volunteers to serve the same functions as CARE Teams did at the bigger cousins of his operation. Three of the pilots offered to be the contact point for the families of the two pilots and flight attendant. In a subsequent phone call, one of HighJet's board members volunteered to be the liaison with Senator Holmes' family and the family of his aides. The board member was a friend of the Holmes family anyway. HighJet did not need the same level of organization as was needed at the airline CARE Team level, but the goals were similar. Mike was very glad that the ERP guided him to assure he did not neglect important details in such stressful times.

One of his pilots and a HighJet mechanic would be en route tomorrow to Cleveland on a commercial flight to be HighJet's party members to the NTSB investigation. He could not afford to send more people right now, and hoped that his two guys could cover it. Though he had planes and pilots available to dispatch to CLE to get them there sooner, the ERP emphatically advised against such action. There was no rush – the wreckage would be there tomorrow. A year ago his bosses had agreed to send the two designees to the NTSB Training Center, a decision he hoped they all remembered right now. If not, he would remind them of it in the next few days. HighJet was as prepared for this as they could reasonably expect to be.

He decided to pick up the phone and call Frank Mills again. It was the second time this evening he had spoken to Frank; in fact, Frank was the first person he had called after he heard about the crash. Frank was a consultant he had hired two years ago, at the beginning of HighJet's move toward building an SMS. The Board had nearly choked when they saw the bill for Frank's services, but he knew that he had made the right decision. In the consultant business you pay for experience, and Frank had that in spades. Six months ago, when he delivered the dog and pony show to the Board describing how HighJet was going to accomplish Phase 4 of SMS implementation, one of the Board members had even commented that it looked like the investment in Frank had paid off. Tonight was proving to be a test of that statement.

Frank answered his cell phone on the first ring. "Hey Frank, me again," Mike said. "Hope I'm not imposing too much."

"Of course not, Mike," said Frank. "What's up?"

"Well, we're in reasonably good shape right now. Working the plan. But there's an aspect to this accident that is just not dealt with in the ERP. Senator Holmes is, correction, *was* one of the most influential people in the US. As cold as it may seem, this is not just your ordinary plane crash with John Q Public aboard. We're not ready for this. And I think it might affect our operation moving forward too. We're going to have to get out in front of this in a couple of ways, or we're going to get eaten up. And Lord help us if Harry and Jim – I'm sorry, our two pilots – screwed up somehow."

"Mike, the whole industry is rethinking what 'phase four' SMS implementation really should be," said Frank. "You guys have done a remarkable job of getting to the continuous improvement phase of implementation. But we're all recognizing that risk management in an operational sense does not happen in a vacuum. There are many kinds of risk. The risk you're dealing with right now, of losing a high-value passenger, has implications for your Board to consider, from insurance, to PR, to marketing. I honestly don't want to make this sound self-serving, but maybe we should accelerate the integrated risk management strategic planning that we've been talking about. No better time than now to sell the Board on the idea."

Mike responded, "Small minds, Frank. That's the reason I called. I need your help formulating an approach that will work, and I think we need to do it quick."

"The other thing I would recommend, Mike, has to do with your comment on the high visibility of this accident and its effect on your ongoing operations. As we discussed a few hours ago, the most important piece of the ERP is the Continuing Operations section. Preach that until nobody can stand you anymore. But you need to reach out in the next few days to your pilot group and reinforce your safety culture. Give your ASAP and FOQA pilot specialists extra time off to focus on their programs." HighJet could not afford dedicated staff for their ASAP and FOQA programs, so two pilots had been assigned the ancillary duties of running these safety initiatives. "Especially encourage ASAP reports right now; it will bring safety to the forefront of everyone's mind. And to counter the inevitable pressure your pilots are going to feel from press scrutiny, get out there and remind them of their strong existing safety culture; that once a passenger is aboard their jet, it doesn't matter who it is, the same safety rules apply. I'd suggest you do spot checks of your preflight risk assessment forms to make sure they are being done correctly, in fact make the fact that you have this system a point of pride when you speak to your pilot group. You guys are ahead of the pack on this."

"Excellent advice as usual, Frank," Mike said. "I'll keep in touch concerning the integrated risk management strategy. I have a feeling that I'll be seeing a lot of the Board in the next few weeks."

After the phone call, Mike reviewed the quick notes he took as Frank made his recommendations. What was remarkable to him was that having an SMS gave them both the vocabulary to discuss and tools to work the problem. Safety culture, preflight risk assessment, ASAP (Aviation Safety Action Program), FOQA (flight operations quality assurance), continuing operations – the decades of work required to develop the discipline of SMS paid off in a few minutes of insightful discussion, with a plan to proactively use the SMS toolkit to place another few barriers between normal operations and tragedy.

As bad as the evening had been, he felt a bit better now. He had heard this paraphrased a couple of times before, and it sure felt right – he could not know the fate of their operation, but what he did know was this: he was not going to be the hunted. He was the hunter.

Event Plus Six Weeks, St. Louis

Ken Tiller briskly walked through the door that connected his office with the spartan board meeting room, saying "Ladies and gentlemen, let's get started." He took his place at the head of the table, awakened his sleeping laptop, and brought up the meeting's agenda. "We have a lot of work ahead of us today." He turned to Sean Werner, his chief administrative assistant, sitting next to him. "Sean, as usual, you're the note taker and time keeper, we'll rely on you to keep us on schedule. How about giving us a quick overview of the agenda?"

As Sean ran through a high level summary of the plan for the monthly exec meeting, Tiller looked around the table. All of the company officers were present: the VPs of Ops, Maintenance, Security, Marketing, Human Resources, Corporate and Government Affairs, the Chief Financial Officer and Chief Information Officer, and the two new additions to his staff, Greg Murphy, the VP of Quality, and Georgia Marks, formerly the Director of Quest's Internal Evaluation Program who had taken the place of Seyfat as VP of Safety.

When Sean was finished, Tiller said "Don," gesturing toward the VP of Ops, "how about giving us a quick summary of where the 242 investigation stands."

"Well, we're still months away from the release of the NTSB factual report, but essentially we know what happened. Our guys were responsible for a runway incursion, entering the protected zone of an active runway without clearance from ATC. But there were just an unbelievable number of happenstances that led up to the incursion. It just shows that if something like this happened to Captain Hargrove, it can happen to anybody."

"First of all, the weather; that storm dumped two feet of snow on Cleveland in 24 hours. It's amazing that they kept up with snow removal as well as they did. By the time of the accident the front had moved through and most of the snow event was over, but with the temp falling almost 30 degrees the last six inches or so of the snow didn't stick to anything; the wind just blew it around. Except for the blowing snow, the visibility wasn't that bad – ¾ of a mile or more. But especially since snow removal was in progress on 6 Left, all that stirred up snow blew right across to where 242 was taxiing."

"And the snow removal – Cleveland airport ops didn't have a good day. Somehow nearly all of their snow removal equipment got bad gas, and broke down. They got good support from the surrounding communities in sending plows and blowers in to help, but those drivers were not trained for snow removal on an airport. One of them took out about 300 feet of runway edge lighting, and – this next is critical – the wig wag lights on the south side of the intersection of taxiway Romeo and runway 6 Right. The driver radioed in and told ops control what he had done, but mistakenly said runway 6 Left, and also didn't mention the wig wags. The Ops One truck was sent out to inspect the damage, but was over on the left side and got stuck in the newly installed EMASS at the end of runway 24 Right." Doppermeyer went on to explain to the non-pilots in the room that the wig wag lights were an additional way to identify hold short lines on taxiways, and EMASS (Engineered Materials Arrestor System) was a soft paving material installed on runway overrun areas to slow aircraft down in case of an impending runway excursion. The soft pavement had done its job by giving way under the weight of the ops truck.

"By the time someone got over to inspect the damaged lighting, so much was happening that no one noticed the mistake in runway identification. So the Notice to Airmen (NOTAM) describing the problem went out saying 6 Left. Actually the NOTAM said the damaged area was east of Golf taxiway, which should have been enough for someone to notice it couldn't be runway 6 Right, but no one did, including our crew."

"Here's the kicker. Snow removal on 6 Left and its taxiways was proceeding in the worst possible place for our aircraft. All that blowing snow made it really tough for our crew to see. What we think happened was that the crew could see the reflection off the falling and blowing snow from the wig wags on the other side of the runway and just mistakenly thought that was where the hold short line was."

"And finally, just as our aircraft was making the turn from Lima to Romeo, they get a call from Load Control. The FO (flying officer) told them to stand by, but they called back again. He said STAND BY, forcefully this time, but by then it was too late, they had crossed the hold short line."

There was silence around the table for a few moments, then Tiller said, "Georgia, this next idea goes beyond the purpose of this meeting, and we don't want to do it until we know we have all the facts. But as soon as the NTSB (National Transportation Safety Board) releases the factual report, schedule a meeting with all the appropriate people to go over our risk management plan. I want to know where our controls went wrong, and where we need to beef them up."

"Will do," the VP of Safety said.

"We've got to get on with the items on the agenda, but I've got to say, one thing kept occurring to me as I heard the story. Memories of that old car company. How about you, Greg?" turning to the new VP of Quality, "Anything ring a bell?"

"You bet. Supply chain management," said Greg.

"Bingo. Nearly all our recalls were due to parts someone else made for us, not because of faulty work on our part. Nevertheless, we paid the penalty and wrote big checks because of it. We finally stopped feeling sorry for ourselves and did something about it. That was right around when you were hired, right Greg?" asked Tiller.

"That was my first assignment in the Quality Department," recalled Greg.

"For us here at Quest, instead of widgets, a lot of our suppliers provide information for us. Georgia, Don – what are our standards for information quality for flight-critical information?" inquired Tiller.

Don and Georgia looked at each other and shook their heads. "Well Ken, we pretty much take what they give us."

"Perry, how about you?" Tiller turned to the CIO. "What are our standards for information delivery? How are we measuring that?"

"We know when it breaks, 'cause someone can't do their job."

"Mr VP of Quality," said Tiller, gesturing toward Greg Murphy, "there's your first quality assignment. Figure out how to apply supply chain management techniques to flight-critical information."

"Got it." Murphy responded.

Tiller turned to the DOS and said, "Georgia, here's another to-do for you. Something that is just so clear from this accident is how great an effect other organizations" processes have on our operation. I know we're pretty proactive in getting out there and engaging, but I want to make sure that we're managing it and doing it in a strategic way. I want to devote a goodly portion of the next QSR to identifying and risk-assessing interfaces with our operation." The QSR (Quarterly Safety Review) was Quest's method of staying on top of safety assurance by means of a thorough management review of the key performance indicators (KPIs), established to continuously monitor the effectiveness of risk controls in the airline's operation. The heads of all operational departments as well as representatives from Quest's FAA Certificate Management Office attended the quarterly meetings.

Georgia responded, "That's a tall order, Ken."

"Yep," was Tiller's reply.

"Okay, next on the agenda is hearing from Greg Murphy. As all of you know, I asked Greg to plug into the Safety Department, and especially to the Event Response Team following the crash. My theory was that Greg would provide a set of fresh eyes, and especially with his strong background in quality management, he could give us an appraisal of how our SMS is working. Greg, you're on."

"Well," said Greg, "my initial observation is related to the fact that this has been my first exposure to airline operations. As Ken mentioned, I came into this role with the perspective of an outsider, so I think I saw things that all of you take for granted. The observation is that we have an incredible level of expertise and professionalism in this company. And as I've been studying the SMS literature and our own core documentation, what strikes me is that this high level of expertise and professionalism is based on Quest's safety culture. Everybody I talked to over the last six weeks, from front line employee to you guys at the officer level, just gets it – professionalism is expected, information flow is encouraged, when weaknesses are found, things are fixed. In fact, one thing we should all be aware of is that many of our people are asking 'How could this have happened to us? We all try so hard.' I think a concerted effort over the next few months is warranted, to emphasize once again that their efforts are worth it, and do make a difference."

"Well put, Greg," said Tiller. "Sean, please include that point for review and action planning at the end of today's session."

Murphy continued. "So let me proceed by dividing up my comments between SRM and SA – safety risk management and safety assurance. I'll start with SA. Georgia, as the former Director of Internal Evaluations and facilitator of the QSR, the rest of us around this table are indebted to you for your internal evaluation program (IEP) work. We have excellent measures of each department's self-auditing processes, good policies to assure self-auditing is done, and evidence of management intervention to make sure nobody slacks off. The IEP audits are consistent and there is evidence of effective follow-up. Todd," Murphy turned to Todd Jacks, the VP of Maintenance, "our Continuous Analysis and Surveillance program is the best audit program I've ever seen. Your auditors are phenomenal."

"That's where we put the cream of the crop," responded Jacks.

"And Perry," Murphy turned to Perry Verner, Chief Information Officer, "your department's integration of applications so that self audits, IEP, CASS (continuing analysis and surveillance system) and external audits are all in one database, that's world-class stuff.

"So once an issue is identified and good KPIs are targeted, the Quest SA processes really work," Murphy summarized.

"I sense a "but" coming," said Tiller.

"But," Murphy paused. "If there is one area I would focus on as a critique of our SMS, it's in the safety risk management area. And it's fundamental."

"In our documentation, in the FAA documentation, indeed in the ICAO documentation, the first step in SRM is 'describe the process'. Every core document, including our own, reflects that SMS is very closely aligned with quality management, and that a focus on process rather than outcome is the fundamental similarity between the two. In my opinion, I don't see much evidence of our having process-based SRM."

"Listen, we haven't done anything any worse than everybody else, in fact not any differently that everyone else," Murphy continued. "We followed the example of the FAA Air Traffic Organization (ATO). As ATO brought their own SMS online, knowing that a full, top to bottom review of all of their processes to risk-assess each one would be a herculean task, what they did was to declare their present processes as having an acceptable target level of safety, absent evidence to the contrary. In addition, they said that when any existing process is modified, a full SRM assessment would be accomplished. Theory was that eventually all processes would be reviewed and risk assessed, and that the ATO could evolve into a complete and rigorous SMS."

"In our documentation, we explicitly say the same thing. And there is evidence of our doing just what we said we would do. Two years ago when we had that embarrassing slide deployment incident, we did a thorough review of gate arrival processes." Jacks, the DOM, couldn't help but let out a cackle as he remembered the incident, and the laughter became infectious around the table. In that incident, as the last passengers were disembarking at the gate in St. Louis a catering guy opened the right front door where somehow the slide had not been disarmed. The inflating slide had pinned a very rotund lady against the bulkhead. Understandably shaken, she was further freaked out when a mechanic started stabbing the slide with a screwdriver to deflate it. The woman had mistakenly perceived that she, not the slide, was the mechanic's target, and as soon as her arm was freed she had cold-cocked the mechanic. No one was hurt in the incident, but the local press had a lot of fun that day.

"So," Murphy tried to bring sobriety back to the meeting, "just as we said we would, when a process changes, or we see it needs to be, we do process-based SRM. The trouble with this approach is that once again we are reactive. Most of the time we react to relatively small things. Sometimes, big things happen."

With everyone around the table knowing that Murphy was referring to 242, a quiet intensity returned to the discussion. Murphy concluded, "In my opinion, we have to get very serious about getting back to fundamentals in SMS, and that means a thorough, documented review of our processes."

"Exactly so," Georgia quietly stated, with heads nodding all around the table.

"That ain't easy, or cheap," remarked Jacks. The VP of Maintenance was well known for his almost curmudgeonly nature. "But – I agree with you."

Tiller leaned forward, hands resting on the table. "This leads right in to the main topic for the day. We've talked about this before. We've now got to do it. These last six weeks have been an object lesson for us all. We have all kinds of risk in this operation. Bob," pointing to the company's Chief Counsel, "your hands are full dealing with our legal risks following 242. We have financial risks. We have marketing risks. Rick," now pointing to the VP of Security, "Murphy's critique of the lack of process-based thinking could probably be leveled at your security risk management too. And Georgia, you now have environmental issues under your purview. Do we have KPIs identified as a result of a process-based environmental risk assessment?"

"Understand, I'm not pointing the finger at any of you – well I guess I was, wasn't I? But I don't mean to say we've been failing at our job. But we haven't, or best said, I haven't, really embraced the truth that we need transformational change."

"So let's try this on for the rest of the day: In two years, we will have a fully functional, process-based integrated risk management system in place at Quest Airlines. Help me think this through. How do we get there?"

By 5 p.m. they were all in agreement that they were nowhere near done figuring out their new strategy. But they all were in agreement that they were heading down the right path.

Introduction

Alan Stolzer, Carl Halford and John Goglia

> There is nothing more difficult to take in hand, more perilous to conduct or more uncertain in its success than to take the lead in the introduction of a new order of things.
>
> Niccolo Machiavelli

Review of Principles of SMS from Safety Management Systems in Aviation

In *Safety Management Systems in Aviation*, we attempt to add depth and context to the excellent Safety Management System (SMS) materials available from the Federal Aviation Administration, the International Civil Aviation Organization, and other sources. We discuss the fundamentals of SMS, including the importance of securing top management support for the SMS initiative, developing a safety culture, performing effective risk assessments, and the importance of moving toward more scientific discovery and analysis methods for those with the capabilities to do so – case studies in probabilistic risk assessment and data mining are presented. Throughout the book we encourage the reader to think about SMS from systems and process-based perspectives, including with safety risk management and safety assurance. Since the development and implementation is at its core, a project, we also offer a very brief overview of project management. We emphasize that SMS must be aligned with and grounded in quality management and we presented – importantly, we believe – the fundamentals of quality management for readers to consider. We discuss issues regarding managing the SMS, and briefly discuss SMS implementation.

It is this latter topic that is the focus of this book. We will examine issues associated with implementing an SMS program in some detail, but before doing so, we'll offer a brief, selective review of some of the important points made in *Safety Management Systems in Aviation*.

Top Management Support

There is arguably nothing more important to the successful implementation of SMS than gaining the support of senior management. Without it, even the most well thought-out program is doomed to fail. The CEO simply must realize that the responsibility for safety of the system cannot be delegated to anyone, no matter how talented or capable that person may be. If the CEO doesn't have that mindset, the SMS champion's work is going nowhere.

CEOs must be personally and actively involved in the SMS. Obviously that does not mean that the CEO must sit through a two-hour risk assessment meeting regarding a potential slip-and-fall hazard in the hangar, but those in the meeting must know that the CEO cares about the outcome of that meeting. What it does mean is that the CEO must be invested in the SMS's success by promoting and supporting the safety policy and, importantly, committing the resources of the organization to it. In short, it must be a top priority in word and deed. Anything less will inevitably result in employees deserting the SMS effort. In the following paragraphs, we will review for emphasis the importance of creating a culture of safety in an organization. This is so important because embracing safety as a core value, and as a business function, is a cultural change for many organizations. We are just beginning to grasp the influence top management has in creating or destroying the cultural climate in an organization. Paradoxically, the CEO with the greatest power to influence the cultural of the organization in a positive way is the one who uses that power to create a more participative and less authoritarian workplace; one who values teamwork and breaks down silos; and one who ties their own fate to the organization's bottom line in the same way that the workers are required to do. That CEO is a must for creating a positive safety culture.

Chapter 3 is dedicated to the subject of obtaining top management support for the SMS effort.

Safety Culture

Safety culture can be described as the values, beliefs and norms that govern how people act and behave with respect to safety. All of us who belong to an organization are affected by the values, beliefs and norms of that organization, whether or not we are conscious of it. Thus, if we want to behave and act differently, that is, be safer, we need to change the culture of the organization.

A positive safety culture is regarded as the fusion of the following cultures within an organization: informed, flexible, reporting, learning and just. Each of these cultures must be cultivated and developed in order to achieve a positive safety culture.

Talking about developing a positive safety culture and doing it are two very different things; it is no easy task. The process should begin with an objective assessment of the existing safety culture in an organization. There are many tools that can be used to assess culture, including self-assessment tools.

Of course, establishing a positive safety culture begins with a sound safety policy, and the enthusiastic promotion of that policy by, again, top management. The statement should unambiguously lay out the organization's expectations regarding safety along with who is responsible for achieving them. The policy statement is not something to be hidden; rather it should be widely disseminated throughout the organization and referred to on a frequent basis.

As stated previously, the policy should state who is responsible for achieving the safety goals. Particularly for a large organization this can be cumbersome, but it is vital. In an SMS, the responsibility for safety resides in people throughout the organization; it is *not* vested in just a few. That being so, there must be alignment between the job descriptions of these workers and the responsibilities they have; it is completely ineffectual to hold people responsible for safety unless that responsibility is explicit in the workers' job description. Any effort to implement SMS must include an audit of job descriptions throughout the organization to assess safety responsibilities and, of course, those with safety responsibilities must be held accountable for results. This alignment of job requirements, safety responsibilities and accountability promotes a strong safety culture because the culture is linked with the organization's operations.

Safety culture is so important to an SMS that we include two chapters (4 and 5) on the topic.

Performing Risk Assessments

Risk assessment is the valuation of the potential loss associated with a risk, as compared with the cost of effectively controlling that risk.

SMS is all about managing risk. To manage risk, we need to understand where our risks exist, and then assess them. Not every organization assesses risk the same way, nor should they. The risk assessment program should make sense for the organization, and should include processes that are well thought-out, rational, and appropriate for the scope of the organization.

It would be surprising to see a Mom and Pop fixed base operator (FBO) using some of the more advanced risk assessment tools such as probabilistic risk assessment (PRA). However, that small FBO would probably benefit considerably from working through its risk exposure using a simple risk matrix. Conversely, a major air carrier or a large airport, both of which have a wide array of exposures, should be using tools beyond simple risk matrices.

There are a variety of tools available to safety personnel and others for risk assessment, including PRA, data mining, simulation, modeling, root cause analysis methods and many others. Several of these tools are discussed in *Safety Management Systems in Aviation*.

Systems Perspective

Our natural tendency today is to break things down into pieces that are simpler to understand, that is, reductionism, and then apply methods to solve the resultant, definable problems. The problem with that thinking is that it fails to take into account the complexities of those various pieces and, importantly, the interaction of those pieces with one another. In our earlier book, we traced this concept to the 1940s to Karl Ludwig von Bertalanffy who, based on his work as a scientist, concluded that the whole is more than the sum of its parts. His discovery was that a system possesses critical properties that none of its individual parts possess, properties that arise from the interaction of those individual parts. The system, then, is the amalgam of the parts and those interactions. When we analyze parts of a system, we lose those critical properties and, thus, our conclusions, whatever they may be, are often erroneous as a result. Systems thinking encourages a special focus on the interaction of the parts in addition to the parts themselves; this is particularly important when some of the parts are social rather than technical in nature.

A thorough study of systems thinking (versus analytical thinking) is recommended for anyone trying to build a system within an organization. It is very different thinking than most of us are accustomed to, and it will produce a different SMS than would be built in its absence. We assert that an SMS built without a systems foundation will not survive in the long term.

Since it is so foundational to an SMS, we lead off the book with a discussion of systems and task analysis in Chapter 1.

Process-based Thinking

The old way of thinking in quality, and in business generally, was to assess the system based on the output of that system. However, for just one illustration of the ineffectiveness of that approach, even 100 percent inspection failed to produce the quality of switches needed by AT&T in the 1920s to support the rapidly expanding transcontinental telephone system. In contrast to a focus on the output, or product of the system, process-based thinking focuses on monitoring performance (quality assurance activities) so that we know that the system is capable of reliably producing an acceptable level of output. If the capability analysis reveals deficiencies in the system, the process can be adjusted.

As we noted in our earlier book, that it is only through the exercise of *producing process descriptions* that the true extent, detail, dependencies, contingencies and outputs of an activity can be understood. The noted quality guru W. Edwards Deming once remarked that 'if you can't describe what you're doing as a process, you don't know what you're doing'.

Good practice dictates that process descriptions contain all of the following: purpose; scope; interfaces; stakeholders; responsibility; authority; procedure/ description/flowchart; materials/equipment/supplies; hazards; controls; process measures; targets; assessment; change management; qualifications required for participants in process; documentation/record-keeping.

If process descriptions don't exist in an organization, it would be an excellent idea to make this a top priority of SMS development.

Project Management

As with the other topics we're discussing in this brief review of SMS, a project management approach is essential for successful SMS development and deployment. A project management approach will cause participants to think in terms of specific objectives, a defined lifespan, multiple units participating and the triple constraints of time, cost and performance.

The triple constraints have been likened to a three-legged stool; all three must be managed to remain in balance according to the objectives of the project. The project manager must have a clear understanding of the organizational resources he or she can manage, what the limitations are, what adjustments can be made, and which constraints have priority and when.

SMS implementation projects pass through four often distinct phases:

1. defining
2. planning
3. executing and
4. delivering.

In the important defining phase, goals, specifications and objectives are established, and major responsibilities determined. In the planning phase, schedules, budgets and resources are set, constantly keeping in mind the objectives of the project. Risks to the project are determined and assessed. In the executing phase, the actual work of the project is accomplished, including status reporting, change management, quality and forecasting. In the delivering phase, the product or service and associated documentation is handed over to the customer, training is provided, project personnel are released and lessons learned are recorded to contribute to organizational learning.

One of the very important aspects of project management is project scope, which is the end result or what is delivered to the customer. Project scope must be clearly documented into an unambiguous scope statement; statements include deliverables, timelines, milestones, technical requirements and exclusions. Given the tendency of project scopes to *creep* over time, it is very important to spend adequate time on this and to get it right.

Project scope statements are important in that they state the objective of the project, but they do not provide a usable map of the project; that is the purpose of the work breakdown structure (WBS). WBSs breaks down the overall objective into greater detail so that individual tasks can be accomplished. Work packages are the lowest level of activity planning in the project.

A team that one of the authors once worked with indicated that they were following a project management approach in completing a project. When asked for the documentation to review, the team provided a printout of a Gantt chart from Microsoft™ Project as the sole artifact, and indicated that the chart constituted their planning. Project management is more than creating a Microsoft™ Project file!

We believe it is well worth the organization's investment to train its personnel in the project management approach.

Quality Management

Stated simply: SMS needs a quality management system because SMS requires documented, repeatable processes. The quality management system provides the strong foundation upon which an SMS is built. Aviation Safety's QMS will be the springboard to our SMS (Sabatini, 2005).

Nick Sabatini, former FAA Associate Administrator for Aviation Safety, recognizes the importance of a quality management underpinning for an SMS. We suspect that recognition lies in the following points.

Quality focuses on *continuous improvement* of products, services, or processes. Continuous improvement is beyond simply taking corrective action. It is about using information and data to reduce the variation of product or process parameters around target value. Quality recognizes that there is a cost to that variability and constantly seeks to reduce it through the application of tools and methods on a continuous basis. Those tools and methods range from the simple plan-do-check-act to more sophisticated methods such as six sigma, lean, and total quality management.

Second, quality brings a focus on *satisfying customers*. If we're in business we have customers, and those customers must be satisfied. But what if we're a service function within a large organization, for example, a safety department within an airline, a manufacturing company, or a governmental organization? Who is our customer? Of course, there are many types of customers, including primary, secondary, indirect, external, consumer, intermediary and others; a quality organization will know which stakeholders fit into which customer categories. But then we must go further; we must determine what our customer wants, and not just assume we know. Voice of the Customer (VOC) provides tools to determine the customer's requirements and expectations regarding products and services.

Quality practitioners understand and employ the concept of Cost of Quality (COQ). Paradoxically, COQ does not refer to the cost of producing a quality product or service; rather it is the cost of *not* producing such a product or service. By focusing on COQ, organizations are able to concentrate their efforts on the issues that prevent them from achieving an acceptable level of quality. Quality costs normally fall into one of the following categories: prevention (all activities designed to prevent poor quality, such as planning, training, education, evaluations, quality improvement projects, etc.), appraisal (measuring, evaluating, or auditing products or services to assure conformance to a standard or requirement), internal failure costs – occur prior to delivery to the customer, and external – occur after delivery to the customer.

Quality focuses on other vital areas such as documentation, problem solving, process thinking (discussed previously), variation and quality assurance. Having guided the Federal Aviation Administration's (FAA) Aviation Safety organization to ISO 9001:2000 certification in 2006, and operating under a Quality Management System (QMS), FAA officials surely recognized the value of basing the development of SMS on a solid QMS.

As we did in *Safety Management Systems in Aviation*, we strongly encourage the student of SMS to delve into the study of quality. SMS development and implementation will be far better accomplished, less frustrating and more sustainable when it is grounded in quality management.

A practical, in-depth discussion of quality as it relates to SMS is offered in Chapter 6.

Conclusion

Much more could be discussed in even a cursory review of SMS, but we trust that the reader who does not possess a firm grasp of the theoretical constructs of the subject will delve into some of the resources currently available, including *Safety Management Systems in Aviation*. Indeed, volumes could be written about some of these topics, including culture, risk management, project management, quality management and others, and many have been.

In our first book we stressed the importance of understanding quality management as the underpinning of SMS. In this book on *implementation*, the authors would like to encourage the reader who is or will be involved in SMS implementation to become very comfortable with the tenets of project management. Having studied this subject for some time, and having discussed SMS implementation with those who have done it and now appreciate its complexity, we conclude that employing a sound project management approach is essential to a successful implementation. Neither the brief discussion above nor the slightly more detailed discussion in the first book will provide adequate preparation for those who do not already possess the requisite skills.

References

Sabatini, N. (2005). 'SMS: I'm a Believer.' Safety Management Systems Symposium, McLean, VA.

Stolzer, A., Halford, C. and Goglia, J. (2008). *Safety Management Systems in Aviation*. Aldershot: Ashgate Publishing.

To Julie – my soul mate. AJS

To Deborah, who still for some reason waits patiently as I finish one last thing. CDH

To the men and women whose conscientious work makes flying as safe as it is today. JJG

Chapter 1
System and Task Analysis

Don Arendt and Anthony Adamski

Much has been said in safety literature about the effects of human error. Causes of accidents involving human error of some type range from an estimated 60 to 90 percent. However, *root causes* of these errors can often be traced to fundamental problems in the design of systems, processes and tasks related to the human activities in which these errors occurred. Sound safety planning, including hazard identification, risk management and safety assurance, must be based on a thorough understanding of the processes and activities of people in the system, and the other components of the systems and environments in which they work. Consequently, the world of aviation safety has accepted the practice of system safety and looked to the concept of safety management systems.

The foundational building blocks of a safety management system (SMS) are referred to as the four components, or the *four pillars*, of safety management. These pillars are *policy, safety risk management, safety assurance*, and *safety promotion* (FAA, 2006a; ICAO, 2009). The processes of safety risk management (SRM) and safety assurance are in turn built upon system and task analysis. Systems analysis, within the context of SMS, is a process for describing and evaluating organizational systems, and evaluating risk and compliance with safety regulations at the system level. Task analysis is a methodology that supports system analysis by documenting the activities and workplace conditions of a specific task to include people, hardware, software and environment. There are a number of objectives of system and task analysis within SMS, including:

- Initial design of processes (describing workflows, attributes, etc.)
- Task and procedure development (including documentation, job aids, etc.)
- Hazard identification (what if things go wrong?)
- Training development (assessment, design, development, implementation and evaluation)
- Shaping *safety assurance* processes (what needs to be monitored, measured and evaluated?)
- Performance assessment (how are we doing?)

The system and task analysis should completely explain the interactions among the hardware, software, people and environment that make up the system in sufficient detail to identify hazards and perform risk analyses. Systems and task analysis (including design) are the primary means of proactively identifying and addressing potential problems before the system or process goes into operation. Such problems consist of hazards that are conditions and not outcomes. We cannot directly manage errors that occur, but we can manage the conditions that caused them. Oftentimes, hazard identification is viewed as a process that starts more or less spontaneously. For example, employees report conditions that they encounter, audits provide findings of new hazards observed, investigations identify failed processes and uncontrolled environmental conditions, etc. While each of these is, of course, a legitimate source of hazard information, many hazards are *built in* to the system. Many more of the hazards are the result of system, process and procedural designs, or the cumulative effects of individually small, but collectively significant changes that have occurred incrementally in the systems or their operational environments. Thus, it pays to look carefully at the factors that can affect human and equipment performance and take proactive steps to avoid problems that can be identified early.

System and task analyses should completely explain the interactions among the hardware, software, people and environment that make up the system in sufficient detail to identify hazards and perform risk analyses (FAA, 2006a). Analysis, as opposed to description, is a process aimed at identifying system problems, performance difficulties and sources of error (Annett, 2003). Additionally, and perhaps most importantly, system and task analysis provides the opportunity to gain a fundamental understanding of the organization's systems, processes and operational environment in which it operates. This understanding allows for *meaningful compliance* as opposed to *perfunctory compliance*. Meaningful compliance is achieved when the organization applies regulations in the context of their operations in a way that accomplishes the safety intent of the regulations. Perfunctory compliance is basically a *check the block* mentality.

It is important to emphasize at this point that the decisions of management which occur during the design phase regarding compliance and risk management both reflect and shape the organizational culture and organizational behavior. The authors' past experiences have shown that when an organization implements and aggressively uses SMS practices such as employee reporting systems, risk management procedures and safety assurance processes early in the organizational lifecycle, we can expect that there will be success in fostering pro-safety attitudes and behaviors in its employees. In the operational arena, we have seen many safety programs comprised of essentially reporting systems that fail to take advantage of system and task analysis. This leads to treating problems in a form of isolation, as each report, finding, occurrence, etc., gets treated to some type of closure, independent of other related systems or operating conditions (see also Sparrow, 2000, 2008). It is safe to say that a truly predictive process, or a robust management-of-change process, cannot reach fruition without comprehensive system and task analyses.

The International Civil Aviation Organization (ICAO) also emphasizes the importance of SRM and safety assurance. ICAO makes clear that the two core operational processes of an SMS are safety risk management and safety assurance (ICAO, 2009). ICAO goes on to say that SRM is a generic term that encompasses the assessment and mitigation of the safety risks of the consequences of hazards that threaten the capabilities of an organization, to a level as low as reasonably practicable.

The Federal Aviation Administration (FAA) states that SRM is a formal process within SMS that describes an organization's systems, identifies hazards, analyzes risks and assesses and controls those risks. The FAA also cautions that the SRM process should be embedded in the organization's operational processes that are used to provide product and services, and not a separate or distinct process (FAA, 2009a). As such, SRM becomes an activity that not only supports the management of safety, but also contributes to other related organizational processes. In essence, the results of the SRM process, beginning with a sound system task analysis, become part of the system rather than an *add on* or an after-the-fact corrective process. Risk management isn't something you do – it's the way you do it that makes the process function properly and safely. It is important to remember that SRM should be part of the entire management domain of the organization. It cannot be just relegated to a safety officer or safety department. It has to be part of the way that the management, and especially senior management, runs the entire business enterprise.

Safety assurance is a formal management process within an SMS that systematically provides confidence that an organization's products or services meet or exceed safety requirements (FAA, 2009b). ICAO (2009) reminds us that safety assurance must be considered as a continuous, ongoing activity. It should be aimed at ensuring that the initial identification of hazards and assessments of the consequences of safety risks, and the defenses that are used as a means of control, remain valid and applicable over time.

As stated earlier, SRM and safety assurance rely on system and task analysis. We have observed that these processes are vastly misunderstood by a great number of operators. We have seen many operators shy away from these processes because they were unfamiliar with them and they, frankly, were scared to delve into what they believed to be a very complex and purely academic area, or an endeavor that requires additional specialist personnel, or processes that are beyond their means. In reality, system/task analysis does not have to be that complicated. While there are some sophisticated software tools and complex methods available, simple lists and descriptions can often suffice. In fact, simple flow charts that we will explore later can pave the way.

One of the primary purposes of system and task analysis is to reach a complete and predictive state within an organization's SMS. This is the type of risk management in which the organization considers activities that are in the future. The organization reviews changes in systems and processes, revisits its mission, evaluates business practices and assesses the operational environment. It then becomes the basis for proactive safety management, where the organization considers its current practices and foresees contingencies – the 'what if...' questions.

System and task analysis are at the forefront of an effective risk management system (see Figure 1.1). Alternately, system and task analysis are termed as 'system analysis' in the systems engineering world, 'systems description' in much of the industrial system safety literature, and 'establishing the context' in the AS/NZS 4360 standard (Standards Australia, 2004). ISO 31000 risk management standards are concepts by which organizations can gain a fundamental understanding of their respective systems that are essential to hazard identification, risk analysis and assessment and risk control. The relationship between system/task analysis and SRM is depicted in Figure 1.1. It reflects that some degree of variability can be expected in *normal* human performance. In addition, workplace conditions can also affect this variability, as well as introducing *triggers* that activate human error (see Dismukes, Berman and Loukopoulos, 2007; Hollnagel, 2004).

As this chapter progresses, we will explore recommended methodologies to conduct both systems analysis and task analysis. Before we delve too deeply into actual analysis methods, it is important to provide a brief review of systems, and of the process approach to system analysis as it relates to SMS.

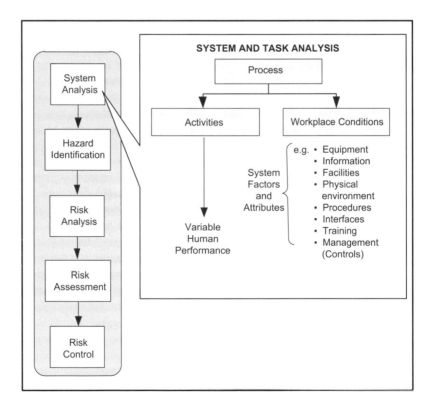

Figure 1.1   Safety risk management and task analysis

Systems Review

Roland and Moriarty (1990) provide an excellent definition of *system* in their text on safety engineering. They define a system as 'a composite of people, procedures and equipment that are integrated to perform a specific operational task of function within a specific environment' (p. 6). Notice particularly the terms *integrated, function* and *environment*. Every system is designed to perform a specific mission, or produce a product or service; consequently, it is the responsibility of management (the owners of the system) to define the goals for the mission, product, or service. These goals represent their expectations for the outcomes of their investment. In order to meet these expectations, an understanding of the system's environment and the processes that make up the system becomes critical. A common error is to only define a system in terms of physical assets (for example, aircraft, computers, facilities), and to refer to organizational components as *organizational factors*. In the context of a safety management system, the organization *is* the system and these assets are resources used to accomplish the organization's mission.

Systems include the processes which consist of activities people do within the systems. Systems also include the tools, facilities, procedures and equipment that are used to perform the processes. In addition, systems have a defined mission, purpose, or objective. This is where the concepts of systems and processes converge. A system has a mission or objective – to produce a product or service (outcome). A process has outputs which are the results of the efforts put forth by the people in the system.

Processes are the essential operations that a system must perform to achieve its outcomes. ISO 9000, the international quality management system standard, defines a *process* as an 'interrelated or interacting set of activities which transforms inputs into outputs'. In every case, inputs are turned into outputs because some kind of activity is completed. Note the active nature of this definition. It does not define the product or service, but rather the things that people in the system do to produce them. A process is *people doing things* rather than *things people have done*. The products or services are the outcomes, what the system owners want out of it, while the processes are the activities that get them there (ISO, 2000).

There is a subtle but important difference between outputs and outcomes. Process outputs should be defined such that they can be measured. Practitioners need to avoid defining outputs without considering process measurements. Additionally, outputs should be defined in a manner to support the desired outcomes. Activities are often defined in terms of outputs – flights made, maintenance procedures accomplished, audits conducted, etc., but the goal of meeting the system's desired outcome has been neglected. It's easy, though, to define things that are easily *countable* and neglect or even ignore outcomes. For example, a common measure in practice is timeliness. While this is a desirable goal in most activities and is usually easy to measure, it does not matter how timely we produce an inferior result. In safety, it does not matter how many audits we conduct, how many reports we process, etc., if we aren't achieving the outcome of enhanced safety.

In some cases, outcomes are difficult to measure directly. This is often (if not always) the case in safety where the desired outcome (absence of accidents) can be intangible and its inverse events (accidents, incidents, etc.) are rare. Process measures are often employed in these cases. Process measures place emphasis on performance shaping and process supporting factors such as equipment, procedures, personnel selection, training, qualification and supervision and facilities. The theory being that, if these process factors are well managed, confidence in the probability of successful outcomes is enhanced (readers are referred to ISO 9001-2000 sub-clause 7.5.2 for an example standard for process verification and validation).

The Process Approach

The process approach is basically a systems approach in that the process owner is identified, procedures are evaluated, outputs to the next process are verified, controls to insure desired output are confirmed and finally, performance measures are reviewed to ensure consistent results. This approach accepts that different types of processes take on different characteristics and have different expectations. For the purpose of our analyses, there are three basic types of processes—operational, management and resource allocation processes.

- *Operational processes* do the real work of the system. Inputs are transformed into outputs in the form of products or services. These may be direct customer services such as providing transportation, manufacturing or repairing aircraft or aircraft components, directing air traffic, etc. They may also be internal processes such as providing employee training.
- *Management processes* serve as controls on operational processes. This may take the long-term form of such activities as establishing policies or authoring and disseminating policy and procedural documents. It may also take the form of more real-time activities such as direct supervision and management of employees' activities.
- *Resource allocation processes* concern procurement and allocation of personnel, facilities, equipment, tools and other resources that are used in operational processes.

Process attributes are important factors to consider in system analysis. These attributes are the same attributes as described in the FAA's Air Transportation Oversight System (ATOS), however, we use a slightly different definition of the attributes as defined in the 2009 ICAO System Safety Manual 9859, section 9.6.3:

- *Responsibility* addresses who is accountable for management of the operational activities (planning, organizing, directing, controlling) and their ultimate accomplishment. Management and individual employee accountability, and therefore responsibility, are fundamental to the management of safety.

- *Authority* addresses who can direct, control or change the procedures, and who cannot, as well as who can make key decisions such as safety risk decisions.
- *Procedures* specify ways to carry out operational activities that translate the *what* (objectives) into *how* (practical activities). Procedures address the who, what, where, when and how of the activities that make up a process.
- *Controls* address the elements of the system including, hardware, software, special procedures or procedural steps and supervisory practices designed to keep operational activities on track. Controls include procedures, supervision and assurance processes.
- *Interface* provides an examination of such things as lines of authority between departments, lines of communication between employees, consistency of procedures and clear delineation of responsibility between organizations, work units and employees.
- *Process measures* provide the means for feedback to responsible parties that verify required actions are taking place, required outputs are being produced and expected outcomes are being achieved.

We have provided a diagram that reflects how these attributes interact with the input, activity, and output system process (see Figure 1.2) and have borrowed from the IDEFØ methodology, which we will discuss later, to construct the diagram. The diagram depicts the performance objectives and expectations of a process and reflects where the process attributes fit into a well-designed process. The ICAO System Safety Manual maintains that these areas (attributes) are necessary to assure safety through safety performance monitoring and measurement. Both SRM and safety assurance depend on reliable system and task analysis. We can use many of the analysis tools for both system analysis and task analysis. We also can use many of the same tools whether we are designing new processes or modifying or revising an existing process. Whether designing new or revising old, the first step is the same – answer the question: 'Who is in charge?'

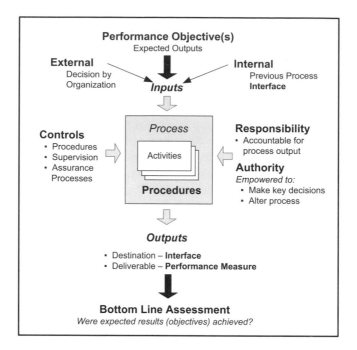

Figure 1.2   Workflow process diagram

Conducting the Analysis: The Organization Chart

Using the process attributes as a guideline, we begin a system/task analysis with a hard look at how the organization is laid out in terms of the major operational processes (for example, flight operations, ground operations, maintenance operations). This is done by comparing the organization's current organization chart with actual operations, or developing a chart if none exist. If you are an organization new to SMS, then this step is absolutely critical. If your organization is in operation with established processes, this step is crucial to confirming exactly how your organization is wired beyond a perfunctory level. We have found that many organizations really don't know how their organization is laid out, who does what, who is truly responsible for what, and how the lines of communication work. Things work because dedicated, diligent people work carefully but, in this environment, critical safety breakdowns can occur unexpectedly, often due to poor communication, uncertain lines of responsibility, or ambiguous lines of control.

This initial step involves designing or reviewing your organizational chart. You know that strange-looking graphic that hangs on a company wall and few people actually look at? The question to answer is, 'Does the organization chart reflect what is actually going on in regards to lines of responsibility and authority?' These attributes tell us not only who is accountable, but who has authority to make key decisions, such as risk acceptance or process revision approval. The chart can also tell us something about lines of communication and chain of command. This, in turn, will introduce us to the *interface* attribute. Here we can check if handoffs of responsibility and authority between interfacing processes are taking place effectively. (Note: This is especially critical where processes are contracted out.) System analysis of processes and associated attributes starts with the organizational chart.

We will use the FAA's definition of *responsibility* and *authority* attributes, as the FAA distinguishes between *responsibility* and *authority* (FAA, 2009a), while other definitions do not. The FAA defines the *responsibility attribute* as 'a clearly identifiable, qualified, and knowledgeable person who is accountable for the quality of a process'. They define the *authority* attribute as 'a clearly identifiable, qualified and knowledgeable person with the authority to make key risk-acceptance decisions and establish or modify a process'. The *responsibility* and *authority* attributes tell us not only who is accountable but who has authority to make key decisions, such as risk acceptance. It also tells us something about lines of communication and chain of command, which leads us into the *interface* attribute. There is really a strong relationship between responsibility, authority and interfaces. One of the key aspects of an interface is the handoffs of responsibility and authority between interfacing processes. Without a clear definition of who holds responsibility and authority, interface problems will definitely emerge.

The attributes of responsibility and authority target what we need to identify during our first step in system analysis. We can validate an existing organizational chart or design a new chart by laying out the organization's processes to diagram lines of communication and authority. This can be accomplished with a simple block diagram, a basic IDEFØ diagram, or a more detailed cross-functional process map.

Laying It Out: Process Mapping

A process map is 'a step-by-step description of the actions or activities taken by workers as they use a specific set of inputs to produce a defined set of outputs' (Marrelli, 2005, p. 40). A process map provides an easy to understand visual depiction of a process that relies on simple flow charting techniques. For our purposes, the map reflects inputs, the performers, the sequence of activities or actions taken by the performers, and the outputs in a flow chart format. A flow chart graphically represents the activities that make up a process much in the same way as an aeronautical chart represents some specific geographic area. It paints a picture that makes it easier to understand the activities and tasks that formulate a process. Flow chart diagramming is one of the clearest ways to organize multifactor information and to detect disconnects in management control, input–output relationships between processes and poor process definitions. However you go about mapping your processes, we recommend keeping it as simple as possible. The following will help you in this process.

Flow charts

Before we begin our discussion of process mapping, let us review the symbols that are commonly used in the construction of a flow chart. The only hard and fast rule is to be clear and consistent in your use of symbols. It is critical that your flow chart translate readily into a clear depiction of the target process. The objective is to have anyone in the organization view a process map created in one department, and have it be easily understood by those in other departments. This becomes the true test for clarity and consistency.

The International Organization for Standards (ISO) has established a standard set of flow chart symbols, and has set out four basic symbols. If you can keep your flow charts (process maps) down to this number, they will be much easier to understand. Aviation processes, at times, require more symbols to adequately map activities and flow. This is perfectly fine, so long as you remain clear and consistent across all of the flow charts (process maps) in your organization. There are ten symbols that are often used in an analysis (see Figure 1.4), however, many simpler analyses can get by with two or three quite nicely. There are four basic symbols that will fulfill most process mapping requirements for system safety analysis purposes. Be aware that the more symbols you use, the harder it will be for everyone in your organization to readily understand your process map. However, beware of oversimplification where important details can be overlooked or neglected. In the words of the famous American journalist H.L. Mencken, 'For every complex problem there is an answer that is clear, simple, and wrong.' Or in the words of a less famous author (D. Arendt): 'Keep it simple, stupid, but don't make it simply stupid.'

We recommend that you establish a standard set of symbols for use within your organization which is published in some manner so that all those involved in process mapping can remain standardized. Standardization can become extremely important when you are addressing interface issues between processes.

The elongated oval or rounded rectangle signifies the starting and ending points of a process (see Figure 1.3). Typically, the word *start/stop* or *begin/finish* is inserted in the oval symbol. The activity rectangle represents a function or an active part of the process. It is the most commonly used symbol in both flow charts and process maps. Use this symbol whenever a change in the process occurs and a new activity takes place. Normally, a brief description of the activity is included in the rectangle, and since the rectangle represents an activity, the descriptor uses verbs or verb phrases.

Symbols are normally connected by a line and arrow. The arrow depicts the direction and order of the process flow. It reflects movement from one symbol to another and signifies interactions among activity rectangles.

The diamond symbol represents a point in the process flow that a decision must be made. There two directions flowing from the decision point and each direction leads to a different series of steps. The steps will vary depending on the type of decision. This symbol is used only where one possible outcome can be chosen. Typical outcomes involve yes/no, approved/rejected, or accept/revise decisions. While activities will be described in terms of what people do (task descriptions, procedures, etc.) in documentation accompanying the flow chart, decisions should be described in terms of responsibility, authority and decision criteria (who can makes the decision and how).

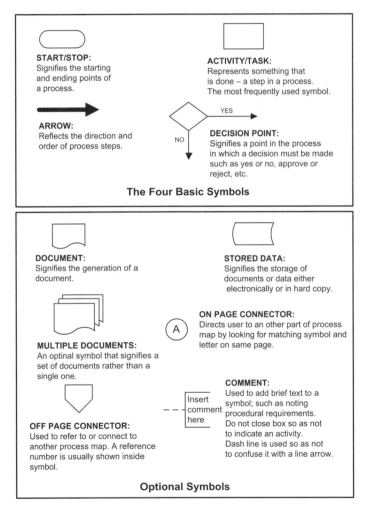

START/STOP:
Signifies the starting and ending points of a process.

ACTIVITY/TASK:
Represents something that is done – a step in a process. The most frequently used symbol.

ARROW:
Reflects the direction and order of process steps.

DECISION POINT:
Signifies a point in the process in which a decision must be made such as yes or no, approve or reject, etc.

The Four Basic Symbols

DOCUMENT:
Signifies the generation of a document.

STORED DATA:
Signifies the storage of documents or data either electronically or in hard copy.

ON PAGE CONNECTOR:
Directs user to an other part of process map by looking for matching symbol and letter on same page.

MULTIPLE DOCUMENTS:
An optinal symbol that signifies a set of documents rather than a single one.

COMMENT:
Used to add brief text to a symbol; such as noting procedural requirements. Do not close box so as not to indicate an activity. Dash line is used so as not to confuse it with a line arrow.

Insert comment here

OFF PAGE CONNECTOR:
Used to refer to or connect to another process map. A reference number is usually shown inside symbol.

Optional Symbols

Figure 1.3   Flow chart symbols

The wiggle-bottomed rectangle represents the generation of a document or the alteration of a document. It is the step in a process in which some type of document (paper or electronic) is to be created or modified such as a hazard report, an action plan, a maintenance record, or a training record. This symbol makes it easy to go through a flow chart and identify documentation and record-keeping requirements. The rounded-ends rectangle signifies the storage of documents or data either electronically or in hard copy (for example, the storage of crewmember training records). The multiple wiggle-bottomed rectangles represent a set of documents rather than a single one (for example, the set of documents that make up an air carrier's maintenance program).

The circle symbol is an on-page connector. It directs users of the diagram to another part of the process map by looking for a matching symbol and letter on the same page. The symbol that looks like a home-plate is an off-page connector. It is used to refer to, or connect to, another process map. A reference number is usually shown inside the symbol. Our last recommended symbol is the open-sided comment box. It is used to add some brief text to a symbol, such as noting procedural requirements. The box is not closed so as not to indicate an activity. A dash line is used so as not to indicate directional flow or confuse it with a line arrow.

Process Maps

Process mapping is an excellent tool for system and task analysis. Both flow charting and process mapping are common practices in quality management systems and other management practices. In SMS, process maps serve two basic purposes. The first is developmental in nature as it is used to diagram a new process in its design or conceptual stage. For this purpose, the map is used to identify as many potential defects and failure points prior to the process design being finalized. This mapping technique makes use of the development team's *requisite imagination*. Requisite imagination refers to the ability to imagine key aspects of the future we are planning (Westrum, 1991; Adamski and Westrum, 2003). Most importantly, it involves anticipating what might go wrong, and how to test for problems before the design is finalized. Requisite imagination often indicates the direction from which trouble is likely to arrive as it takes advantage of the development team's collective experiences, decision-making skills, design abilities, planning skills and more.

Our second type of process map is confirmative in nature. Its purpose is to analyze an existing process to determine if the process remains effective and continues to meet its objectives. It ensures that the design remains effective. The primary objectives of both types of process maps are the same; that is, to identify defects and potential failure points (hazards) within the process, and to ensure that the objectives are being achieved. Each type uses the same diagramming methods, and although they resemble each other, they do serve different purposes. The developmental process map and the confirmative process map may be constructed using various types of flowcharts, but we recommend three basic types: the block flow diagram, the block flow procedural diagram, and the cross-functional process diagram. Additionally, we will discuss the use of our adaptation of an IDEF∅ model, which has also been used by Erik Hollnagel (2004), who has also applied this technique to safety. This model is typically used for high-level process design.

The Block Flow Diagram

The block flow process map is a basic flow chart which serves as the basic process map. It provides a quick and uncomplicated overview of a process. It is the simplest and most common type of flow chart, and is used to simplify large, complex processes or to document process-related tasks. The block flow diagram is our starting point in the development of a more complex process map, whether its purpose is developmental or confirmative. Figure 1.4 provides an example of a block flow process map, which depicts an air carrier's manual revision process. It is a high-level map, meaning that it reflects the overall process and major activities. The following discussion provides a brief overview of the process map that is depicted. It is important to note at this time that this example and the following examples are just that – examples, and should not be construed as an FAA-approved process.

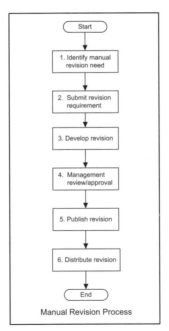

Figure 1.4   Simple block flow diagram

Our block flow process map begins with a start symbol that establishes the beginning of the manual revision process. The start/stop symbols establish the boundaries for the process. This may not seem to be an important step, but it is. Without establishing the boundaries of the process, where a process begins (inputs) and where it ends (outputs, which in turn may serve as an input and interface into another process) can easily become a confusing, gray area. The FAA emphasizes that describing a system involves the act of 'bounding the system' (FAA, 2008).

In our example, we have numbered each of the activity rectangles, but this is not necessary. We have done so for ease of explanation and discussion purposes. In fact, numbering the activities could lead to confusion when revising steps. Note that some of the activities could be further decomposed into individual task flow charts to further describe the process. Of course the question arises: 'How detailed do we need to be?' This depends on your needs. For example: Activity 3 (Develop Revision) most likely involves a number of activities that make up that specific task (e.g., develop draft, write revision and edit revision). The level of decomposition will largely depend on the purpose of your process map. A developmental map will most likely not need the level of detail as a confirmative map. We suggest that when your level of detail meets your needs, stop there. If you are still lacking information and the process map does not provide you with what you need, then go back and decompose the various activities again to more fully reflect the process. Additional detail can also be addressed in a narrative form to accompany the block flow diagram and provide procedural guidance, for example, manuals, standard operating procedures (SOPs), user guides, job aids. The process ends with a start/stop symbol to indicate the end or finish of the process, and establishes the ending boundary.

When in a design phase, keep in mind the process attributes, which were defined earlier, and where they fit into your design. We recommend that you capture your work, both in initial design and in subsequent work, in the form of some simple, spreadsheet-like forms. Such documentation will aid you in determining and considering workplace conditions, possible hazards, procedural requirements, training needs, etc. Often, a hazard is a workplace condition with an adjective (that is, poor lighting)! By documenting your findings, elements will not slip through the cracks or be unintentionally ignored.

Block Flow Procedural Process Map

The block flow procedural process map is a block flow process map that also provides the procedural requirements for each activity (see Figure 1.5). The procedural requirements can be annotated alongside the activities by means of a comment box. The comment box, which is open, is connected by a dashed line to the associated activity. A dashed line is used to signify that the process does not flow along that path. Additionally, the comment box is open so as not to imply that it is an activity symbol. This type of process map is most suitable in process development. It can provide you with a flight plan for procedural requirements or other additional information. Note that we do not write procedures just yet; at this point we simply determine what procedures appear to be needed.

This type of map can also be used for confirmative evaluation purposes in that we can match existing procedures to process activities to confirm that the procedures follow a logical sequence that are found in the approved process design. Confirmative evaluation is a marriage of evaluation and continuous improvement (Dessinger and Mosely, 2003). Unlike other types of evaluation which are used during the design process, a confirmative evaluation follows a period of time (for example one year) after the process or program is in place. It tests the endurance of outcomes, the return on investment, and establishes the effectiveness, efficiency, impact and value of the original design over time. In the confirmative process, we are conducting a type of gap analysis to determine if there is a difference (gap) between 'what is' (procedures as practiced) and 'what should be' (procedures as written).

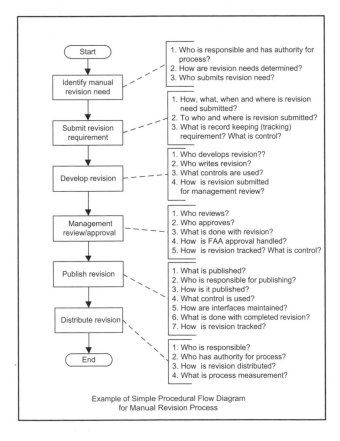

Example of Simple Procedural Flow Diagram
for Manual Revision Process

Figure 1.5   Procedural process map

Cross-functional Process Map

The cross-functional process map is perhaps one of the more useful tools for systems and task analysis. Not only does it depict the activities, decision points and directional flows, it also reflects who completes the various activities, and who is responsible for the various activities. Figure 1.6 presents the manual revision process example in a cross-functional format. The top row reflects the various managers, people, departments, regulatory agencies, etc., that are involved in our example process. In your map, columns may be added or deleted as necessary. Activities are then placed in the appropriate column to reflect sequence and to indicate who is responsible for an activity, or what entity completes a specific activity (see Figure 1.6).

Note that as we diagrammed the manual revision process in the cross-functional format, a number of additional activities and decision points were needed to make the process more meaningful and logical. This is typical: as we start assigning responsibilities for various activities, the need for more detail becomes apparent. Also, take note that as an activity arrow crosses from one column into another,

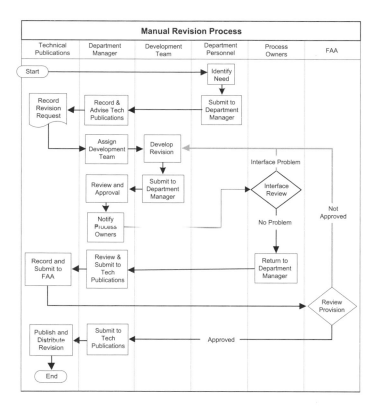

Figure 1.6   Cross-functional process map

it indicates a potential for a communication problem. We needed to add a document generation symbol to indicate that a document is required at this point in the process. The example also depicts feedback loops which emit from the two decision points. The feedback shows what actions are taken when a revision has an interface problem, or when it does not receive FAA approval.

When one constructs such a process map, documentation and procedural requirements quickly become apparent. We suggest that you construct a procedural diagram and a cross-functional process map when designing or revising any operational process. Such maps will identify procedural requirements and assist you in transforming these requirements into actual procedures.

Process Design: The IDEFØ Diagram

At times when the organization deems it necessary, you may wish to engage in a more sophisticated analysis, such as in the design and develop of a process that is complex and involves multiple interfaces. One good, and rather simple, way to begin this type of design process is the use of some type of model to serve as a focal point around which you can center the ensuing design-team discussion. The model we suggest is an adaptation of IDEFØ diagram, which was mentioned earlier (see Figure 1.7). This adaptation provides a number of things that the basic block flow process map does not. Using IDEFØ diagramming techniques, the design team can more easily discuss the process attributes and types of inputs that will or should occur, the controls that will or should affect the process, the mechanisms and resources that are necessary for the process to take place, what types of outputs should result and what types of output–input relationships (interfaces) may occur.

IDEFØ may be used to model a wide variety of processes and systems. For new processes or systems, it may be used to define process requirements and specify functions, and then to formulate a process design. For existing processes or systems, it can be used to analyze the functions the system performs and to record the mechanisms (means) by which these are done. IDEFØ is a formal method of describing processes or systems, using several techniques to avoid the overly complex diagrams which often result from using other methods.

Our adaption of the IDEFØ diagram consists of two fundamental symbols, the activity box and arrows. The activity box is used much the same way as in our other process maps. The box represents a function or an active part of the process; consequently, the boxes should be named with the use of verbs or verb phrases. Unlike our previous process map arrows, the IDEFØ convention calls for a slightly different application for arrows. Not only do the arrows represent direction of flow, they represent things and interconnections among activity boxes (McGowan and Marca, 1988). The letters 'IDEF' represent *ICOM DEF*initions. ICOM means input, control, output and mechanism – the initials for the four arrows which surround the basic box symbol of an IDEFØ diagram (see Figure

1.7). The zero indicates that the model is a basic function model, which is the most commonly used.

There are four possible relationships that things may have with an activity box; consequently, arrows may connect with any of the box's four sides that represent inputs, controls, outputs and mechanisms (or resources). The arrows signify the things that input or output from or to the appropriate sides of an activity box, and may also indicate interconnections among other activities (for example an output from one activity may serve as an input into another activity, such as output into control).

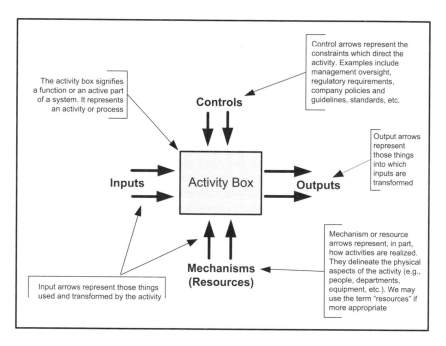

Figure 1.7   A basic IDEFØ chart

McGowan and Marca (1988) explain that input arrows represent those things which are used and transformed by the activity. For example, safety reports, audit findings, employee hazard reports, risk statements, FAA inquiries, manual revisions, or action plans are some of the things that may be represented by input arrows. Control arrows represent the things that influence, direct, or constrain how the process works. For example, management supervision, CFRs, company standards, safety standards and FAA requirements are some of the things that may be represented by control arrows. Output arrows represent those things into which inputs are transformed. For example, a hazard report (input) may be transformed into an action plan (transformation of input into output). Mechanism (or Resource) arrows represent how activities come about. They depict what resources (hardware, people, information, etc.) are required to complete the transformation of input into output.

Our adaptation of IDEFØ, however, uses only one activity box to represent a process. The one box serves as our focal point for discussing inputs, outputs, controls and mechanisms. We may of course use more than one activity box to further our discussion, which is more in line with actual IDEFØ modeling. Figure 1.8 presents an example of an IDEFØ diagram that has been constructed to serve as a discussion tool for designing a manual revision process. If you notice some things or items missing from the diagram, then our example model is working. The idea is that your design team will collectively identify the various tasks and attributes necessary for the process to function effectively.

We have found IDEFØ to be most valuable in discussions about different types of processes, such as the interactions among management processes, operational management processes, resource allocation processes and the relationships of attributes such as the process attributes defined earlier. Additionally, the model allows us to explore the differences between inputs, controls and resources (we have found that *resources* rather than *mechanisms* seem to communicate better). In standard flow charts, all activities look the same and it is harder to show these concepts. Everything goes in as an input and comes out as an output. Thus, we have found that IDEFØ is the tool to facilitate the understanding of these various factors and interrelationships using simple, conceptual diagrams.

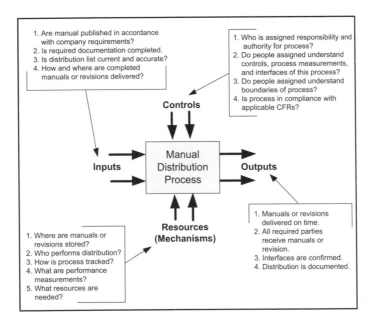

Figure 1.8   An IDEFØ discussion example

Be aware, however, that an IDEFØ model can get very complex, very fast. Therefore, we recommend confining IDEFØ to conceptual applications and discussions. An alternative is to use an IDEFØ breakdown to establish high-level relationships in the macro system and then shift to simpler tools (process maps) once it's broken down.

Once things have been narrowed down to a specific process or task flow, the use of the process maps we discussed previously is probably clearer and easier to understand for most audiences. Organizing the flows in some manner, such as in the cross-functional process map, to assign responsibilities is an example of controlling the complexity and avoiding getting lost in it.

The goal of the final product is to lay out the process or system to the degree feasible so that we have a firm grasp of the things necessary for an effective operation. Additionally, the goal includes integrating the process attributes into your new or revised design. In system safety, we try (emphasis on try) to make clear that the purpose of the system and task analysis is to identify potential hazards in the way that we do business. Thus it is not necessary to make these analyses masterworks of human factors engineering or systems design as long as they are clear enough to facilitate essential understanding of these processes.

Hazard Identification

System and task analysis are one of our most effective means for identifying hazards. As Clif Ericson, the past president of the System Safety Society reminds us, 'Hazards, mishaps and risks are a reality of life, and are often a by-product of man-made systems' (2009, p. 28). As such, hazard identification becomes one of our primary tasks in SRM and safety assurance. In order to conduct effective hazard identification, one must first understand the nature of hazards.

There are basically two viewpoints regarding the perception of a hazard. One is that a hazard is merely a hazard source such as jet fuel, high voltage, toxic chemicals, freezing rain, etc. The second viewpoint, which encompasses the FAA's perspective, looks at a hazard as a potential condition that could lead to an accident or incident. The FAA defines *hazard* as, 'Any existing or potential condition that can lead to injury, illness, or death to people; damage to or loss of a system, equipment, or property; or damage to the environment' (FAA, 2006c, p. 2). It is a condition that is a prerequisite to an accident or incident, which is referred to as a *mishap,* per system safety definitions (Ericson, 2005; DoD, 2000). Similarly, ICAO (2009, p. 4-1, para. 4.2.3) defines a hazard as 'a condition or an object with the potential to cause injuries to personnel, damage to equipment, or structures, loss of material, or reduction of ability to perform a prescribed function'.

Principles of system safety emphasize that accidents and incidents don't just happen – they are a result of actuated hazards. Accidents and incidents (mishaps) are random chance events that are, in most cases, predictable and controllable and can be predicted via hazard identification and prevented or controlled via hazard elimination or hazard control (Ericson, 2009). This starts with the evaluation of the system's design – interfaces, tools, procedures, documents, training and organizational structures – to identify possible error-producing scenarios and the potential consequences of errors and failures. Such analysis can be a test of those structures, relationships, adequacy of procedures, etc. Many times, *holes* may be found in these system elements and attributes where potential error-producing conditions are the result of incomplete system design. System hazards are often man-made! This emphasizes the importance of well developed safety risk management and safety assurance programs which rely heavily on effective safety analysis and task analysis.

A hazard and a mishap are linked by a transition referred to as a *state transition* (Ericson, 2005). The *state transition* is an event or factor that changes the hazard (conditional *before* state) to an actualized mishap (*after* state). The change is based on two factors, the set of hazard components involved and the risk presented by the hazard components. The hazard components are the things that constitute a hazard. The risk factor is the probability of the mishap occurring, and the severity of the resulting mishap loss (Ericson, 2005).

There are three components that make up a hazard: (1) the hazard source, (2) the hazard initiating mechanism and (3) the hazard target and threat (Ericson, 2008). In the field of system safety, they create what is called the hazard triangle (see Figure 1.9).

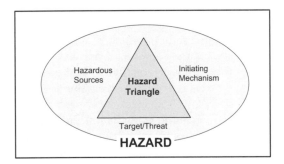

Figure 1.9   The hazard triangle

The *hazard source* is the rudimentary element of a hazard. It provides the impetus for the hazard to exist. As such, every system that contains a hazard source will have an associated set of unique hazards. Hazard sources are generally energy sources or safety critical functions (e.g., electricity, fuel, compressed gas, rotating machinery, aircraft velocity, etc.). The *initiating mechanism* is the *causal* factor, or set of *causal* factors, that transform the hazard into a mishap. They transform a dormant conditional state (latent failure) to an actualized hazard (mishap). Most often, the initiating mechanism is a combination of two or more factors (hardware failure, human errors, environmental factors, etc.). The *hazard target and threat* is the potential target and threat of the potential mishap event. The target is the person or thing that is potentially vulnerable to injury and damage. The threat is the severity of the potential damage or loss expected to result from the mishap event.

We have adapted an FAA model (see Figure 1.10) that is used in safety management training – the Integrated Risk Management Model. It is a somewhat eclectic blend of material taken from a number of safety experts and the traditional severity and likelihood engineering models. James Reason (1997) argued that while human error reduction and containment is a worthwhile goal, human error is a fundamental part of the human condition and is inevitable. Erik Hollnagel (2004) points out that reasons for system failure are more often due to the variability of the context and conditions in which operators work rather than the variability of their actions. Dismukes, Berman, and Loukopoulos (2007) in their analysis of 19 air carrier accidents maintained that, even though clear errors were made by the crew in each instance, a case could be made that the crew's performance was within range of potentially expected performance for a typical crew given the conditions.

The model is divided into three levels. The first is System/Task Analysis and represents the collection of data and analysis that addresses processes, activities, operational environment and system components which should be factual (this part of the model was previously depicted in Figure 1.1). Gathering facts and not opinions is not only important in auditing, evaluation and other safety assurance activities, it is also extremely important to hazard identification. The next level is Hazard Identification, which involves inferences. It represents the transformation of raw data into useful information, and making inferences about potential outcomes (success, failure, error, accidents, etc.) of operations under the given conditions (noted as 'workplace conditions' in the models in Figures 1.1 and 1.10). The remaining levels represent Risk Analysis, Risk Assessment and Risk Control. These levels involve judgments that apply a value system to the previous inferences. We caution not to apply judgments too early in the process as they represent opinions which could possibly cloud any corrective process. We maintain that one has to understand the entire loop in context, as the system will be altered if the analysis shows that the original design, plan, program, etc. has features that leave unattended hazards or unacceptable risk.

The system/task analysis level of the model, which is our area of interest, reflects processes, activities and workplace conditions. Different types of processes take on different characteristics and have different expectations. For the purpose of our analyses, there are three basic types of processes: operational, management and resource allocation processes. Each can harbor hazards. Operational processes do the real work of the system. Inputs are transformed into outputs in the form of products or services. A key element of analysis is examining the interfaces between processes. Frequently, problems are found where one or more interfaces are not well organized, or poorly documented (often a case when key functions are contracted out). Ineffective interfaces are a source of hazards.

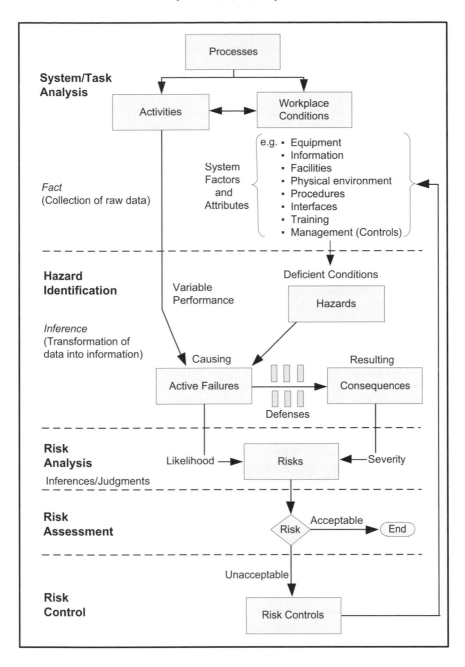

Figure 1.10   Integrated risk management model

Workplace conditions are often a key factor towards hazard development. For a workplace condition to be identified as a hazard, there must be potential for harm. It is important to emphasize at this point that *a hazard is a condition, not an event, not an error, and not a consequence of either* (see also ICAO, 2009, paragraph 4.3.1, pp. 4–2).

The integrated model reflects the essential elements of systems analysis and the analysis that provides a proactive/predictive identification of hazards. The bridge between system description, including operational processes, activities, workplace factors and hazards, requires an imaginative process. That is, we have to analyze the system and environmental factors and imagine how these elements could combine to create hazards, and what system failures or errors could result, as well as the potential consequences of those failures or errors. We will discuss the use of imagination in our discussion of gap analysis.

A fundamental difference between reactive, forensic methods (for example accident investigation and analysis) and predictive, forward-looking hazard identification, based on system analysis, is that in forensic analyses, the outcomes or consequences are known and the objective is to work back upstream to determine the root cause. In many cases, the failures, errors, and underlying conditions are relatively self-evident. However, forward looking inferential or imaginative analyses using system analysis techniques are the only way to identify hazards *before* new systems, processes, procedures, or changes are implemented. Moreover, system analysis techniques are also a highly effective way to proactively identify hazards in existing systems before the occurrence of unwanted outcomes.

When searching for hazards during your systems or task analysis, the following considerations may help trigger identification and recognition (Ericson, 2005):

- evaluate the hazard triangle components;
- utilize past knowledge from experiences and past lessons learned;
- make use of good design analysis practices;
- review design safety criteria, regulatory requirements and current safety practices;
- make use of good system analysis and task analysis techniques.

Although the purpose of this chapter is to explore system and task analysis and the major components of each, we believe we should end this discussion on hazard identification by briefly mentioning ways to deal with identified hazards. The potential for error/failure typically corresponds with the likelihood element of risk, and the types of consequences correspond more closely with severity. Risk analysis and risk assessment often talk about worst credible outcomes, but in practice, it is advisable to run separate risk analyses and risk assessments for different outcomes. Some hazard outcomes can have a limited range of severities, some hazards can have a number of different types of outcomes, and some hazards can have a single type of outcome. Thus, an initial analysis may need to have several exploratory sub-analyses to fully complete the system analysis and deal with the identified hazards.

Hazards are primarily eradicated by eliminating the source component (for example, clean up spilled fuel on ramp). Hazards (risks) are primarily mitigated by reducing probability of the initiating mechanism to occur (reduce probability of fuel spills by properly training line employees). Hazard targets and threats can be mitigated by reducing the effective danger of source, by protecting target or reducing threat (provide warning signage in fuel areas). One last point which the system safety experts emphasize: *when a hazard (risk) is mitigated, it is NOT eliminated.*

Training

Although this chapter focuses on system and task analysis, we have included this brief discussion on training development as it is a critical component of a successful SMS. Training provides the means for organizations to prepare their people to practice SRM and safety assurance programs, and it is an integral part of safety promotion (FAA, 2009a). Training, however, isn't just for the SMS functions, but is a fundamental part of a system and its operational functions. System and task analysis should include a determination of the tasks that are necessary for the processes and activities that are being designed or revised to effectively achieve output. Risk assessment will tell us how critical the tasks are.

The first DC-10 that was lost was a Turkish Airlines DC-10 departing from Paris. It was brought down due to an improperly secured cargo door. The employee who closed the door was not provided proper training and did not read English, which was the language that the placards presented. The criticality of the task may also drive other choices in training – do we just need to give people general familiarization training or does the task criticality demand testing and possibly an experience component?

Training evaluation is also part of the safety assurance function. We should not only look at the quality of the training itself, but also check to see that it isn't having an impact on system performance during operational audits. These audits may take us back to redesign (back to the 'analyze' step) if we find that the training program isn't hitting the mark.

While training shouldn't be considered a substitute for sound equipment, process, or procedure design, it is a key component of overall system design. During initial system analysis, one of the factors that should be considered is the competency demands of the personnel performing the operational tasks. In fact, the ability to adequately assure the necessary competencies through selection, training and supervision, as well as the characteristics of the target pool of employees, may drive key system decisions. At any rate, training shouldn't be relegated to an afterthought, or a correction for design deficiencies.

One of the first undertakings for effective SRM and safety assurance programs is to ensure that the people involved in managing and conducting these programs have a thorough understanding of the configuration and structure of the organization's systems and processes. Thorough training is usually how this is accomplished. Additionally, training is a key to organizational learning, which provides information about risk controls and lessons learned. Training, itself, also serves as a risk control by providing instruction in how to avoid hazards and reduce the likelihood of associated risks (FAA, 2006a).

If you are new to training development, or you have been charged with putting together some type of training program, it is recommended to use an instructional system design model to guide you through the process. The most basic instructional system design (ISD) model used by professional trainers is called the ADDIE model – analyse, design, develop, implement and evaluate. You will often hear ADDIE referred to as ISD or instructional design (ID). This model is essentially a generic, systematic, step-by-step framework used to ensure effective course design and development. Its purpose is to ensure that instruction is not delivered in a haphazard, unstructured way. Figure 1.11 illustrates the classic waterfall ADDIE model.

Clark (2004) tells us that when the model was first introduced, it was strictly a waterfall method as each step progressed in a waterfall approach except for evaluation. Evaluation is performed throughout the entire process. This was a criticism of the original ADDIE model in that it appeared to be a purely linear process. Experienced users, however, knew that reality was not quite so orderly. Most instructional designers found that the model's steps overlapped and that it was often necessary to go back to do more analysis and evaluation. As time went on, the model evolved into a more dynamic process in which the user could go back to the other phases as needed and revise steps as determined by evaluation. Figure 1.12 illustrates the dynamic ADDIE model.

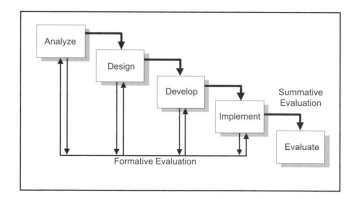

Figure 1.11   The ADDIE model: classic 'waterfall' representation

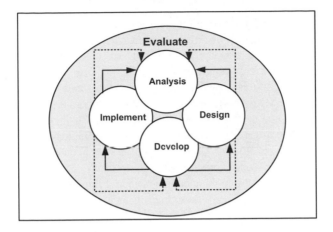

Figure 1.12   The dynamic ADDIE model

The Dynamic ADDIE model reflects the importance of evaluation and feedback throughout the entire training development process. The model also reflects the process as an iterative flow of activities (dynamic) through the use of solid and dashed arrows going in opposite directions between the phases. The model shows that each activity includes a formative evaluation which is used to refine goals and evolve strategies for achieving goals throughout the entire duration of the ISD process. Clark (2004) maintains that this integration of formative evaluation produces skilled and knowledgeable performers by improving the learning program while it is still fluid.

A suggestion offered by Haig and Adison (2009), respected instructional designers, recommend that the most time in training program development should be spent in the first two phases of the ADDIE model — analyze and design. This, they maintain, yields the best results in ISD. They also warn that the *analyze* and *design* phases are the two most neglected phases in the ISD process. Design deserves extra effort as it is the key to the instructional system, as well as process and procedural development for organizational systems. If you find yourself in the training arena, we recommend that you review some texts and websites to gain a full understanding of ISD. We have provided some suggested sources at the end of this section.

The ADDIE Phases

The literature on ADDIE estimates that there are well over 100 different ISD variations in use today, with almost all being based on the classic ADDIE model. The following provides a brief overview of the phases of the basic ADDIE model. The *analysis* phase is the most important phase of ADDIE. In regards to SMS, it is during this phase that we determine who is responsible for each safety-related task. This is further refined during our system and task analysis, with criticality of the tasks being set during the hazard and risk analysis steps. The analysis phase also identifies areas requiring training taking into account the results of our system and task analysis, performance assessments and regulatory mandates. Additionally, it defines the target population to develop an understanding of their needs, constraints, and existing knowledge and skills. Lastly, the analysis defines the desired outcome of the training and describes the ultimate objectives and goals of the training.

The *design* phase is the process of planning and determining the course specifications and criteria to include course objectives, lesson plans, topic content, training methods, course delivery time, training materials, learner exercises, costs and assessment and evaluation criteria. Often, detailed prototypes are developed at this time in order to conduct test runs on selected parts of a complex training program.

The *development* phase is the actual production and assembly of the training materials that were specified in the design phase. At this point, it is important to address time schedules, deadlines and production runs. All course materials and audio/visual media are prepared during this phase. Note that in some instructional design circles, development is lumped together with the design phase. This is often the case in smaller organizations where a person assigned to the training project is also responsible for the entire project. In many larger organizations, actual instructional designers create the design while others such as artists, technical writers, programmers and experienced trainers actually build the various components as specified by the design. The key point is to keep system design, risk management and training design in sync.

During the *implementation* phase the course is actually put into action, and the final product is presented to the target population. This is where training design and system design also have to be kept in sync – we can't deploy systems, new procedures, etc., ahead of required training. Conversely, it can be a mistake to begin training so far in advance that the knowledge has already eroded by the time that these new or changed systems come on line. Again, training must be an integral part of the system, including deployment and implementation planning, as well as change management. After delivery, the effectiveness of the training materials and course is typically evaluated.

The *evaluation* phase consists of (1) formative, (2) summative and (3) confirmative evaluations. Formative evaluation is conducted during each phase of the ADDIE process. It judges the value or worth of a particular phase and focuses on the processes or activities that make up that phase. Summative evaluation is conducted at the end of the ISD process. It focuses on the outcome (summation), and provides opportunities for feedback from the users. Confirmative evaluation assesses the transfer of learning to the real world after a period of time, perhaps months or even years. Its purpose is to determine whether the training program remains current and effective by enabling trainees to meet the original program objectives. Training evaluation is part of the safety assurance process. It not only is part of what the personnel or organizations responsible for training do, but also should be a subject of emphasis on internal audit and internal evaluation functions. Training should be considered part of the overall operational system in all respects. This means that safety assurance isn't complete unless training is continually monitored, measured and evaluated.

Many in our industry approach training in an ad hoc method that often fails to identify the issues that are covered by the analysis (A), implementation (I), and evaluation (E) components of the ADDIE model. That is, we often don't have a good handle on what really needs to be trained, the characteristics of the target audience and the operational environment that will exist for the tasks that are the subject of training. Likewise, we often have poor control over training *as designed* versus *as delivered*. Moreover, there is often no real evaluation of both the direct training outcome (did the people learn what was intended) and subsequent on-the-job performance. There seems to be a lot of synergy between the system and task analysis in the SRM part of system design and the first stages of ADDIE, as well as synergy between the safety assurance part of the SMS and the later stages of ISD. We maintain that training should be considered to be part of the system, and therefore, training design is to be part of the system design.

Some Suggested Sources to explore ISD

A good place to start is Don Clark's website: http://nwlink.com/~donclark/hrd/ahold/isd.html.
Gagne, R.M., Wagner, W.W., Golas, K. and Keller, J.M. (2004). *Principles of Instructional Design*. Belmont, CA: Wadsworth.
Dick, W., Carey, L. and Carey, J. (2009). *The Systematic Design of Instruction*, 7th edn. Upper Saddle River, NJ: Prentice Hall.
Leshin, C.B., Pollock, J. and Reigeluth, C.M. (1992). *Instructional Design Strategies and Tactics*. Englewood Cliffs, NJ: Education Technology Publications.
Jonassen, D.H., Tessmer, M. and Hannum, W.H. (1999). *Task Analysis Methods for Instructional Design*. Mahwah, NJ: Lawrence Erlbaum Associates.

We would like to end this section with a quote from a professional airline trainer that sums up what this section is all about (Brian Wilson, personal communication, August 12, 2009):

> Being a professional trainer who has spent the time to educate myself as to the current research, opinions, studies etc. as to how adults learn best, I have come to the conclusion that the general approach of ALL US airlines to training pilots leaves a lot to be desired ... I believe the general consensus in the industry is to do the minimum training necessary to satisfy FAA requirements and ... any additional training to achieve a higher level of pilot expertise is generally seen

as wasteful. Beyond that if we want to convey expertise – where is the standard that defines a higher level of expertise than the FAA Part 121 or AQP standards? Where are the ISD-developed task analyses that frame that body of knowledge and to what level is that knowledge held? Who gets to develop them? Who gets to administer them?

Shaping Safety Assurance Processes and Performance Assessment

To complete our discussion of system and task analysis, we end this chapter with an examination of the roles of system and task analysis in shaping safety assurance and performance assessment. As defined earlier, safety assurance is a continuous, ongoing formal management process that ensures that the processes within an organization's SRM remain valid over a period of time. Performance assessment, for the purposes of SMS, is the means by which an organization periodically examines its systems' performance objectives and design expectations to ensure that all SMS outputs are recorded, monitored, measured and analyzed (FAA, 2009b; ICAO, 2009). The *ICAO Safety Management Manual* (2009) states, with reference to Element 3.1 of the ICAO SMS Framework, 'The [organization] shall develop and maintain the means to verify the safety performance of the organization and to validate the effectiveness of safety risk controls.'

Figure 1.13 presents the major activities that make up the safety risk management and safety assurance processes. The FAA SMS Assurance Guide (2009b) and the soon to be published AC 120-92A define the activities of safety assurance as follows.

- *System operation* is the monitoring and management of risk controls. This activity is one of the most important activities in safety assurance. This is the day-to-day, flight-by-flight, task-by-task monitoring of operations and other safety-related activities to assure that risk controls are in place and being used as intended. For example, if an airplane is being operated with known inoperative equipment under the authority of an approved minimum equipment list (MEL), any required maintenance and operational limitations must be monitored each time the aircraft is dispatched or otherwise released for flight.
- *Data acquisition* is the collection of a variety of data to provide information and output measurements for analysis. The data ranges from continuous monitoring of processes, to periodic auditing, to employee reporting systems and to investigations that examine failures.
- *Analysis* is the process by which an organization analyzes the collected data. As in SRM, it addresses the data in terms of performance objectives and seeks to determine root causes of any shortfalls. Additionally, the analysis looks for any new conditions that have not been seen before, and unexpected results of system performance.

- *System assessment* is the process in which decisions based on analyses are made. In this process, we answer whether any identified variation to a process is *common cause* (that which normally occurs and is acceptable) or *special cause* (that which is not normal and is unacceptable). The decision boils down to whether the process in question needs revision or total redesign.
- *Corrective action* involves the development of an action plan that will resolve the problem. The plan addresses how to correct the system.

Sometimes, even though everyone is doing everything that is expected, but it just isn't working, expected outcomes and risk controls are not being accomplished and things seem to be awry (possibly the conditions have changed so that the original controls are no longer appropriate). Systems change over a period of time and it is important to remember that every system and process involves some risk. Safety experts tell us that there is always some residual risk even after all safety measures have been employed (Bedford and Cooke, 2001; Ericson, 2005, Mol, 2003; Stolzer, Halford, and Goglia, 2008). Residual safety risk is defined as 'the safety risk that exists after all controls have been implemented or exhausted and verified' (FAA, 2009a, p. 9). Sidney Dekker (2005), in his text on human error, reminds us that 'the greatest residual risk in every sociotechnical system is the drift into failure' (p. 18). He defines *drift into failure* as 'the slow incremental

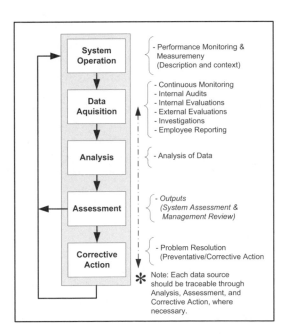

Figure 1.13   Safety assurance flow and activities

movement of system operations towards the edge of its safety envelope' (p. 18). Drift is one of the greatest enemies of safety assurance. Two types of drift can occur; drift due to incremental and accumulative changes in systems, and changes in the operational environment.

Incremental changes are referred to as *system creep*. Operational systems in aviation operate in a constant state of change. While some operational environments are relatively stable and others are very dynamic, nearly all organizations make at least minor changes to their systems as time progresses. An accumulation of minor changes can result in the same effects as a single significant change. However, these changes can go unnoticed, with unexpected results, due to the fact that no single increment of change was considered significant (Dekker et al., 2009).

The process of day-to-day, change-by-change monitoring should be supplemented by periodic system audits to ensure that the system has not drifted into an unsafe condition. The technique of system auditing is frequently practiced in quality management systems, and is a requirement in other aviation-oriented audit programs. Periodic audits and evaluations should compare original process objectives and conditions with current process status. Assurance system planning should identify critical operations, technical parameters and risk controls that need to be included in monitoring, periodic audits and evaluations. Again, the follow-up audits and evaluations should check not only on the technical parameters but also on the monitoring program itself.

Along with changes in systems, changes also occur in operational environments. These changes can include changes in demographic characteristics of pools of potential new employees (for example differences in experience levels of potential job candidates due to changes in economic conditions), changes in supplier or support contractor practices and regulatory changes, among others. These changes should also be evaluated carefully to determine if system changes are needed to respond to them.

At any rate, when changes of any nature take place within an organization, it is necessary to be cognizant of the potential for system creep or drift. Adhering to safety assurance processes and continuing performance assessments should give warning of drift creeping into operations. Now the question becomes, when our systems and task analyses have shown that something is askew, what is our next step? This is the point at which we enter into performance assessment.

Performance Assessment

Assessment is the method by which worth and importance are judged. Performance assessment is the process of identifying performance and organizational problems, prioritizing the problems, determining causal factors and gathering data to select alternate courses of action. It incorporates a number of overlapping methods including system analysis, task analysis, needs assessment and evaluation. Performance assessment verifies that the current system is still in line with regulations, design requirements or expectations and objectives of the organization. Periodic assessments should also include analyses of the operational environment versus the existing system to assure that the system is still appropriate for the current situation. The practice of reviewing system design baselines is common in standardized quality, environmental and occupational health and safety management systems where comprehensive system audits are often accomplished on the order of every three to five years or more often if major changes to the system or environment have occurred.

Although performance assessment is often linked with evaluating learning outcomes in an educational setting, here we are looking at performance assessment in terms of process outcomes (performance objectives and expectations) within an organizational setting. Evaluation, however, does play a role in this setting as well. Unfortunately, a jargon jungle exists in regards to performance assessment. Terms such as needs assessment, organizational analysis and behavioral analysis are used in various organizational settings by a host of practitioners all seeking answers to similar problems. What we are really trying to do is exemplified by the title of Zemke and Kramlinger's text *Figuring Things Out* (1982). In other words, we are attempting to find the gap or flaw in a process to determine what went wrong, why it went wrong, and then follow up with corrective action.

Gap Analysis

Figuring things out involves comparing the actual outcomes (reality) with the desired outcomes (performance objectives and functional expectations) of each process. This is done by means of a *gap analysis*. The analysis makes use of established standards to determine the difference between what is and what should be. It might be better thought of as *gap recognition*. This is the type of gap analysis that the *FAA SMS Safety Assurance Guide* employs.

The most common usage in the SMS literature refers to a *design gap analysis*. This analysis is actually an up-front activity that is used to determine differences between the current status of a company's systems, programs and practices, with a standard or other set of expectations for a system that they desire or are required to conform to. This type of gap analysis is more associated with design criteria, requirements, or expectations than with performance. It is usually the first task in implementing an SMS.

To conduct an effective gap analysis, one must establish two benchmarks. The first benchmark is represented by the question 'What is'? The second is represented by the question 'What should be'? The Gap Analysis Model (see Figure 1.14) reflects these benchmarks as *actual* and *optimal*.

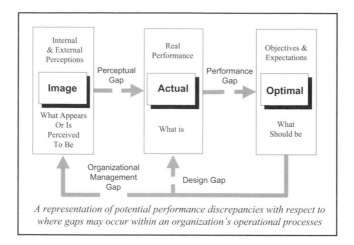

A representation of potential performance discrepancies with respect to where gaps may occur within an organization's operational processes

Figure 1.14   Gap analysis model

The model illustrates three types of performance gaps and four kinds of gaps. The performance types are termed *image, actual* and *optimal*, and the gaps are termed *performance gaps, perceptual gaps, design gaps* and *organizational/ management gaps*.

Performance Types

Image represents the internal and external perceptions of the organization's safety status. It is exhibited by physical manifestations, including the organization's mission statement, goals, objectives, documentation, environment and working conditions. Image is reflected through both an internal image (how the organization perceives itself) and through an external image (how those outside the company perceive the organization). Perceptual gaps exist when management perceives that the organization's standards (for example operation manuals, checklists) are optimal, but they are actually lacking. Such gaps can lead to lack of management support, resources, etc. The corrective action for a perceptual gap is found in the strategic planning process in which the mission, goals, and objectives are revisited and realigned with SMS criteria. This can be a very dangerous gap as the lack of management support and resources can kill any safety program.

- *Actual* and *optimal* have previously been discussed, and are basically the *what is state* and the *desired state*. The difference between the actual and the optimal establishes the performance gap. Corrective action calls for the determination of the cause of the gap. Basically, the cause can reside in the process itself (a design gap), hardware (tools, equipment), software (manuals, checklists, training), live-ware (worker skills and knowledge), or environment (work conditions, facilities, etc.). Each cause will require a different approach to corrective action. Suffice to say, the only cause that training will correct is certain problems in live-ware (for example skill and knowledge). The others require a much more in-depth look at causation.

Four Types of Gaps

A *performance gap* is a discrepancy between an actual and an optimal; it is a difference in performance that is indicated by an inconsistency with the desired performance. It is identified by symptoms and indicators that a problem exists, and is typically linked to results. Performance gaps are usually described in quantitative and specific behavioral terms. For example, we find that 75 percent of pilots do not have current (quantitative) operation manuals, as the last two revisions have not been inserted by the required date. Additionally, performance gaps may be negative or positive. A *negative gap* results in an outcome that does not meet standards. It is a liability to the company (for example unapproved work-a-rounds). A *positive gap* is one that results in an outcome which is actually better or of a higher standard than the predetermined optimal. This performance is an asset to the company and should be incorporated into operating procedure once approved by management.

A *perceptual gap* is a difference in perception between the image and the optimal. For example, management may believe that the organization's operations manual revision process is more than adequate (the internal image), and they may argue that manual distribution and revision procedures are adequately published in the company policy manual (the external image), but the outcome is that a number of crew members consistently have out-of-date manuals. It is a gap rooted in management's perception of what is claimed to exist and what actually exists.

A *design gap* is a result of failing to properly correct for the difference between *what should be* (optimal) and *what is* (actual) in a corrective action, or a gap created in the initial design of a process.

The *organizational/management gap* is a deficiency that exists within the planning, organizing, directing and controlling functions of management that affect the safety status of the organization. It is a gap that is rooted in one or more management functions which inhibit or block the achievement of optimal performance. Perhaps the best description of the importance of management's role in safety is found in the words of H.W. Heinrich, whose text *Industrial Accident Prevention,* first published in 1931, revolutionized industrial safety. Heinrich (1959) argued that the methods of most value in accident prevention were analogous with the methods for the control of quality, cost and quantity of production.

The desired state is a *zero gap* in which the actual performance matches the optimal performance. Within established operating systems, the intent is to determine the deviation (or error) between an expectation or requirement and the current state. In the design of new processes, the intent is to foresee as best as can be done any gaps that may occur because of the design. Subsequent gap analysis is then undertaken to determine the tolerability of an identified gap, determine its cause, and to identify potential avenues for corrective action.

Within SMS, gap analysis generally refers to the activity of studying the differences between standards and the delivery of those standards. For our purpose, we are determining if there are differences between practices and standards. Typically, the gap analysis is done against a template or model, such as the *Safety Management System Assurance Guide* (FAA, 2009b). Organizations often use gap analysis for initiating efforts to register with an established standard such as ISO 9000, IATA Operational Safety Audit (IOSA), International Standard – Business Aircraft Operations (IS-BAO), etc. In these cases, the organization compares its current practices to those required to register to the desired standard. When Transport Canada developed its regulations that addressed SMS, the operators' gap analyses were used to determine what actions were needed to be taken in order to comply with the new SMS. The critical point in gap analysis is to be sure a clear understanding of the standard or requirement is established. We have also found that the use of the FAA's Air Transportation Oversight System (ATOS) Data Collection Tools serve air carriers as an excellent template for identification of gaps in systems associated with regulatory compliance. At its core are two basic questions: (1) Where are we? (2) Where do we want to be?

We suggest beginning with a comparison of documented safety and quality management processes with SMS expectations, and whatever standard has been established. This checks the design. Common problems at this stage are often found in documentation. As we mentioned earlier, the initial examination of the company's organization chart, and the first iteration of looking at the *as is* process/ system flow charts will often uncover problems in organization and management documentation even before getting to the actual gap analysis. Thus, we have found that many of the gap analysis type questions that accompany a standard such as the FAA SMS Assurance Guide are addressed as follows: Does the organization... [name the requirement]?, answered by Don't know.

Once we are past this initial stage, then we can get into the actual gap analysis process, that is identifying gaps in actual performance. If a gap is detected, then it must be determined whether the practices (performances) are not appropriate, the procedures are lacking, the performers lack competence, the work conditions or equipment leads to substandard performance, or if the process itself is flawed. Gap analysis can be a powerful tool for most performance assessments. However, it does have the limitation of not being easily applied to circumstances within which one of the benchmarks (what is, or what should be) is missing. When the *what is* is missing, this means that we have inadequate data regarding the current state scenario. Usually, a more rigorous analysis process, including additional data collection, documentation, verification and examination, will correct this circumstance.

But what do we do when the missing benchmark is *what should be*? The answer is that we must undertake an alternate form of performance assessment that focuses on the determination of *what should be*. The determination of *what should be* may be found in two unique but related concepts: *conscious inquiry* (Westrum and Adamski, 1999) and *requisite imagination* (Adamski and Westrum, 2003).

Conscious inquiry is a characteristic of a high-integrity organization that encourages its members to investigate, query and challenge the status quo. In this type of organization, it is expected that its members will think and let their thoughts be known. Westrum (1993) explains that this is a 'license to think.' Consequently, conscious inquiry makes the SRM and safety assurance processes much smoother as members of the organization feel they can freely share their knowledge. With a license to think, members of the organization will not only identify faults within the organization, they will determine *what should be* by use of requisite imagination.

Requisite imagination is a means by which the people involved identify problems and use their individual and collective intellectual resources to describe future states of what, when, how and where things can go wrong. It establishes standards when no standards exist. It allows for the issues identified through conscious inquiry to set the standards (optimals) for gap analysis.

Evaluation

Once the performance assessment is complete and gaps have been identified, the next step involves determination of cause. This step often requires an evaluation to determine the adequacy of some specific thing (for example a checklist, a procedure, a training program, a trouble-shooting guide). For example, numerous pre-flight errors are occurring and an evaluation of the aircraft checklist reveals that a number of incomplete steps are leading to errors. Evaluation is used to determine the adequacy of processes, products, programs, training, etc. It is the process of gathering valid and reliable evidence to help management make arduous decisions. In the performance assessment process, it can help determine the causal factor to a gap and set the stage for corrective action. Be sure to document all evaluation findings. Without good documentation you run the risk of repeating steps or losing data.

It is important to point out that evaluation is not decision making; rather, it is the process that collects the information about a specific procedure, thing, or process that provides the basis for a decision. Consequently, it is important not only to identify and prioritize gaps during the performance assessment, it is also necessary to analyze the causal factors and evaluate applicable components that lead to gaps.

Almost anything can be the object of an evaluation. The evaluation process for safety assurance concerns (gaps) incorporates a number of possible focal points for evaluation that are detailed in the *FAA SMS Assurance Guide* (2009b). The guide provides a step-by-step process to assess the objectives and expectations of a SMS, as well as identifying things that may require evaluation. Figure 1.15 provides examples of gaps discussed in this chapter and examples of analysis, evaluation, and possible corrective actions.

Examples of Gap Analysis				
Expectations	Performance Gap Analysis	Perceptual Gap Analysis	Organizational Gap Analysis	Evaluation Findings
Do the organization's SMS procedures apply to the development of operational procedures?	A number of procedures in flight operations manual are outdated and ignored. ———— Conduct in-depth evaluation of manual to determine currency and validity of procedures.	Management believes current manual is more than adequate. ———— The internal and external images do not match. An evaluation of manual is required and findings reported.	Sufficient resources are not allocated to maintain operations manual. ———— Identify inhibitors and present findings for management review.	The manual has numerous outdated and confusing procedures. Insufficient resources are allocated for manual upkeep. Technical writing is poor. Distribution process is lacking.
Does the organization clearly define the levels of management that can make safety risk acceptance decisions?	Employees report that some decisions are confusing and in conflict between certain managers. ———— Interview employees and managers. Review company policy manual.	Senior management believes that current management structure and policy manual are adequate. ———— Present findings of interviews and manual review to management.	Senior management provides insufficient direction to middle management. ———— Present findings of interviews and manual review to management.	Some middle management does not understand its responsibilities or scope of operations. Policy manual and job descriptions need revision. Company organization chart is outdated.
Does the organization's line management ensure regular audits are conducted of their safety related departmental functions which are performed by subcontractors?	The Director of Safety reports that audits of subcontractors have not been conducted. ———— Identify contractors, review contracts and agreements. Review Internal Evaluation Program.	Senior management believes that current Internal Evaluation Program is adequate. ———— Present findings of analysis to senior management. Request review of contracts and Internal Evaluation Program.	Senior management Is unaware of number of subcontractors being used. ———— Present findings of analysis to senior management. Request review of contracts and Internal Evaluation Program.	The Internal Evaluation Program fails to address subcontractors that have recently been employed by certain departments. Feedback to Director of Safety is lacking.

Figure 1.15   Examples of gap analysis

System/task Analysis Example

Remember that systems and task analyses consider how people work, how they coordinate and work with each other, how they're supervised, what tools they use and the conditions under which the work is done. As an example, let us look at an aircraft fueling process, which is common in aviation operations, and conduct a sample system/task analysis. For the purpose of our example, we'll limit the discussion to a simple operation; but first, let us review the recommended steps to complete a system/task analysis (see Figure 1.16).

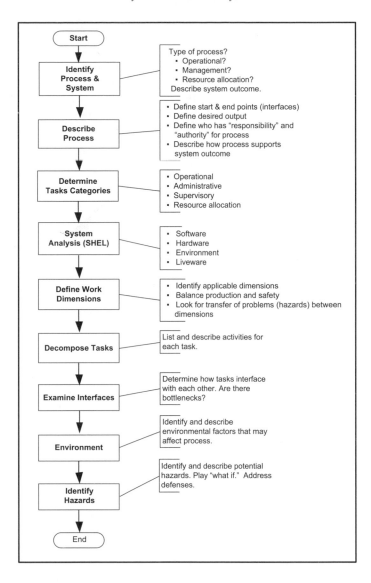

Figure 1.16   Steps to system/task analysis

Identify Process

The first step is to clearly identify the process to be analyzed and the system in which it resides. Here we should define the type of process, be it operational, management, or resource allocation. The process we will examine is aircraft fueling which is part of the Line Operations system.

Process and Tasks

Our second step is to clearly describe the process to include *responsibility* and *authority*, the beginning and end of the process, and its desired output. Additionally, it needs to describe how the process supports the system outcome(s). Our example begins with the request for a fuel order and ends with the proper fuel load uploaded, with the output being a properly fueled aircraft. Our next step is to determine the basic task categories. This step will help us determine the various objectives that are part of the process. The tasks that support our example fueling process can be divided into four basic task categories, each with its own principal objective:

- operational-service: deliver fuel to aircraft
- operational-administrative: receive and coordinate orders for fuel
- supervisory: oversee fueling operations
- resource management: provide bulk fuel to fuel tanker(s).

System

Next, we need to examine the system in which the process exists. Here, using the *software, hardware, environment and 'liveware'* (SHEL) model, we define and describe the components of the system to identify any system factors that might affect task performance. Remember that systems include the people, their work instructions, and the necessary equipment to perform the tasks. Our analysis should define the system's desired outcome. Some system models, such as the SHEL, also include the aspects of the environment in its consideration of a system. We believe it is important to include task-environmental factors somewhere in your analysis as they can easily be a source of hazards. Thus, using the SHEL model, a breakdown of system and environmental factors might reflect as follows.

- Software, such as the fueling manual, employee checklists or other job aids, fuel orders, fuel records and accounting software.
- Hardware, such as fuel storage, fuel trucks, fuel sampling equipment, other miscellaneous equipment (chocks, ladders, tools) and personnel protective equipment.
- Environment, such as airport ramp (concrete/asphalt, parked and moving aircraft, other vehicles, pedestrians), day/night operations, dry/rain/snow conditions.
- Liveware, such as fueler(s), one per shift in our example, clerk/dispatcher, line supervisor, pilots, customers and other employees (for examplemechanics, bulk fuel handlers).

Work Dimensions

In a SMS, we might easily think that we are only interested in operational safety. While this is one of our primary objectives, a useful analysis should also consider other dimensions such as quality of service, employee occupational safety and health, customer service and environmental protection. These dimensions should be considered for two primary reasons. First, managers of real operations must constantly balance the requirements between safety and production; consequently, they must consider all of the dimensions of the work at hand in order to make appropriate decisions about such areas as resource allocation, process revision, etc. Second, none of these dimensions operate in a vacuum; risk controls for potential problems in one dimension may introduce hazards in another. For example, hazards in these other dimensions include inaccurate fueling with the wrong type of fuel (safety – operational), vehicle traffic (safety – occupational), fuel spillage (environmental), customer satisfaction (quality of service), or lost labor hours (efficiency of service), which can all transgress into other dimensions.

Different dimensions may have compatible goals, so solutions to one problem may solve problems in other areas. Conversely, goals may be incompatible, so conflicts may have to be resolved using more than one solution. For example, if a customer discovers that an incorrect type or grade of fuel has been dispensed to their aircraft, the customer will understandably be dissatisfied. This will also lead to wasted labor-hours involved in defueling the aircraft, the possibility of fuel spillage in the process and delays in fueling other aircraft. If the pilot does not discover the problem, a serious operational risk and, potentially, a serious accident could occur. Preventing this type of problem would obviously be beneficial in all of these areas. The goals of all of these dimensions are compatible in this case. In another scenario, however, customer pressure for a quick turnaround (rapid fueling operation to get the aircraft on its way with minimum delay) could pit customer satisfaction against operational and employee safety concerns. In this case, the goal of satisfying the customer may run counter to safety goals.

Task Decomposition: Listing the Activities

Our next step is to prepare a descriptive list of the activities in each of the tasks identified earlier. A simple list will suffice to start the process (a whiteboard can be the most effective analysis tool here). It is also beneficial to include the employees who typically do the work in these discussions. Including them will provide important insight into how they interpret the task and how they understand the potential problems. It can also engender a greater sense of involvement and empowerment on the part of the employees. From there, a basic flow chart (process map) is developed as discussed earlier. The chart should reflect each task sequence, which will help to ensure that all activities have been accounted for and that they flow smoothly from one activity to another. If needed, more complex process maps may be developed at this time, such as a procedural map or cross-functional map. In any task analysis, it is essential to consider the background and capability of the employees who will actually perform the work to ensure that we do not generate invalid expectations. Oftentimes, this will necessitate additional training, supervisory practices, hiring criteria, or allocation of tasks to other personnel with more experience or higher levels of training.

Attending to the Interfaces

Once we have completed our process map, we should examine how the tasks interrelate. That is, how does the clerk/dispatcher communicate with the fueler? How much and what type of supervision is expected? How is the tanker refueled? Are there other companies or employees that must be considered? The IDEFØ model, which was described earlier, may be used if the processes are simple or, if necessary, in-depth process maps may be used. Here again, the simplest method to begin may involve a simple list on a whiteboard of the interfaces between tasks. We should also consider interfaces between shifts, if this is the way the operation is organized.

Effects of the Environment

Next, we need to look at environmental factors. Are there any differences in the way that our employees need to work in night vs day operations? What are the impacts of the various types of inclement weather that may exist in the area? These factors may have potential impacts on the way that the work must be adapted to perform safely and efficiently under changing conditions.

The operational environment is not confined just to the physical environment. It includes the complete situation in which work is done. The operational environment includes system design factors, procedures, training, selection, training and supervisory practices, tools, equipment and facilities. The elements of the SHEL model and the six system attributes discussed earlier all make up the set of organizational and workplace factors that may affect task performance (Reason, 1997, 2008).

Hazard Identification

The next process that we will consider will be hazard identification. In this process, we will look at the system/task analysis and play *what if*. We will attempt to break the system in such a way as to uncover potential hazards or risk factors and design defenses for them.

Examples of an Analysis

The scenario for our example problem, as stated earlier, is a fueling operation run by a fixed-base operator at a municipal airport and provides services between 6:00 a.m. and 10:00 p.m. The company also runs a flight school, an air taxi operation and sells fuel to non-company aircraft based at the airport, as well as transient aircraft. They handle two types of fuel, aviation gasoline and jet fuel (which are not interchangeable in the serviced aircraft).

The company uses a two-shift operation with separate personnel assigned for customer service (administrative process) and for fuel delivery (operational or technical process). A single employee is used in each position for each shift. The shifts overlap by one-half hour exits and each employee is allowed a half-hour lunch break. The administrative person is supervised by the company secretary/ treasurer, and fuelers are supervised by the company director of maintenance, who is also the ramp supervisor. Bulk fuel is obtained from permanently installed tanks in a fuel farm located on the airport. All airport users purchase fuel from this facility, which is operated by a separate contractor. A general outline of the fueling operation and its key process relationships are depicted in Figure 1.17, along with procedural requirements and associated document needs, controls, and interfaces.

Figure 1.17 reflects a simplified representation of two successive fueling operations and their associated management and resource allocation processes. Each individual process could be analyzed using the techniques we have discussed. The circled letters refer to the process attributes; procedures (P), controls (C), and interfaces (I). Note that each of the processes carries a 'P' designation. This

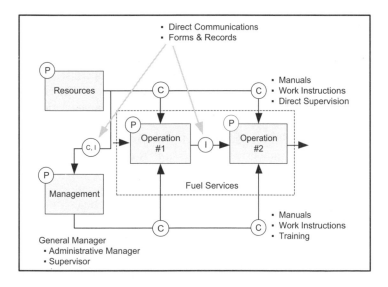

Figure 1.17   Fuel service: process relationships

indicates that a documented procedure is necessary. The administrative service and resource management processes reflect an 'I' designation which identifies important points of interface. Finally, the supervisory function supplies controls to all of the processes, which points are designated with a 'C'.

This latter point bears some discussion. The attributes of *authority* and *responsibility* are applied to each of the processes, but they have special significance to management and supervisory functions. The outputs of these functions are controls on other processes, but these controls reflect the responsibility of management to provide organizational controls as well as their authority over these functions. In order for controls to be effective, managers and supervisors in the system must be given appropriate levels of authority and their span of control must be commensurate with the responsibilities given to them. If managers and supervisors are not given ample training, direction, ability to observe and reasonable workloads, they will not be able to control functions placed under them effectively.

Likewise, controls may be of several varieties. Management has the responsibility to design processes and develop procedures. Management needs to publish work instructions, checklists, manuals and training syllabi so that employees know what is expected of them, and are trained to a level of competence commensurate with those expectations. This level of control is usually delivered in the form of written documents and in training programs.

Controls are also delivered in the form of direct supervision and communication. In our fueling process example, supervisors apply direct control over employees based on communication and direct observation. They may supply direct instructions to ensure conformity to procedures as well as to assist employees with prioritization of work. Managers and supervisors also provide information, instructions and feedback to resource suppliers so that they can deliver the right kind and amount of raw resources to the process. Managers are also responsible for controlling quality of incoming goods and services, both directly and through supervision of subordinate employees. In our example, supervisors interact with the fuel farm contractor to coordinate delivery of bulk fuel. They also supervise fuelers to ensure that they perform required quality and contamination checks on the bulk fuel received.

One of the more difficult attributes to address and examine is *interface*. It would be easy to assume that we could look at each process in isolation and draw the simplistic conclusion that, if everybody did their job, the process would function adequately. However, experience has shown that this is not always (if ever) the case. Many processes have broken down, sometimes catastrophically, where each individual involved was diligently doing their job according to what they thought was an appropriate course of action. Thus, it is important not only to look at individual task descriptions and sequences, but in the process as a whole, pay special attention to how employees communicate with each other and how work passes from one to another participant in the process. Interfaces also include the maintenance of feedback loops so that each person involved remains aware of what is going on in the process.

One final area that we should address before looking at our process example in more detail is that of *process measurement*, which can take several forms. As a process is ongoing, managers, supervisors and other participants in the operation must remain aware of the status of the ongoing operation. Process measurements are those elements that we can measure and determine if our output is meeting expectations. In these cases, process measures may take the form of simple verbal reports, computer inputs, or paper forms. Over time, managers must also look at their operations to see that they consistently produce the performance desired. Managers must ask, are we consistently providing the level of service that we intend to, are we conserving the company's resources, and are essential risk controls being applied and working as intended?

Figure 1.18 provides process maps for our example fueling operation. We have outlined the tasks that make up the four process categories which we identified earlier. The circled 'I' indicates points of interface. Such diagramming and analysis can easily point to interfaces and where possible information flow problems may occur. Such problems include information bottlenecks, distortion and even avoidance. Additionally, if you were to take such mapping to a deeper level, one could easily identify points where process measurements should be employed.

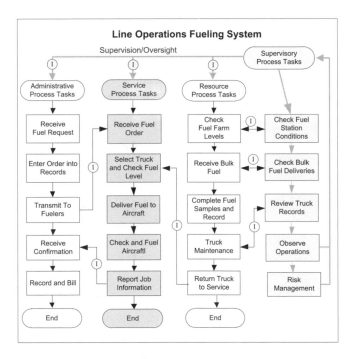

Figure 1.18   Example: fueling operation processes

Our example (see Figure 1.18) begins with a request for fuel as a task in the administrative process. Before we are finished, however, we need to consider the entire scope of each process's tasks, their relationships and their associated attributes. For example, under the service process tasks, the lineman must select a truck with the proper type of fuel, ensure that the truck has sufficient fuel to meet the request, deliver fuel to the aircraft, sample the fuel, upload gallons requested and complete the necessary paperwork. As our fueling operation diagram reflects, there are numerous interface points (I), procedure points (P) and control points (C) that relate to more than one process.

Following the steps outlined in Figure 1.16 (steps for analysis), our next step involves addressing environmental factors and hazard identification. Our emphasis should be on identifying elements of the system, and the operational environment that could affect task performance in ways that could cause safety-related errors. We have provided a sample form that can be used to document findings (see Figure 1.19). Our sample depiction shows part of the fueling process. It reflects the flow from processes and activities, how workplace conditions become hazards and what types of errors could result. An additional column could be added to show unsafe states as well. In our example, however, since the error due to the fueler is the introduction of an incorrect type of fuel into an aircraft, the resulting unsafe state of an incorrectly fueled aircraft flows from this error in a straightforward manner. You can use our format or design your own. The point is that you must document your findings in order to complete and record your analysis. Our sample form depicts sample tasks, conditions, hazards and potential errors.

As our example reflects, there are numerous factors that can adversely affect a fueling operation. For simplicity and discussion purposes, we only show two of the many tasks that make up a fuel service process. Additionally, we identify the contributing component of the identified hazards. Identifying the hazard components can often lead to a more effective resolution of the conditions.

Different environments can affect operations in different ways. The analysis should consider elements of the operational environment to include not only physical conditions such as weather, time of day, terrain, but also such things as the backgrounds of employees, relationships to customers, economic conditions and regulatory factors.

Once the tasks and environments have been determined, hazards can be identified via a *what if* analysis. This is similar to a failure mode effects analysis (FMEA), often used in major systems design and analysis. It is using your requisite imagination to project what hazardous conditions might occur. While this might seem like a daunting task, it's really just a matter of working through a set of job tasks and activities, and asking ourselves a couple simple questions such as, what kinds of things could go wrong (failure modes), and what would happen as a result (effects – these are the consequences of the failure)?

The types of hazards we might identify include hazards in operational safety, occupational safety, environmental elements, quality concerns (customer service/satisfaction) and efficiency (waste of resources, damage to equipment). It is

Fuel Service Sample Tasks, Conditions, Hazards, & Errors				
Process	Task/Activity	Workplace Conditions	Hazard	Potential Error

Administrative	Process fuel request	• Office Environment
		• Non-technical staff
		• Entry-Level employees
		• OJT training
		• Fuel orders received by:
		  - Radio
		  - Telephone
		  - Walk-in
			Employee experience is minimal	Wrong type of fuel uploaded to aircraft

Operational – Fuel Delivery	Deliver fuel to aircraft per fuel order	• Ramp environment
		  - Outdoors
		  - Night/Day
		  - Variable
		  -Weather
		• Entry-level staff
		• OJT training
		• 12-hour shifts
		• Two types of aircraft (i.e. piston/turbine)
		• Aircraft sensitive to type of fuel
			• Effects of light (Source)
			• Effects of adverse weather (Source)
			• Minimal employee experience (Mechanism: mistake prone)
			• Long shift length (Mechanism: slip prone)
			• Two types of fuel (Source)
			• Aircraft are fuel sensitive (Target)
				Wrong type of fuel uploaded to aircraft

Figure 1.19   Sample tasks, conditions, etc.

important to consider workplace conditions when identifying hazards. The point is that action can be taken regarding workplace conditions but we cannot directly fix an event, error, or consequence. It is easy to slip into the trap of classifying events, error, or consequences as hazards if we are not careful. Thus, in our example, the hazard is not the misfueled airplane but rather the factors that led to the misfueling. The misfueling itself is the error and the misfueled airplane is the outcome of the error. In the words of Dan Maurino, a well-respected human factors expert, hazards are something that existed 'while the pilots were at breakfast' (Maurino et al., 1995.

While we're going through this analysis, we should also use this as an opportunity to identify all types of potential problems; those related to delivering safe, high-quality products and services with a minimum of waste. Even though the emphasis of this chapter is on system and task analysis, we should also look at all types of hazards, those to customers, employees, equipment and the environment. Managers are responsible for controlling risk in all of these areas, so doing a more comprehensive analysis will prove beneficial in conducting business and setting priorities.

Summary

Recent research has confirmed the importance of system and task analysis. Benner and Rimson (2009) have found that in at least five fatal air carrier accidents, it appears that the accidents were a form of repeats of prior accidents. All the accidents resulted in fatalities and enough factual data has emerged to establish that not one of the accidents was historically unique. Design-related system safety issues were cited in two of the accidents, and system safety implications relating to training and operational issues were identified in the others. Each case repeated well-documented behaviors of prior occurrences from which attainable lessons were not recognized or learned. As such, we argue that system and task analysis are key factors to hazard identification and a successful safety program. Risk management and safety assurance provide us the tools to conduct and maintain an effective SMS, and both of these processes rely on system and task analyses that can identify the various gaps which create performance problems. System and task analyses also rely on performance assessment and gap analysis. The *FAA SMS Assurance Guide* can provide both the certificated operator and the un-certificated operator guidelines to maintain an effective SMS. Additionally, there are numerous materials available that describe the theories, methods and techniques for system analysis, task analysis and performance assessment. Remember, just as in training design, system and task analysis needs a top-down approach. You cannot analyze a system or design a process by ignoring the whole system. Start at the top, and work your way through the various components of the system to examine the processes and tasks that comprise your operation.

The objective in system and task analysis is to better understand the processes that an organization will use to produce its products and services, the activities that the employees in the organization will accomplish in those processes, elements of the environment that could cause those activities to go awry, and what can be done to eliminate or mitigate the risk of those problems. While many sources of hazard information exist, the first line of defense in system safety is to provide a robust design with integral risk controls. Sound system analysis is an essential activity that provides decision makers with the information they need to direct these essential design actions.

References

Adamski, A.J. and Westrum, R. (2003). Requisite imagination: The fine art of anticipating what might go wrong. In Erik Hollnagel (ed.), *Handbook of Cognitive Task Design* (pp. 193–220). Hillsdale, NJ: Lawrence Erlbaum Associates.

Annett, J. (2003). Hierarchical task analysis. In Erik Hollnagel (ed.), *Handbook of Cognitive Task Design* (pp. 17–35). Hillsdale, NJ: Lawrence Erlbaum Associates.

Bedford, T. and Cooke, R. (2001). *Probabilistic Risk Analysis: Foundations and methods*. Cambridge: Cambridge University Press.

Benner, L. and Rimson, I.J. (2009). The curse of the retros. *Journal of System Safety*, *45*, (4), 4–7.

Clark, D.R. (2004). *ADDIE – 1975*. Retrieved September 30, 2007 from http://nwlink.com/~donclark/hrd/ahold/isd.html.

Dekker, S. (2005). *Ten Questions about Human Error: A new view of human factors and system safety*. Mahwah, NJ: Lawrence Erlbaum.

Dekker, S., Hollnagel, E. and Nemeth, C. (2009). *Resilience Engineering Perspectives*. Burlington, VT: Ashgate Publishing Company.

Dessinger, J.C. and Moseley, J.L. (2003). *Confirmative Evaluation: Practical strategies for valuing continuous improvement*. Hoboken, NJ: Pfeifer, John Wiley and Sons.

DoD (Department of Defense) (2000). *MIL-STD-882D: Standard Practice for System Safety*. Washington, DC: Department of Defense.

Dismukes, R.K., Berman, B.A. and Loukopoulos, L.D. (2007). *Limits of Expertise: Rethinking pilot error*. Burlington, VT: Ashgate.

Ericson, C.A. (2005). *Hazard Analysis Techniques for System Safety*. Hoboken, NJ: John Wiley and Sons.

Ericson, C.A. (2008). The four laws of safety: Part 3. *Journal of System Safety*, *44*, (1), 6–9.

Ericson, C.A. (2009). The first principles of system safety. *Journal of System Safety*, *45*, (2), 28–30.

FAA (Federal Aviation Administration) (2009a). *Safety Management System Framework: Safety management system (SMS); Pilot project participants and voluntary implementation of organization SMS programs (rev. 2)*. Washington, DC: Federal Aviation Administration, Flight Standards Service – SMS Program Office.

FAA (Federal Aviation Administration) (2009b). *Safety Management System Assurance Guide: Pilot project participants and voluntary implementation of organization SMS programs (rev. 2)*. Washington, DC: Federal Aviation Administration, Flight Standards Service – SMS Program Office.

FAA (Federal Aviation Administration) (2008). *Aviation Safety System Requirements*. FAA Order VS 8000.367. Washington, DC: US Department of Transportation, Federal Aviation Administration.

FAA (Federal Aviation Administration) (2006a). *Introduction to Safety Management Systems for Air Operators*. Advisory Circular 120–92. Retrieved July 18, 2009 from http://www.faa.gov/regulations_policies/advisory_circulars/index.cfm/go/document.information/documentID/22480.

FAA (Federal Aviation Administration) (2006b). *Advanced Qualification Program*. Advisory Circular 120—54A. Retrieved July 18, 2009 from http://www.faa.gov/regulations_policies/advisory_circulars/index.cfm/go/document.information/documentID/23190.

FAA (Federal Aviation Administration) (2006c). *Introduction to Safety Management Systems for Air Operators*. Advisory Circular 120—92, Appendix 1. Retrieved July 18, 2009 from http://www.faa.gov/regulations_policies/advisory_circulars/index.cfm/go/document.information/documentID/22480.

Haig, C. and Addison, R. (2009). *Instructional System Design Model. Performance Express*. Retrieved from http://www.performanceexpress.org/0905.

Heinrich, H.W. (1959). *Industrial Accident Prevention*, 4th edn. New York: McGraw-Hill.

Hollnagel, E. (2004). *Barriers and Accident Prevention*. Burlington, VT: Ashgate.

ICAO (International Civil Aviation Organization) (2009). *Safety Management Manual*. Document 9859 AN/474. Retrieved July 10, 2009 from http://www.icao.int/anb/safetymanagement/Documents.html

ISO (International Standards Organization) (2000). ISO 9000-2000. *Quality Management Systems: Fundamentals and Vocabulary*. Geneva: International Standards Organization.

McGowan, D. and Marca, D. (1998). *SADT: Structured Analysis and Design Technique*. New York: McGraw Hill.

Marrelli, A.F. (2005). Process mapping. *Performance Improvement Journal, 44*, (5), 40–44.

Maurino, D.E., Reason, J., Johnston, N. and Lee, R.B. (1995). *Beyond Aviation Human Factors*. Brookfield, VT: Ashgate.

Mol, T. (2003). *Productive Safety Management*. Burlington, MA: Butterworth-Heinemann.

Reason, J. (2008). *The Human Contribution: Unsafe acts, accidents, and heroic recoveries*. Brookfield, VT: Ashgate.

Reason, J. (1997). *Managing the Risks of Organizational Accidents*. Brookfield, VT: Ashgate.

Roland, H.E. and Moriarty, B. (1990). *System Safety Engineering and Management*, 2nd edn. Mississauga, Ontario: John Wiley and Sons.

Sparrow, M.K. (2000). *The Regulatory Craft*. Washington, DC: Brookings Institution.

Sparrow, M.K. (2008). *The Character of Harms*. Cambridge: Cambridge University Press.

Standards Australia (2004). *Standard AS/NZS-4360. Risk Management*. Sydney: Standards Australia.

Stolzer, A.J., Halford, C.D. and Goglia, J.J. (2008). *Safety Management Systems in Aviation*. Burlington, VT: Ashgate.

Westrum, R. (1991). *Technologies and Society: The shaping of people and things*. Belmont, CA: Wadsworth.

Westrum, R. (1993). Cultures with requisite imagination. In J. Wise and D. Hopkin (eds), *Verification and Validation: Human factors aspects* (pp. 401–416). New York: Springer.

Westrum R., and Adamski, A.J. (1999). Organizational factors associated with safety and mission success in aviation environments. In D.J. Garland, J.A. Wise and V.D. Hopkins (eds), *Handbook of Aviation Human Factors* (pp. 67–104). Mawah, NJ: Lawrence Erlbaum Associates.

Zemke, R. and Kramlinger, T. (1982). *Figuring Things Out: A trainer's guide to needs and task analysis*. Reading, MA: Addison-Wesley.

Chapter 2
Perspectives on Information Sharing

Part 1: Information Sharing – Types and Principles

Michelle Harper

For any organization striving to implement SMS, it quickly becomes clear that information management is a critical component of the effort. Information concerning the existence of hazards, the monitoring of controls in place to cope with those hazards, and measurements of the effectiveness of controls all must be collected, analyzed, assessed and made available as actionable intelligence. Those responsible for SMS implementation must 'right-size' this effort to assure essential information is obtained and managed, while not adding undue burden to the organization's operations. In general, this right-sizing can be accomplished by asking a simple question – 'are we getting the information we need to manage risk?'

As a SMS matures, it becomes obvious that there are sources of information that are essential in understanding and managing organizational risk that are outside an individual organization's control. Examples are easy to find. An airport SMS depends on safety reports from those who conduct flight operations at the field, yet the airport authority cannot directly control reporting from that operator's employees. An airline needs information about a new route it is considering, but cannot demand relevant safety information about that route from another airline presently flying it. The national-level regulator supporting an Information-Sharing System (ISS) needs information concerning routine operations to detect hazards that occur at rates too small for individual operators or airports to detect, but from a practical perspective cannot mandate that each operator surrender every detail of those operations.

Every SMS has a sphere of direct operational control, and then surrounding it, an entire environment where critical information resides, information necessary for the effective management of risk. A mature SMS requires knowledge of both direct operational control and information about the environment in which the operation occurs. This is true of both an operator-supported SMS and an industry-focused SMS. To establish and grow an SMS an organization must understand this fact, and engage in programs that have access to information about the environment they operate in. Critical to the maturity of both an industry's SMS and a national-level SMS is the sharing of all available information.

Even during the earlier days of aviation safety, information sharing was recognized as a key element of accident prevention. As accidents become increasingly rare and more importantly, technology for collecting data and governance for protecting data have expanded, there has been a shift in the aviation industry. This shift is characterized by a clear movement towards a proactive approach to safety, focused on use of data to identify and mediate emerging hazards.

The shift towards a proactive safety approach has taken on several forms in the last decade or so, each with a specific focus on types of data and methods of sharing that data. Woven into all of these efforts are common elements that have been increasingly recognized as the essential ingredients of the new vision of proactive safety. It's worthwhile to quickly review some of these initiatives in order to identify those elements. As we review them, it will be helpful to keep in mind the complementary and contrasting natures of information-sharing forums, programs and systems.

An information-sharing forum is a method of information sharing that has evolved to meet the needs of its constituent members to exchange important information, but does not itself have formalized methods of data collection, analysis, or distribution of results. Forums can, and frequently do, focus on one particular area of risk, and often this focus includes the sharing of limited sets of data from its stakeholders in order to accomplish assessments of those risks. In general, though, forums are structured to allow stakeholders to share the analyses of their own risks, and their attempts to mitigate identified hazards. The strength of information-sharing forums is that member organizations are able to share methods, insights and solutions with each other, in an unrestricted, informal manner.

In contrast, an information-sharing program supports the explicit goal of the standardization of a target dataset, methods of analysis for that data, and tools for the distribution of results to its stakeholders. An information-sharing program is, by its nature, more formalized, and therefore harder to construct, govern and manage. In the field of aviation, information-sharing programs have excelled at collecting single sources of information and operators have in turn put this information to use in supporting their safety objectives.

In contrast to forums and programs, an ISS is a planned infrastructure consisting of interrelated and interdependent data sources, information processing and analysis tools and information distribution capabilities. These attributes continually influence one another (directly or indirectly) to maintain their impact on the system. As with any functional system, an ISS should support four basic capabilities:

1. Inputs, processing, outputs, and feedback mechanisms
2. Maintenance of internal steady-state architecture
3. Display properties that are used to represent the whole system
4. Defined boundaries that are usually set by the system's stakeholders.

Unlike information-sharing forums and programs, ISSs provide an opportunity to derive a coherent and unified way of viewing and interpreting the operations and the environment supported by an individual or an organization. The aviation ISS is an open system (see http://www.businessdictionary.com/definition/open-system.html), meaning that the system itself supports an integrated and unified view of attributes that are consistently changing. The aviation ISS has 'permeable boundaries' through which a continual exchange of information, needs and feedback must be incorporated. In this way, a successful ISS must support a level of scalability that is driven by an ever changing industry.

A serious challenge for any information-sharing forum, program or system is the same as any business process – to maintain operational relevance to its stakeholders. Since it takes time and resources to construct such initiatives, there is a constant need for attention to assure that the questions the initiative was designed to answer have not been overcome by events, and that the results of its analyses are continually provided to the important decision-makers, the list of which is in constant flux.

Information-sharing forums, programs and systems have complementary and critically important functions. In fact, a mature ISS (in addition to its formalized methods of data collection, analysis and information distribution), has built into it connections to information-sharing forums and programs. These connections serve as conduits for the intelligence gained from access to domain knowledge, data and feedback on the products of the system. Such influences guide the evolution of formalized structures and direction of analytic efforts, all of which drive the ability of the system to maintain relevance.

In the following section is a review of some of the safety initiatives that have played an important role in aviation safety information sharing over the last few decades. You are invited to categorize the distinctions, merits, challenges and ultimately the potential impact of each information-sharing forum, program and system.

Information-sharing Forums

Inter-airline Information Sharing

Virtually every airline in the world belongs to an industry organization that represents the member airlines' interests and communicates those interests to the public, the regulator and to legislators. Organizations such as the Air Transport Association (ATA), the Regional Airline Association (RAA) and the International Air Transport Association (IATA) have grown to become a significant and critical part of the operation.

Typical in all of these organizations is a structure and process that regularly brings together the directors of safety from their member airlines. The safety department of the industry organization is charged with the responsibility of facilitating information sharing between its constituent airlines, and accomplishing aggregate analysis of shared information in ways that the individual airlines cannot do. An example of such an information sharing structure is the ATA's Safety Council. A new attendee at a Safety Council meeting is immediately struck by the remarkable openness and candor of the discussions. The fierce competition that characterizes the airline industry is nowhere to be found.

Such inter-airline sharing is not confined to the United States. Some airlines are organized into 'pods', one of which, for example, consists of Air France, British Airways, Cathay, Emirates, Lufthansa, South African, Swiss, Qantas and Qatar, in which they share the results of their individual FOQA analyses. A similar group that shares a wider range of safety information is LABSKI – Lufthansa, Air France, BA, Swiss, KLM and Iberia.

Inter-airline forums are excellent examples of an essential element of successful information sharing – collaboration. These industry organizations have consistently encouraged safety collaboration so effectively that collaboration is now a norm in the worldwide airline safety culture.

Inter-union Information Sharing

The airline industry is a mature one, and as such many of its employee groups are well represented by unions. To the general public, union representation is generally associated with wage, benefit and working condition issues. But in the airline world, from the very beginning unions have made safety a core concern. Airline employee unions invest a remarkable proportion of their resources to support work safety, and similar to their companies, union organizations have institutionalized safety information sharing. That is not surprising, since front line union members are the ones exposed to risk every working day. They have a personal stake in safety.

Perhaps the most interesting and definitive element that unions have contributed to the maturation of safety information-sharing forums is that of 'assertive engagement'. No one presently involved in safety information-sharing forums would ever consider leaving union representation out of any effort, but union representation in that effort does not depend upon the goodwill of others. They see to it that they are represented. This willingness to step up to the plate and become involved is another core element of success.

InfoShare

Undoubtedly one of the most successful activities promoting the proactive assessment of safety-related incidents is regularly scheduled meetings in which airlines openly present and discuss high-risk events. These meetings, referred to as InfoShare, are held twice a year and are hosted by the FAA. InfoShare meetings are closed to the general public and open only to registered members of the aviation community. The primary purpose of these meetings is to enable airlines to share information on events that they encountered and best practices they have adopted to address identified hazards.

InfoShare meetings have been largely successful due to the growing participation of airline safety representatives and, on average, meetings include over 300 attendees. The latest expansion of InfoShare includes operators representing cabin, dispatch and maintenance operations. Future proposed enhancements to the meetings will include working groups representing a range of aviation service operators which will focus on identified safety concerns. The large presence of airline representatives at InfoShare demonstrates the growing need that operators have for sharing information on systemic safety issues. InfoShare meetings demonstrate two elements of information sharing, protection for the information shared by organizations and support for collaboration across the industry.

The limitation of InfoShare has been that the sharing is completed only after an airline's recognition of a close call or experience with a high-risk event. Additional limitations are based on the fact that this information is shared only twice a year, and although InfoShare meetings include two days of presentations, informal discussions during this time often prove to be of most value to operators. Sharing of safety information through InfoShare has provided and continues to provide a proactive way for airlines to learn from their peers, but this type of sharing does not by itself support a systemic approach to addressing and mediating emerging hazards.

A recent advancement for InfoShare is the establishment of the connection between this forum and a new ISS. Consistent with the relationship between forums and systems discussed above, InfoShare is increasingly serving a critically important role by providing a conduit for information and feedback to this new system. This ISS, referred to as the Aviation Safety Information Analysis and Sharing (ASIAS), will be discussed later in this chapter.

Information-sharing Programs

ASRS

On the morning of December 1, 1974, TWA Flight 514 was inbound to Washington's Dulles Airport. TWA Flight 514 originated in Indianapolis, with an intermediate stop in Columbus, Ohio, and had originally been scheduled to land at Washington National Airport. The flight had to divert to Dulles because of high winds at National. The crew received an approach clearance for a VOR/DME runway 12 at Dulles 44 miles from the airport. At 11:09 a.m., the Boeing 727 crashed into Mount Weather, Virginia, destroying the aircraft and killing all 92 people aboard.

The NTSB investigation revealed that the aircraft had descended below the initial approach altitude for that approach. The primary factors contributing to this error were found to be that the procedures and terminology for issuing approach clearances were not well defined. But the most significant fact that the NTSB uncovered during this investigation was that this was not the first time this type of event had occurred. Six weeks before the TWA accident, a United Airlines flight had made the same mistake on the same approach to Dulles, but fortunately realized the mistake and corrected it. The critical difference between the two situations was that United had implemented an anonymous safety awareness program a few months before, and the crew of that United flight had reported the error. United immediately issued a notice to its flight crews to preclude the recurrence of a near-fatal misinterpretation of an approach clearance. Unfortunately, this information was not made available to any other airline outside of United.

In the aftermath of TWA 514, the FAA was quick to accept the need for the creation of some sort of national safety reporting system, even before the publication of the NTSB final report on the accident. But as the drafts of the program were developed, problems were immediately apparent. After decades of adversarial dealings with the FAA, pilots were extremely wary of submitting any report that could potentially be used against them or their colleagues. It became clear that the only way the system would work was if a neutral, trusted broker of voluntarily submitted information could be found.

In 1976, the FAA entered into a memorandum of agreement with NASA that created the Aviation Safety Reporting System (ASRS). NASA would be responsible for the design, data collection, report de-identification, and aggregate analysis of submitted safety reports. NASA would also be responsible for the dissemination of intelligence gathered from the aggregate analysis.

The critical element incorporated into ASRS, one that is now at the heart of all successful information sharing, is its non-punitive nature. Pilots are encouraged to report threats or errors with the promise that the report will not be used by the FAA in certificate action against the pilot.

Another element that has been critical to the success of ASRS is the concept of trusted broker. NASA, being a neutral third party with no responsibility or capability to assign penalties, is the entity that creates systems to guard against the misuse of identifying information, while passing along the critical safety intelligence contained in the reports. Over the years, the trusted broker concept has matured into very robust systems that guard against the abuse of information.

LOSA

The Line Operations Safety Audit (LOSA) is a proactive safety program that includes the observation and recording of flight crew performance in a normal operating environment. LOSA observations are completed by trained observers sitting in the jump seat and recording flight crew behavior. Audits are conducted under a non-jeopardy agreement – traditionally signed by the participating airline, pilot union representatives and the research organization or entity completing the observations. During the completion of a LOSA, the trained observer records potential threats to safety, how the crew managed these threats, associated errors committed by the flight crew, and a range of other types of *behavioral markers* that characterize the flight crew's performance. The information collected through the completion of a LOSA is retained by the airline and is often used for measuring and monitoring identified hazards through the completion of repeated audits. As an alternative to internal completion of a LOSA, airlines can opt to complete a LOSA with support from the LOSA Collaborative. This option enables the airline LOSA data to be part of a repository of audits completed by a large and diverse number of airlines, which enables airlines to be provided with benchmarking information, or information on how their audit results fall in comparison to other airlines represented in the repository.

A few words on the term 'audit'. Although the term can have negative connotations, a LOSA is only conducted under the voluntary cooperation of the airline safety department and representative pilots union. Furthermore, any crew approached to be part of the LOSA project can opt out – meaning the participation of the crew itself is voluntary. And, of most importance, information collected during a LOSA can only be used to support proactive safety initiatives and cannot be used in any way to penalize a crew. In this way, LOSA supports the non-punitive, non-jeopardy information sharing element.

LOSA was initiated in 1991 through the University of Texas at Austin Human Factors Research Project under the guidance of Dr. Robert Helmreich. The program was initially funded by the FAA through the Human Factors Division with the goal of establishing a methodology that could be used to identify and record crew resource management (CRM) practices. The primary focus of this research was to develop a process to identify the ways in which CRM was used by crews during normal line operations, as opposed to the training environment, where previous research on CRM had been completed. The result of moving the observations of flight crews out of a training environment and into normal operations expanded an airline's knowledge of factors that impact crew performance, and most importantly, how flight crews manage these factors. This expanded knowledge includes information regarding how pilots manage or mismanage threats in their operating environment, as well as how they manage the errors they commit as a crew.

During the development and refinement of LOSA, a model for collecting and categorizing information related to the LOSA observations was developed. This model is referred to as the Threat and Error Management (TEM) model. The TEM model supports both a taxonomy and a framework that provides a relational structure for the concepts of threats, errors and their management. The model is based on the premise that threats and errors are an integral part of ongoing flight operations and must be managed by the flight crews. Collection and categorization of information related to threat and error management provides a level of detailed information that can be used to identify countermeasures used by a flight crew to address threats and their own errors. The TEM model has received widespread support throughout the aviation community due to its practical approach to categorizing both positive and negative characteristics of the performance of an operator during normal operations.

In 2002, ICAO published the LOSA Manual and endorsed LOSA as a primary tool to be used by airline operators to develop countermeasures to human error. The number of operators supporting a LOSA program and the completion of reoccurring audits as part of their proactive safety approach has grown constantly since this time. This trend has also expanded to a large number of major international operators from different parts of the world and across diverse cultures.

LOSA supports several elements of information sharing; use of a third party or trusted broker to collect and protect data, non-punitive use of data, access to previously unknown issues and support for benchmark findings. These elements enable airlines to identify areas of potential risk in their own operations. LOSA provides extremely important insight into these essential elements of a mature safety ISS, and especially the relationship of such systems to SMS. First of all, LOSA is process-based. Before any observations are made by LOSA auditors, an exhaustive study is completed concerning what processes should be observed in normal operations, what threats might be observed, and what errors might take place. This *process-based SRM/SA* (i.e., safety risk management/safety assurance) is exactly as SMS would have it. Second, LOSA has at its roots the TEM model. Far from being an ivory tower concept, the TEM model provides a guide for how to support a data-driven approach to safety by providing a framework for identifying what is important to observe and what those observations mean. Bob Helmreich liked to quote Curt Lewin, one of the giants of social science theory, by stating, 'There is nothing more practical than a good theory.' The success of LOSA across the aviation industry demonstrates that a theoretical foundation can guide the work of the information-sharing effort and make it more efficient and effective.

This leads to perhaps the most important contribution that LOSA has given to the safety information-sharing revolution now underway. The airline professionals involved in the creation of LOSA did not try to do it alone. They understood that their area of expertise was in flying airplanes and operating an airline, not in constructing data collection systems, building databases and analyzing complex data. They reached out to academia and the research community, collaborating outside the silos of the airline world, and found the expertise needed to succeed in establishing an ISS. This critical insight – *that standard processes for collection and sharing of information outside of an airline's operation can be used to generate information for improvements to internal safety practices* – has greatly contributed to the proactive safety advances for both the airline and industry.

FOQA

Flight operations quality assurance (FOQA) is considered to be the most objective of the proactive safety programs, in that it collects data recorded by an aircraft's flight data acquisition and recording systems and aggregates it into a database that incorporates part or all of an airline's fleet. Routine flight operations are then analyzed by sophisticated software that identifies exceedances and trends. The evolution of FOQA programs has been associated with both information-sharing forums, programs and ISSs. Indeed, a discussion of the events surrounding the publication of 14 CFR 13.401, commonly referred to as the 'FOQA rule', illuminates many of the problems involved with information sharing, as well as the solutions that have been created through joint efforts by government, industry and labor.

The following discussion is excerpted from the previous book in this series, *Safety Management Systems in Aviation* (Stolzer, Halford and Goglia, 2008.)

From the birth of proactive safety programs in the 1990s, there has been the assumption that information and intelligence gained from one operator's program would somehow be integrated with that of others, and that the result would be a greater understanding of the issues and risks associated with the question under study. The exact method that this sharing should take, though, has been perhaps the most controversial issue in the short life of these programs, and more than once has threatened their very existence.

The FAA has been consistent in its explanation for desiring the submission of aggregate data from proactive programs. In the preamble to 14 CFR 13.401, the FOQA rule, the FAA described one of its motives (FAA, 2001):

> One of the principal benefits to the FAA and to public safety of aggregate FOQA data submission will be the opportunity it affords to target the limited resources available for FAA surveillance to those areas where it is most needed.

In the years leading up to the publication of the FOQA rule, both industry and labor expressed concern over submitting data or information from these proactive safety programs to the FAA. There were two aspects to this concern. First, there was worry that the FAA might use such information in a punitive action against the reporter. Second, the legal staffs of both airlines and unions were especially concerned about the possibility that any information submitted to the federal government would be available to the general population under the Freedom of Information Act (FOIA).

Part 193

To answer these concerns, in 2001 the FAA enacted 14 CFR Part 193, Protection of Voluntarily Submitted Information. This regulation provides protection for safety information voluntarily submitted to the FAA from both misuse by regulators and, in most instances, from release to third parties under FOIA.

The concerns of the industry and unions regarding misuse of safety data were understandable. Many industry professionals felt that the FAA had demonstrated a propensity to use punitive measures such as fines and certificate action as the primary levers by which it attempted to ensure safe operations. But in the years since the publication of the FOQA rule, this suspicion has generally been replaced with the realization that the leadership of the agency had made a serious long-term commitment to changing its approach, wherever possible, from punishment to partnership. This commitment has now been continuously demonstrated through multiple changes of leadership at the FAA, and the reasonable observer has concluded that this enlightened partnership approach to safety has been firmly institutionalized.

With protective regulation such as Part 193 in place, many of the original objections to information sharing on the part of industry and the unions were answered, but there was still considerable disagreement as to the appropriate method in which sharing should take place. The FOQA rule was purposely vague about sharing methodology, stating only that the sharing must take place 'in a form and manner acceptable to the Administrator'. This vagueness resulted in a somewhat disjointed evolution of the national data-sharing initiative, but there was no better way to proceed. That very vagueness was necessary to allow industry, labor and regulator the time to develop dialog and trust, and time to explore which methods would prove most likely to result in meaningful data-sharing.

Directly after the FOQA rule was put into place, an Aviation Rulemaking Committee (ARC) was established to determine how best to create that *form and manner* of sharing. During months of discussion and negotiation within the ARC, sharing was accomplished in an information-sharing forum – regular InfoShare meetings (discussed above), in which members of the community would present the results of studies from their proactive programs, primarily from FOQA and ASAP. The ARC process, being charged with the creation of policy recommendations, was frequently contentious and difficult, but informations shared by airlines through InfoShare was consistently remarkably open. It was InfoShare that better represented the true desires and motivations of industry, union and regulator.

By 2004, the FOQA/ASAP ARC had completed all the homework necessary to begin an ISS – the Distributed National Archive (DNA). ASAP and FOQA data, voluntarily provided by airline partners, would be shared in a protected system, analyzed collaboratively and managed by a trusted broker. The distributed nature of the network allowed proprietary data to remain on company property, while still allowing researchers access. NASA was invited to build the program due to its experience in FOQA research, and in turn invited the University of Texas at Austin's Human Factors Research Project to assist in the creation of the ASAP archive. Purposely limited to a two-year demonstration project, NASA successfully demonstrated that the DNA could function and deliver meaningful studies on national safety issues. More important than the specific conclusions of those studies was the experience of the various stakeholders in the project as they worked together to produce the studies. The FAA continued to demonstrate its enlightened approach to partnership and non-punitive action, and the airlines continued to share sensitive and proprietary data and fully engage in the conduct of the studies. By the time the demonstration project was complete, important precedents had been set, and trust had been earned.

Aviation Safety Action Program

The purpose of the Aviation Safety Action Program (ASAP) is to provide airline employees and other aviation service operator employees with a way to report safety critical issues, including errors they have made, to their safety departments without the risk of incurring punitive consequences. The first handful of airlines to start ASAP programs did so with the belief that further safety enhancements to their operation would require the identification of infrequent yet hazardous or high risk issues that only their employees, operating on a daily basis in the national airspace system (NAS), would be able to identify. There was further belief that not only could operators provide the most immediate information regarding potential safety risks, but that the airlines themselves were best suited to develop corrective actions to address these issues. At the time that the first ASAP programs were initiated, the only sources of this type of information were mandatory reports, required if the event included a violation of a Federal Aviation Regulation (FAR), and voluntarily disclosed reports provided by employees to the airline, if the reporter felt it important enough to place themselves at risk of potential punitive action.

The powerful response the airlines and their employees have had to supporting ASAP is driven by the fact that the program offers an alternative to the standard FAA legal enforcement policies and internal company disciplinary action. The alternative ASAP offers is in the form of an assessment of the reported event completed by the Event Review Committee (ERC). This committee is composed of a member of the airline safety department, the representative employee union and a member of the FAA. The ERC reviews every ASAP report and determines the appropriate 'corrective action' to be completed by the reporter. Based on the consensus-derived decision of the ERC, when a reporter meets specific reporting criteria (the reporter must not have intentionally disregarded safety), and the reporter completes the corrective action, the reported event is closed with no further action.

As such, ASAP represents a significant departure from both the normal FAA enforcement actions and programs like ASRS. Previous to ASAP, ASRS was the only reporting program to offer limited protection to reporters who provide information on errors they committed that were associated with an FAR. ASRS offers protection from suspension of a pilot's certificate but not from potential federal repercussions of the violation of the FAR. Under ASAP, a report will be accepted and the reporter provided with protection from federal punitive action, if provided within 24 hours of the time the employee became aware of the event and if the reporter did not commit actions that implicate an intentional disregard for safety. This decision is reached by the ERC, which fully reviews all available information to determine why the event occurred. If consensus is achieved by the ERC on acceptance of the report and corrective action is completed by the reporter, the report will be closed and any record of the FAR deviation kept from the employee's record.

Beyond protection from potential FAA punitive action, ASAP provides an additional incentive to reporters. The program provides reporters with an established method to report safety concerns that they may have encountered from a single incident or ongoing problems that they believe could lead to a more serious incident or potential accident. The program enables reporters to communicate these issues directly to their airline safety department with the knowledge that it is required that the report be reviewed and the reporter be provided with feedback. As a result, the types of issues reported through ASAP not only include events where the operator was in error but also safety concerns that are encountered and could lead to more serious events.

Approximately ten years following the release of the initial FAA Advisory Circular on ASAP, there are 91 aviation service providers that have active FAA-approved ASAP programs. Over the ten years of development much has been said about ASAP. One of the simplest statements of the impact that ASAP has had is 'ASAP is a fast, pragmatic approach to proactive safety'. ASAP has been so successful that it has expanded to include additional employee groups, including dispatchers, flight attendants, maintenance employees, load controllers, flight following support and ground personnel. And, as discussed below, March 2008 marked the much awaited start of an ASAP program for air traffic personnel.

Air Traffic Safety Action Program

In March 2008 the ATO, NATCA, and FAA signed a memorandum of understanding that launched the Air Traffic Safety Action Program (ATSAP). ATSAP has been established as a system for all air traffic personnel to voluntarily identify and report safety and operational concerns. ATSAP is modeled after the Aviation Safety Action Program and supports similar voluntary, non-punitive and cooperative policies and protection for reporters, in that no decertification or negative credential actions can be taken against a reporter who provides information to ATSAP. The goal of ATSAP is to encourage self-reporting of safety events and to use this information to derive appropriate actions to improve flight safety.

The establishment of ATSAP fills a large gap in the information available from the air traffic segment of the NAS, which is information on what types of threats and what types of errors are occurring at the level of the air traffic controller. This type of information, provided directly by the air traffic personnel, has historically not been available, and even established safety reporting programs like the ASRS receive a very minimal number of reports from air traffic personnel. Reports from air traffic personnel make up less than 5 percent of the ASRS database, with the majority of reports coming from pilots.

Following the model established by existing ASAP programs, ATSAP data is currently being used to identify hazards and support the development of mitigation strategies. Even in the first year of the program's initiation, stories of proactive use of the data are being reported. These include intervention strategies aimed at improving air traffic personnel training and equipment, mitigation strategies to address reported issues occurring at the terminal level and outreach and coordination with the airline community. These initial success stories from ATSAP demonstrate the proactive power of a non-punitive, voluntary safety program. At the time of the publication of this book, the FAA's Air Traffic Organization and the controller's union the National Air Traffic Controllers Association (NATCA) have agreed to support the integration of ATSAP into ASIAS (discussed later in this chapter), further demonstrating a desire for industry collaboration.

The Principles and Structure of Information-sharing Systems

An ISS enables a proactive approach to safety by securing a reliable and continuous stream of data representing a range of data sources, the processing and fusion of these data sources and the analysis of this data to enable hazard identification and the distribution of information to establish and monitor future safety enhancements.

An ISS should ideally support an organization in the completion of the following objectives:

1. Access to a continuous stream of system-wide or systemic sources of data representing routine operations and reported incidents.
2. Processing and fusion of this data in preparation for analytic activities.
3. Analysis of fused data sources to derive results on potential hazards.
4. Confirmation and assessment of the severity associated with the identified hazards.
5. Development and implementation of safety enhancements to address hazards.
6. Monitoring of enhancements to ensure identified hazards have been addressed and the continued monitoring of known hazards.

Six Principles for Managing Data in an Information-sharing System

The primary reason aviation safety has been able to advance from a reactive model of accident and known incident assessment to a proactive approach to improving aviation safety is the growth in availability of data. As previously mentioned, this growth of availability of data is the result of advances in technology and establishment of legal protections for previously unavailable, proprietary data. These advances have in turn been fueled by the continued endorsement by the industry to support information-sharing forums, programs and systems. An increase in the availability of data has resulted in previously unknown operator errors and threats contributing to these errors to be accessible for analysis. It is reported by airlines with ASAP programs that over 90 percent of the events reported through their programs would not otherwise be known to their safety departments. As additional types of information are made available through advances in technology and legal protection, the potential to provide a near complete picture of the NAS is growing. The long history of industry and regulatory groups working together to negotiate the availability of these data sources provides this opportunity. The current task now faced by the industry as a whole and the individual operator is to take this opportunity and turn individual silos of data into integrated components of an ISS.

The following six principles support the generation of reliable, valid, reproducible and traceable results from the utilization of shared data sources. The principles provide guidelines for establishing a set of objectives for the input, processing and output of an individual data source as an integrated part of an ISS, as well as guidelines for the maintenance and growth of a system as a whole.

These principles include:

1. Data source management
2. Data model and taxonomy standards
3. Reliability framework
4. Data quality standards
5. Data fusion
6. Configuration management.

As discussed in the previous sections, information-sharing programs support the availability of data and the forums provide direct insight into the needs of the operators. An ISS provides the architecture for the input, processing and output of information derived from the information-sharing programs and a reliable infrastructure to incorporate the industry's feedback and needs for expanded capabilities.

It should be acknowledged that the means for establishing and maintaining the principles listed above are ongoing activities. The objectives outlined for the establishment and maintenance of each principle may need to be customized for each data source as an integrated part of a system. This customization may be based on the proposed utilization of a data source as well as the resources available to an organization. One additional note: these six principles provide a set of objectives to be strived for whether an ISS is supported by a single carrier or a larger information sharing system, such as ASIAS, focused on system-wide NAS operations.

Data Source Management

There are two requirements for managing a data source that will result in an established and continuously contributing source of information for an ISS. First each data source should be assessed for its utility and role in the ISS. Second each data source should be maintained as an integrated part of an ISS.

Data Source Assessment

Each potential data source should be evaluated for inclusion into an ISS through a data source assessment. The purpose of the completion of a data source assessment, in very basic terms, is to identify how a data source will expand the capability of an organization's SMS and how the data source will become part of the organization's ISS. In other words, a data source assessment identifies the data source's utilization and the requirements for the data source's input, process and output. This assessment should include identification of the gap that the data source will fill in the organization's knowledge of its operations and goals that will be accomplished through the utilization of the data source. This information can then be used to drive the requirements for the processes needed to support the integration of the data source, the analytic processes to be applied to the data source and the proposed output that the data source will provide to the organization's ISS.

The objectives of a data source assessment include:

1. Establish the purpose of the utilization of the data source in terms of the gap that the data source will fill in an organization's knowledge of its operations.
2. Assess the requirements needed to support the input, processing and output of the data source as part of the organization's ISS.

The produce of a data source assessment is a living document that is made available to analysts or any other user who will be accessing or using the data source to complete analytic activities. Ongoing support for maintaining accurate information on the data source is ideally supported by feedback from analysts and other users and monitored for accuracy by an individual identified as an expert on the data source, or a *data steward*.

Data stewardship The second component of data management is data stewardship. Each data source included in an ISS should be managed by a set of data stewardship responsibilities. These responsibilites include initial support for assessment of the data source, and once the data source becomes an operational component of the ISS, the data steward is responsible for the management of the data source. Data stewards are responsible for identifying and distributing information to analysts and other users who count on a reliable feed of the data, and to identify and document any information that they deem necessary to complete data sharing and analysis activities. It is their responsibility to provide the technical support to enable the data source to be an active part of an ISS, to support analytic use of the data source and to identify and help prioritize expansion activities.

The following is a list of responsibilities to be maintained by a data steward:

1. Establishment and maintenance of data source assessment document.
2. Collaboration and coordination with the data source provider.
3. Generation of requirements and maintenance of the data source's taxonomy and database model. This task should include identification and support for maintaining industry data standards applicable to the data source.
4. Requirements development and management of the data source's reliability framework.
5. Requirements development and accountability for data quality.
6. Support and monitoring of a data source's metadata, including completion and output of the data-processing services supported by the reliability framework.
7. Ensure all changes that may impact the integration, analysis or utilization of the data source are documented and made available to analysts or users.
8. Requirements development and management of fusion with other data sources.
9. Requirements development and oversight of governance and information security requirements for access and use of the data source. This information may include de-identification and retention requirements by the data source provider.
10. Development of documentation of the history of the generation of the data source, including configuration management information on the input, processes and output of the data source.
11. Aid in the prioritization and expansion activities associated with the use of the data.

The data stewardship activities can be supported by a selected individual or groups who are responsible for maintaining, monitoring and documenting data processing information and expansion, as well as coordination with analysts and the organization's users. It should be noted that responsibilities of a data steward(s) is not to complete the integrations tasks, monitoring, or maintenance of a data source, but coordination and oversight of these tasks. In this way, the data steward(s) supports the utilization of the data source as a stable and reliable component of an ISS and ensures that the data source is being utilized effectively and to its full potential.

Data Model and Taxonomy Standards

This data management principle supports two sets of objectives; one regarding an individual data source's database structure, and the second regarding the specific data standard values that are applicable to the data source. The purpose for the support of these objectives is to maximize the utility of a data source for fusion with other data sources and for use in analytic activities. The following guidelines should be used to support the implementation of an appropriate data model and taxonomy standards for a data source:

1. A relational database model should be established for each data source in an ISS. A data model should comply with an established table structure and individual record format that has been identified to support analytic and fusion activities.
2. The database should support the three forms of normalization.
3. Key fields should be used to support fused connections to multiple data sources supporting the same taxonomy standard value.
4. Standard taxonomy values should be made available for querying an individual database and across multiple databases in the ISS.
5. Data model and taxonomy standards are used as a reference for analysts and other users. These components of an ISS are resources for referencing a data source, supporting analytic activities and for fusion with other data sources.

Data Model   A data model should support a database structure that will enable data loading, data fusion and querying needs. A relational database structure is the most commonly used database model supporting information similar to that collected in operational settings like aviation as well as manufacturing industries. A relational database model provides a logical and consistent structure for variables and their relationships between tables supported in the database model. A properly designed relational database model should support the first three levels of normalization. A database that has no repeating elements, no partial dependencies on key fields and no dependencies on non-key attributes is considered to support the three forms of normalization. Structuring a database to support these normalized forms eliminates anomalies associated with updates, insertions and deletions of information from a table. A full review of the levels of normalization that can be achieved in a relational database model is beyond the scope of this book, but based on an assessment of data sources common to the aviation industry, a relational database structure that includes the first three forms of normalization will support the basic needs for querying and analysis of information from a data source.

Taxonomy Standards   Taxonomy is a set of concepts that have been established to provide a set of references and a framework to be used for the retrieval assessment and analysis of data from a database. The technical and semantic references used to define the concepts represented in taxonomy are normally maintained in a data dictionary. A data dictionary should include both technical references and semantic definitions describing the taxonomy. The characteristics of a reputable taxonomy include a framework that separates concepts into groups and subgroups that are mutually exclusive, comprehensive and clearly defined. As previously discussed, the TEM model is an example of a framework that aids the user in understanding the concepts identified by the taxonomy. Taxonomies can be derived by an organization internally or an organization may choose to adopt an established taxonomy. The key determinant in this decision should be based on the identified purpose of the data source as part of the ISS. The success of a taxonomy in supporting the identified purposes of the data source is based on the extent to which the data includes features that can be identified in the taxonomy or the application of the taxonomy to the data source.

A complex taxonomy may be impressive to reference as part of an ISS or when describing an organization's SMS, but the true measure of utility of taxonomy is based on the extent to which the fields in the taxonomy can be easily defined and used to identify concepts in a data source.

The second measure of success of taxonomy is the extent to which it can comply with industry, national and international taxonomy standards. Established taxonomy standards support common naming conventions, formats and definitions. These standards include commonly used references such as aircraft type, phase of flight and occurrence categories such as hazards or human factor issues. The establishment and support of taxonomy standard values as part of a data source's relational database model enables the standard value to be used as a common association or key field across multiple data sources.

In 1999, the CAST/ICAO Common Taxonomy Team (CICTT) was formed with the support of the International Civil Aviation Organization (ICAO) and the Commercial Aviation Safety Team (CAST). The CICTT supports the development of common standards for taxonomies to be used across the aviation industry.

See Table 2.1 for an example of the CICTT data standard model for aircraft make/model/series.

Table 2.1   Use of CICTT data standard model for aircraft make/model/series

Current system	Make/manufacturer	Model series	
Accident/incident data reporting	Airbus Industries	A300-B2/B4	
International register of civil aircraft	Airbus Industries	Airbus A300B4-605R	
CICTT valid value	**Make manufacturer**	**Model**	**Series**
	Airbus	A300	B4 605R

Source: CICTT (2009).

The CICTT includes experts from several air carriers, aircraft manufacturers, engine manufacturers, pilot associations, regulatory authorities, transportation safety boards, ICAO and members from Canada, the European Union, France, Italy, the Netherlands, the United Kingdom and the United States. The unifying position maintained by this group is that a lack of international standards for identifying common descriptors used through the aviation community was severely impeding the ability to integrate data from multiple sources. The inability to integrate or fuse data sources was resulting in missed opportunities for the assessment and identification of common safety issues.

The following list includes the current CICTT standards for taxonomies that have been established or are currently under development:

Taxonomies standards – current

1. Aircraft make/model/series
2. Aircraft engine make/model/submodel
3. Phases of flight
4. Aviation occurrence categories
5. Engine occurrence subcategory

Taxonomy standards – under development

1. Hazards
2. Human factor issues
3. Positive taxonomy
4. Aerodrome taxonomy

The CICTT provides an international industry resource of standards that can be used to support an existing taxonomy or aid an organization in the development of a new taxonomy. The CICTT does not provide the taxonomy itself. The adoption, or creation, of taxonomy is the responsibility of the organization supporting the ISS.

Incorporating taxonomy standards as part of a data model enables data sources to be linked not only across an organization's internal ISS, but also across ISSs supported at the industry level. Creating standards that operators around the world can support helps remove one barrier to information sharing. Through the use of taxonomy standards, the aviation community's capacity to identify rare but high-risk safety incidents and to better understand common factors contributing to hazardous operating environments is greatly enhanced.

Reliability Framework

Each data source should be supported by a multi-tiered and configuration-managed data acquisition and processing system referred to as a reliability framework. The purpose of the reliability framework is to provide a reliable and flexible architecture that can be used to support a range of tasks required for data to be made available in a database that is integrated into the ISS. The objective of a reliability framework is to capture, map, transcribe and load a data source into an established production level database. In this way, a reliability framework enables the monitoring of data quality, the importance of which will be discussed in the next section.

Reliability frameworks are unique to each data source, but all support a set of data processes that are independent, yet linked to support the flow of data from its raw state to the post-processed, or production state, where it is made available for analysis. In general, each data source should be supported by a reliability framework that includes the following processing and monitoring services:

1. Data capturing services that access, tag and integrate raw data from the data source.
2. Data transformation services that support the reformatting of the data's file format and table structure to support the data model requirements.
3. Data parsing services, including data transcription and loading to support the coding of data to meet the data standard values.
4. Post parsing services, including data quality assessments, including anomaly filters and when applicable, automated fixes applied to the identified data quality issues.
5. Data expansion services including data fusion with other data sources.
6. Data survey services including failure detection, monitoring, and alerting at each data process level.

The result of the successful flow of data through the steps of the reliability framework is a feed of newly processed data to the data source production database. This database is consistently populated with a feed of transcribed, reformatted and filtered data. A production database in an ISS should include only data that meets the required data standards and passes through data quality filters as supported by the reliability framework. This data is, therefore, ready to be accessed by an analyst, fed into an automated monitoring system or fused with other data sources. The completion of these steps supports the principle of data quality.

The reliability framework enables each individual data processing service to be executed, monitored and the status of the completion of this service logged. Any failures in the data processing services or records that are identified as including bad data through detection and filtering processes are pulled from the reliability framework work flow process and stored in separate anomaly databases. Anomaly databases are then available for review and the data assessed for potential corrections. The collection of information or metadata at each step in a reliability framework supports the principle of configuration management.

Data Quality

Data quality can be broadly defined as the fitness of a data source to support a set of identified uses. Data quality standards are unique for each data source and should be based on the identified purposes and use of the data source as outlined in the data source assessment. Data quality standards include requirements that a data source must support before the data source is *production ready*.

Across all data sources, three types of data quality values should be identified: *stability, completeness* and *accuracy*.

1. *Stability* of a data source can be measured by the extent to which the data source supports the required and scheduled data feed.
2. *Completeness* of a data source can be measured by the extent to which data fields identified as required in the records are present.
3. *Accuracy* of a data source can be measured by the extent to which the data fields represent the targeted concept.

These three types of data quality can be monitored based on results of data source's processing services as supported through reliability framework. The data quality processes supported in a reliability framework can be designed to provide information about anomalies associated with stability, completeness and accuracy as tracked during the ongoing integration of data. Each type of data quality should be measured with the result being identified ranges of acceptance. Any data quality issues that fall outside this range are considered anomalous. An anomaly is defined as any measured parameter or characteristic of a data set that is outside an established set of identified data quality requirements. The occurrence of anomalies is inevitable in any source of data and the reliability framework simply provides the means to create metadata on the integration of a data source and the ability to monitor data to ensure it meets a minimum set of requirements. The establishment of data quality processes is an ongoing task that should be supported by feedback from users, interpretation of this feedback into data quality processing services and the monitoring of metadata resulting from the processing of data through the reliability framework.

For an ISS, use of a data source will continually expand – if in fact the system is seen as a trusted source of information. As a result, adjustments and expansion to data quality requirements and the monitoring processes as supported through the reliability framework must also be expanded.

The achievement of data quality is an ongoing task for any organization managing data sources utilized in an ISS. The goal for achieving an acceptable level of data quality is not based on high levels of stability, completeness, or accuracy, but the extent to which these qualities are measurable and, therefore, known to analysts and other users who will be interpreting the results of the use of the data source.

Data Fusion

Data fusion is one of the most important components of an ISS. It is the ultimate goal of the sharing effort itself in that it combines the information obtained from all the stakeholders in the sharing system. Data fusion enables analysis of the combined dataset in ways unavailable to individual stakeholders operating alone. It is the most promising information-sharing technique available and, at the same time, the most sensitive, since it can potentially link information contained in sensitive and proprietary datasets with information from open, non-proprietary datasets, thus potentially identifying the proprietary data. The entire aviation community recognizes the value of fusion, and also recognizes that care must be exercised in constructing data fusion processes.

The fusion of data sources can be completed through two types of processes. First, it is possible that a study completed using information from an ISS can be enhanced by the insight gained from additional sets of information that are available from other data sources. In this type of fusion, records from one dataset are not tightly linked to the other (on a flight-by-flight basis, for example), but rather are analyzed in such a way as to allow each dataset to illuminate one aspect of the question at hand. This type of fusion is completed at the analyst level and is not a standardized or automated process supported by the system. This type of fusion also requires participation and careful review by subject matter experts to make sure that the subsequent understanding resulting from this fusion makes sense. The usefulness of this type of fusion comes from the fact that, once established, this domain level knowledge can help drive the second type of fusion or automated links between data sources.

The second type of data fusion is created through the use of data standard values, which are used consistently and in an automated fashion to establish relationships across multiple datasets. The integration of data standard values into the data source's database model enables the standard data value to be used as a key field or linked value to other data sources. The common naming conventions supported by the data standard values can then be incorporated into query and search interfaces. This type of data fusion expands a data source's data model. The ultimate key field for such fusion is the flight record itself, which allows analysts to fuse data from multiple datasets in order to construct a complete description of that flight, for example, the exact weather at touchdown (from weather sources), the traffic density during departure (from radar sources), the software version used by the ground proximity warning system (GPWS) box during a terrain awareness and warning system (TAWS) event (from equipage databases), etc. Fused datasets, such as these, allow the analyst to explore correlations and contributing factors in new ways.

The complete view of the flight record ultimately provides an in-depth understanding of the operational environment and operator role in the occurrence of hazard. It is this level of detailed information that is strived for in an ISS.

While such fusion of datasets through linkage of flight records is highly desirable from the analyst's point of view, from the larger perspective, it is important to keep in mind that some of the richest datasets are proprietary. The owner of that dataset has every right to control access to that dataset, and there are legitimate concerns about misuse of such data. This tension between the productive use of data fusion and the legitimate concerns of data providers regarding anonymity is simply an inherent component of data sharing. It clearly accentuates the importance of governance processes in data sharing, and most importantly, it emphasizes the absolutely critical element in any data sharing program – non-punitive use of data. Data fusion holds tremendous promise to become perhaps the most important technique in an ISS's support of SMS information analysis, and also requires that stakeholders supporting data sharing initiatives take very seriously their role in governance, and their responsibilities to use shared information for proactive safety incentives.

Configuration Management

Configuration management can be defined as the establishment and maintenance of the consistency of an ISS. The primary objective for configuration management in an ISS is the assurance that information resulting from the system is reproducible and the methodology traceable. This objective can be achieved through the monitoring and documentation of changes made to the system's data sources, analytic tools and processing and infrastructure for information distribution. The focus of these monitoring and documentation tasks can include changes to hardware, software and other physical attributes of the system. An additional focus of configuration management can include a focus on non-physical attributes of the system including methodology used for completion of studies, design and requirements changes and revisions to governance and data use requirements.

Supporting configuration management requirements of an ISS can be a labor-intensive and ongoing process, but defining the scope of these requirements can minimize the needed resources. For an ISS, the scope of configuration management should be based on two of the primary objectives of the system: (1) to produce information that is reproducible, and (2) to establish methodology that is traceable. Supporting the reproducibility of information, or findings, generated from an ISS can be accomplished through the generation of metadata identifying the data sources and individual records used in a study. The generation of monitoring and documentation of analytic processes, hardware and software support the ability to trace the methodology used to generate the findings.

Reproducibility

In order to reproduce findings generated by an ISS, two primary sources of information must be available. First, references to the individual data sources and the specific records should be retained. This information may not include the full database or records due to storage or governance constraints, but at a minimum references to the unique record identifiers should be retained. The source of information needed to support reproducibility includes a full description of the methodology used to generate the findings. This second source of information is available if the system includes configuration management information related to traceability. This component of configuration management is discussed overleaf.

Traceability

In order to support the requirements that methodology used to produce results from ISS be traceable, configuration management information can be collected on the state and functional capability of the system at the time the data was accessed for analysis. This information should include the hardware configuration, the versions of software used for integration and the descriptions of the analytic processes used during a study.

Configuration management, like data quality, can be a labor-intensive process to establish and maintain, and may not be considered vital or even important for an ISS to support. But it should be noted that if the data records and the methodology used to generate findings comes into question, access to configuration management information can very quickly become critical to the perceived trust and resulting use of the ISS.

Metadata

Before summarizing this section, a few words about metadata are in order. In very simple terms metadata is data about data. But metadata can also be generated from almost any hardware, technical, analytic process and from the data source itself. Metadata can be generated on the analytic processes used to derive findings, including individual data source processing and the use of these data sources to populate individual, fused and derived databases. Metadata can also be collected on processes that are applied to subsets of data sources pulled from the databases and used to support analytic activities. And finally, all processes, including fusion techniques, data quality filters and analytic processes applied to the derived databases can be also tracked through the generation of metadata databases.

One of the standard ways to generate metadata can be through the data source's reliability framework. Metadata from this component of an ISS can include historical data load and data provider information, data quality filtering, anomaly detection and anomaly fixes. Also, through the reliability framework changes to data transcription, loading and parsing processes can be retained. Collecting this type of metadata not only supports configuration management, but also enables more efficient troubleshooting and recovering from failures associated with data integration, and increases knowledge of emerging data quality issues.

It is essential to collect metadata in an ISS to ensure knowledge of the functional state of the system and ensure this state supports the use of the information derived from the system. In order to determine what type and level of detail of metadata to collect, a focus should be made on deriving information on data integration and analytic processes, as these components of an ISS have direct impact generation of and interpretation of findings.

Summary

Establishing and maintaining each of the principles outlined in the previous sections can create a large drain on an organization's resources. Therefore, it is not uncommon for the resources needed to support these principles to be prioritized as less important than generating a continuous stream of results or supporting a consistent request for the completion of studies. The reasons for this are clear, in that the simplest way to show a return on investment of an ISS is to generate findings that can be used to identify potential precursors to high-risk events. However, of equal importance is an organization's confidence that the results generated by the system are reliable, valid, reproducible and traceable. The inability to demonstrate these objectives can have a negative and possibly irreversible impact on the perceived return on investment of an ISS.

Information-sharing System Architecture

In order for an ISS to be a reliable resource of safety information, integration processes, analysis capabilities and data access platforms must work as part of an integrated system. The following section provides an overview of the basic architecture of an ISS. The architecture of an ISS supports the data management processes that are applied to individual, derived and fused data sources and the processes used to complete analytic tasks and generation of safety information. The primary objectives of this ISS architecture are:

1. Provide analysts and users with a stable and reliable environment for accessing and utilizing data and distributing information.
2. Provide analysts with a production and testing environment that enable the development of analytic capabilities to support the generation of reliable, valid and reproducible results.
3. Provide a stable yet flexible infrastructure for expansion through integration of additional data sources and increased access and distribution of safety information to appropriate stakeholders.

Underlying these objectives is the requirement to assimilate and utilize vast amounts of aviation data from many different sources, often stored at different locations, provided in varying intervals and subject to many different usage and governance constraints. Each data source also presents a unique set of data quality issues, data model standards and configuration management requirements. Also, as an ISS becomes an established resource in the aviation community, study requirements and support for analytic capability required to extract relevant information from supported data sources continues to expand.

The wide variety of expanding data sources and the growing requirements for the analytic uses of the data, management and expansion of a system present several challenges. These challenges are addressed through the processes supported in each component of the information sharing architecture.

There are three main components of an ISS discussed in this section:

1. Data acquisition and management
2. Analytic capabilities
3. Information access and reporting.

These three components support data acquisition and transformation workflow processes, a platform for the development and implementation of analytic services and an information access layer supporting analyst and stakeholder interfaces. Figure 2.1 provides an overview of this architecture.

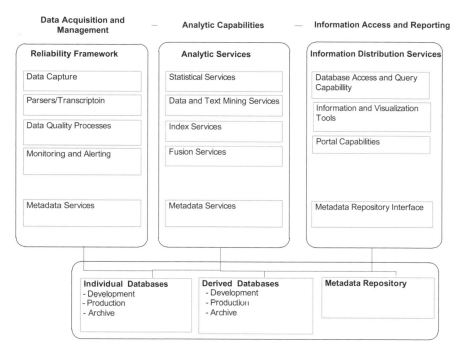

Figure 2.1   An information-sharing system architecture

Data Acquisition and Management

ISS data sources can be provided from a range of industry and government sources, in varying formats and intervals, and through several types of capturing processes. In order to maintain reliable feeds of data required to support analytic capabilities, each data source supports a reliability framework.

The objective of the reliability framework is to support the processes required to make an individual data source easily accessible in an established standardized format and a stable database structure. The reliability framework also provides information in the form of metadata to ensure that analytic services applied to the data result in reliable, valid and reproducible results. The reliability framework supports a set of data-processing services that feed data to a development, production, or archive database.

Analytic Capabilities

The analytic capability component of the ISS supports a platform for analysts and developers to develop, test, validate and implement various types of assessments to be applied to data sources. These capabilities can include a range of tools like data parsing and transformation, data mining, derived data parameters, data fusion techniques and visualization methodologies. The analytic platform enables analytic capabilities to be developed into data services, and these services to be tested and validated before being fully integrated into the ISS. The establishment of these services is required for the initial and periodic production of results, including the generation of metrics, benchmarks and other analysis processes. It is at this level of the information-sharing architecture that research and development tasks can be completed, integration of new analytic tools takes place and the capabilities of the system to provide a feed of information are enhanced.

Information from the use of the data services can be fed to a metadata database that retains historical information on the data sources used to support the study results, the queries and data processes applied to the data source and configuration management information, including versioning information associated with the data service.

For example, in the ASIAS system benchmark results are provided to airlines on a quarterly basis. The production of these results requires data to be pulled from a set of databases, generated by the FOQA analysis system and stored as an analytic database that is separate from the consistently expanding FOQA database. The data residing on the analytic database are processed through a set of analytic data services. These services include the data quality filters, data transformations and various algorithms that support the defined requirements for generating information on the FOQA benchmark topic. Following the development, testing and validation of these analytic services, the services are integrated into the ISS as part of an automated process that is used to generate benchmark results each quarter. The associated analytic databases are then archived, along with the metadata generated by the use of the analytic services, and made available for future use.

Information Access and Reporting

The third component of an ISS supports the requirement that the stakeholder community is provided with access to safety information generated by analytic activities, and that the processes supporting the analysis and generation of study results are made available for review. This level of the ISS is the *face* of the system.

Capabilities at this level of the architecture can include access to processes used to generate study results and interfaces to metadata repositories, including access to production level data sources, metadata repositories, analytic and visualization tools and, most importantly, access to summarized information. The distribution of results is the focus of information access. However, access to information from the metadata repository systems enable a more complete understanding of the processes and data used to generate the distributed information.

The *data access and reporting* component of an ISS can be supported by a portal. A portal is a Web-based interface that enables restricted-use access to reports, data, and additional information that is required to support information distribution. An information-sharing portal provides centralized and user-restricted access to summarized information from studies, views of identified data sources and additional information.

These three components of information-sharing architecture provides an example of how an ISS can be organized to support the establishment and monitoring of a basic set of data management principles as outlined in the previous section. It should be noted that the success of an ISS may not be based on a rigid architecture and static set of requirements for supporting data management principles. Instead, an organization will want to ensure that the architecture and the requirements supporting each principle of the data management of an ISS be scalable.

Summary

As discussed at the beginning of this chapter, an aviation ISS is an open system, meaning that the system itself supports an integrated and unified view of objectives that are consistently changing. To ensure continued use and longevity, an ISS will need to support one last feature – scalability. It is inevitable that an organization will make changes to policies governing access and use of the data, advances in analytic processes and methods for information distribution will create opportunity for further insight into data and changes in the prioritization of resources for supporting a proactive approach to safety may increase (or in most cases will fluctuate). These changes should be welcomed, not dreaded, as they may increase the utility of the system to support SMS.

The aviation industry is not the only industry motivated to manage large amounts of information and use this information to produce actionable intelligence. These needs are present in almost any industry producing, monitoring, or generating information. As a result, there have been huge improvements in the types and capabilities of information management and assessment software tools. These tools support a range of capabilities – from processing services as described in the reliability framework, query and data mining capabilities to support analytic services, and visualization and reporting tools for supporting the communication and distribution of information. As a result of these advancements, technical resources previously required to build these capabilities can now be placed on maintaining and supporting expansion of the ISS and for analyzing data to be distributed to stakeholders. It is important that these tools provide a level of transparency, meaning that the processes and logic used by the tools are available to an organization. In other words, a *black box* software tool will not easily support an open system. Therefore it is recommended that organizations looking to start an ISS or to enhance an existing system complete a market assessment of information management tools.

Part 2: The Aviation Safety Information Analysis and Sharing System

Don Gunther and Jay Pardee

Aviation System Safety: Context and Purpose

The aviation safety community has made significant progress in advancing aviation safety through a forensic approach, decreasing safety rates to the lowest levels in history. This approach to safety incidents has successfully led to addressing known safety risks by identifying safety enhancements and lowering of the rates of fatal accidents and hull losses to the lowest levels in history. The challenge before the Federal Aviation Administration (FAA), and the aviation safety community it oversees, is to drive the commercial aviation fatality rate lower in parallel with the evolution to the Next Generation Air Transportation System (NextGen).

The FAA is consistently and methodically moving towards an overhaul of the NAS. The research, development activities, new technology and capabilities established to support this overhaul are referred to as NextGen. The primary goal for the support and implementation of NextGen capabilities is to enable air travel to be more efficient, convenient and dependable while maintaining and increasing the high level of safety that currently characterizes the NAS.

The anticipated growth and increasing complexity of the air transportation system as it evolves into NextGen are likely to present challenges not only for capacity and efficiency, but also for aviation safety. As new Air Traffic Management (ATM) procedures and technologies are introduced into the NAS, all aviation stakeholders will need to adapt to this changing ATM system. These changes will create a dynamic environment with a focus on continued safety enhancements of the NAS, but will also require focused monitoring, adjustment and responsiveness that current aviation safety systems are not capable of maintaining.

The FAA has responded with an objective in the 'FAA Flight Plan 2009–2013' of a 50 percent reduction by 2025 in commercial aviation fatalities per 100 million on board over the baseline metric in 2007. In support of this objective one of the strategies is to 'promote and expand safety information-sharing efforts, including FAA-industry partnerships and data-driven safety programs that identify, prioritize, and address risks before they lead to accidents'.

The FAA and the aviation community have determined that in order to continue to improve their safety record and achieve the FAA's Flight Plan objective, a different approach is required. This approach is one that utilizes a proactive data-driven approach to aviation safety, focusing on vulnerability discovery and risk management through the leveraging of vast amounts of safety data available across the aviation community.

To be statistically representative, a proactive data-driven approach to aviation safety requires the management, analysis and monitoring of large volumes of data shared by the users and operators of the NAS. The objective of such an approach is to identify potential vulnerabilities before an accident or severe incident occurs. Such vulnerabilities lie embedded in data collected through multiple venues, including automated recording systems generating immense volumes of digital data, self-reported incidents and high-risk occurrences generated directly from operators supporting the NAS, and monitoring data collected directly from audits of normal operations. These sources of data provide insight into initiating events that can become consequential when occurring together and when not properly identified, managed and mitigated.

With a focus on the goal of establishing a collaborative information-sharing and analysis system involving industry and government, the ASIAS initiative was conceived, with the collaborative endorsement of the industry and government for the sharing of data. The FAA, in collaboration with the aviation community, has established a governance process, implemented an information-sharing architecture and integrated data from an initial set of participant stakeholders, including all major commercial FAR Part 121 operators and a growing number of regional operators. With the knowledge that this data holds information on accident and incident precursors, ASIAS has started the process of actively analyzing, and monitoring the data for events contributing to potential high-risk safety issues.

A primary objective of this proactive approach to aviation safety is to leverage and intelligently mine growing data sources in a way that provides new insights into safety issues that may not be identified at the individual stakeholder level. This approach requires the establishment of governance policies, endorsed by both government and industry, to protect safety data, much of which includes proprietary and sensitive information. These governance polices establish the requirement for collaborative agreements between groups sharing data. At the center of these collaborative agreements is the requirement that data shared must only be used to enable safety-driven objectives and not for punitive or regulatory activities. Through these careful and considered processes, data becomes available that holds information on vulnerabilities that might otherwise remain hidden until uncovered in post-accident investigations.

ASIAS and Safety Management Systems

ASIAS is a cornerstone of the SMS. An SMS provides a coordinated approach to use information-sharing systems like ASIAS to proactively identify risks and a systems approach to mitigating and monitoring identified operational risks. As mandated by the ICAO, an SMS is comprised of four principal functions: safety policy, safety promotion, safety risk management and safety assurance.

In the past, aviation safety programs have been based on a forensic approach to safety management which involves the investigation of an event, the determination of the root cause and the development and implementation of corrective action(s). A forensic approach is purely reactive: an event or incident occurs, root causes are investigated and changes are made to prevent a recurrence.

The SRM and SA elements of SMS supplement the traditional, forensic approach to safety management through introduction of a proactive and ultimately preventative approach. Under SMS, multiple data sources are used to identify safety hazards (those circumstances or conditions which, if not corrected, may result in a safety event). Data analysis is continually used to assess risk, identified contributing factors are determined and corrective actions are developed and implemented to prevent a safety event from occurring.

The aggregation, analysis and monitoring of data from multiple sources in ASIAS will enrich the SMS process by enabling comprehensive data fusion capabilities required for comprehensive analyses. These analyses, along with the ability to perform vulnerability discovery, will lead to a more complete understanding of contributing factors and extend industry and government knowledge of initiating events beyond the traditional search for single root causes. The analysis results will also support the development and measured effectiveness of mitigation strategies. ASIAS provides risk-assessment capabilities that can identify emerging safety issues that may otherwise be undetectable through individual data sources or may be missed by operators who lack insight into hazards that are occurring across the NAS.

The focus of the use of data in ASIAS is to aggregate data coming from numerous operators and operations across the aviation community. In this *aggregate* form, the data may present statistically significant trends across a wide range of operated aircraft types, equipment types and airports operations as well as unmonitored areas of the NAS. ASIAS data aggregation and analytic capabilities provide ASIAS participants, the FAA and other government agencies and those entities responsible for the planning and advancement of NextGen capabilities with substantially more robust data upon which to base safety risk-management decisions. In addition, the fusion of data sources that represent the current operational status of the NAS enables multidimensional assessments and monitoring of operations that can be used to predict potential precursors to accidents and serious incidents.

Through analysis of safety data from both industry and government, ASIAS data supports NextGen with an in-depth and comprehensive perspective of operational risks that currently exist and operational risks that could be introduced through changes in ATM procedures, airspace design changes (i.e., sectors and routes), area navigation (RNAV) procedures, airport use, avionics and fleet mixes. ASIAS capabilities further support the development, implementation and monitoring of mitigation strategies implemented to proactively address identified emerging safety issues as the NAS evolves to NextGen. The mission of ASIAS is to provide the data, technology and actionable results to the FAA and ASIAS stakeholders to enable them to optimize SMS risk management performance.

Answering the Call with ASIAS

In response to these challenges, the FAA has endorsed ASIAS as the primary system for the collaboration of the federal government and the aviation community to share data and proactively advance aviation safety. Participation has expanded rapidly to include 34 FAR Part 121 operators by January 2011, representing a majority of the total number of domestic US commercial flights. Many other potential stakeholder participants representing the national and global aviation community have expressed an interest in participating in ASIAS. The current objective for ASIAS expansion is to include all members of the aviation community operating in the NAS and to continue to expand the capabilities supporting ASIAS information sharing, analysis and monitoring technologies as ASIAS participation increases.

Information Analysis

ASIAS conducts various types of studies and analyses under the direction of an ASIAS Executive Board (AEB), of which the authors are co-chairs. The AEB includes representatives from government and industry, and by consensus determines which areas of study are the highest priority in terms of risk and which studies are most likely to result in actionable intelligence. ASIAS analyses include:

1. Directed studies: in-depth assessments of special topics of interest to the ASIAS partners.

2. Known-risk monitoring: a set of analyses that are continuously performed to monitor known safety risks of interest to the ASIAS users.
3. Safety enhancement assessments: development of safety metrics to identify and continuously monitor safety hazards and the effectiveness of changes endorsed by industry and government stakeholders to address known risks.
4. Vulnerability discovery: identification and validated assessments of previously unknown issues or accident precursors.
5. Benchmarking: development of industry-wide metrics applied to aggregated, national data sets to create a point of reference for ASIAS partners to perform safety assessments of their own operations.

In the interest of enhancing aviation safety, the results of these analyses are shared with the ASIAS community and, as approved by the AEB, the aviation community beyond ASIAS.

Information Access

A key element of the ASIAS concept is the information-sharing within the aviation safety community. This includes both data considered proprietary by ASIAS stakeholders, results of ASIAS analyses and the tools developed for ASIAS.

Currently ASIAS retains access to a wide variety of data sources, each of which provide information from various components of the aviation airspace system. ASIAS's data sources include both public and non-public aviation data. Public data sources include multiple sources of information collected and maintained by government agencies, such as data related to ATC traffic, procedures and weather. Non-public sources include de-identified data from aircraft operators, including digital flight data and safety reports submitted by flight crews.

Operating within ASIAS, safety analysts can query millions of flight data records and de-identified textual reports via a secure communications network that connects servers throughout the country. This data is fused with a large number of additional FAA, industry and other government data sources to further understand the context of safety reports.

A key data source for ASIAS is the airline provided FOQA data, which captures digital data from on-board flight recorders. FOQA characterizes both routine and non-routine operations in an objective, quantitative manner. Another significant source of safety information is the Aviation Safety Action Program (ASAP) instituted by the FAA and the airlines over the past several years. This is a source of voluntary reports that describe in detail the circumstances and information surrounding an incident.

ASIAS has established access to proprietary FOQA and ASAP data through governance agreements with participating operators and owners of these databases which enable access to this data under strict controls. The MITRE Corporation's

Center for Advanced Aviation and System Development (CAASD) – a federally funded research and development center (FFRDC) – maintains the ASIAS databases and provides expert analytic support to AEB-approved studies, while assuring that proprietary data is protected as defined by ASIAS and FAA rules.[1]

In July 2010, the FAA's Air Traffic Safety Action Program (ATSAP) data was incorporated into the ASIAS data set. The fusion of this data with the airline FOQA and ASAP data along with other data sources containing information about air traffic, aircraft performance, weather, etc., is providing opportunities that have never before been achieved. Agencies that participate in NextGen through the Joint Planning and Development Office (JPDO) are also providing relevant safety data for incorporation into ASIAS.

As new aviation communities, for exmple helicopters and general aviation (GA), are included in ASIAS, they will bring additional and new types of data which will substantially increase the value of analyses using this data fused with the airline data, particularly with respect to the identification of contributing factors.

ASIAS Principles

The success of ASIAS relies on the continued voluntary participation of the aviation community and willingness of stakeholders (like the commercial aviation operators) to provide information that is proprietary, sensitive and which historically has been unavailable. With the knowledge of the critical importance of continued access to such data, the FAA and the aviation community established key principles for the access and use of ASIAS data. These principles are grounded in a common goal of improving aviation safety in the US and around the world. The key principles of ASIAS include the following:

1. Information will be used for the advancement of aviation safety only and never for punitive or regulatory actions.
2. Procedures and policies will be established through collaborative governance based on industry and government joint decision making.
3. The interests of all stakeholders will be balanced.
4. Results generated will be made available to stakeholders, government entities and industry safety groups with the goal of supporting industry-level safety initiatives and advancements in safety management and risk mitigation strategies.

---

1   In 2007 MITRE was selected by the FAA to develop ASIAS, by leveraging a large number of government-provided data sources, its existing expertise in data management and analytic capabilities and the legal protections it provides to data sources as a Federally Funded Research and Development Center.

Benefits of ASIAS

The ASIAS vision is to serve as a central conduit for the exchange of aviation safety information and analytical capabilities not only among its participants but across the global aviation community. The benefit to the aviation community lies in the capability of ASIAS to provide insight into emerging risks that may not have been detected through the assessment of an individual data source. The key benefits of ASIAS to the aviation community are:

- Benefit 1 – aviation data sharing. Data is shared and aggregated among ASIAS users to more clearly see precursors to accidents. ASIAS will aggregate disparate aviation safety data sources in a central repository, increasing its potential value for analysis-based insight and providing insights that would not be available if data is not shared.
- Benefit 2 – advanced safety analytical capabilities and analyses. Analyses using advanced safety analytical capabilities can be performed that would not be available to individual stakeholders performing similar analysis. These advanced safety capabilities will support analysis of comprehensive data which will unlock new insight about potential safety risks in both the current NAS and as the NAS evolves to NextGen.
- Benefit 3 – knowledge sharing of safety insights. Safety insights from ASIAS analyses will be communicated to the ASIAS users. Stakeholders will leverage insight to identify risk-reducing alternatives or changes to operations or processes. Implemented changes will prevent would-be accidents. Safety insights from ASIAS will be applicable to a broad range of aviation communities (for example, commercial, GA, helicopters) and JPDO and other civilian agencies involved with aviation operations (for example airport operators, airport authorities).

Achieving the ASIAS Vision

The vision for ASIAS is that the aviation community will be able to:

1. Identify systemic risks
2. Establish safety baselines of current operations
3. Identify known and newly emerging system vulnerabilities
4. Monitor safety trends
5. Evaluate identified risks
6. Estimate probability
7. Assess severity
8. Uncover event precursors
9. Diagnose event causation

10. Evaluate safety enhancements of NAS operations and as it evolves to NextGen
11. Assess the probable effects of safety enhancements through studies
12. Monitor intervention effects
13. Assess the effectiveness of safety enhancements in accordance with metrics established by the CAST and other safety organizations.

Achievement of the ASIAS vision has required a multi-year cooperative effort involving numerous industry representatives and FAA support. As previously mentioned the number of participating airlines has grown to 34, as of January 2011, demonstrating an increase of more than five times the number of airlines that were participating less than three years ago. ASIAS participation is now focused on including all segments of the aviation community and providing proactive data driven analyses to enhance aviation safety. By the middle of this decade, ASIAS will be instantiated within the US aviation community as a critical component for identifying vulnerabilities, monitoring NAS-wide operations and assessing current and future safety improvements. This goal will be achieved through analysis of aggregated and fused data, both proprietary and government provided, through the establishment and growth of information sharing systems and through the sharing of results within the aviation community for the implementation of safety improvements.

References

Federal Aviation Administration [FAA] (2001). *Flight Operational Quality Assurance Program, Federal Register, Volume 66, Number 211*. Retrieved December 3, 2007 from http://www.epa.gov/fedrgstr/EPA-IMPACT/2001/ October/Day-31/ i27273.htm.

The CAST ICAO Common Taxonomy Team (2009). Aircraft Make/Model/Series Standards. Retrieved November 2, 2010 from www.intlaviationstandards.org/ MakeModelSeries.html.

Ganter, J. H., Craig, D. D. and Cloer, B. K. (2000). *Sandia Report - Fast Pragmatic Safety Decisions: Analysis of an Event Review Team of the Aviation Safety Action Partnership* (No. SAND (2000-1134)). California.

Stolzer, A.J., Halford, C.D. and Goglia, J.J. (2008). *Safety Management Systems in Aviation*. Burlington, VT: Ashgate.

Chapter 3
Top Management Support

Mark A. Friend

Introduction

John Safer reviewed the airline accident records for the last five years. By looking at accident reports and workers' compensation claim records, he learned that his company had suffered 27 'accidents' and had incurred direct medical costs of just over US$100,000 due to injuries incurred while handling baggage. From his education and experience, John was aware of a number of intervention techniques that could take place, but all would cost money. He has no reason to believe that, unless some sort of intervention takes place, the numbers of injuries and the associated costs will continue to rise. The big question in John's mind is, 'How do I convince management to spend the money needed to intervene?'

The Value of Safety

According to the renowned economist Milton Friedman, 'The business of business is business' (Davis 2005). Management expert Peter Drucker, following years of experience working with top American corporations, said, 'It is the first duty of business to survive. The guiding principle is not the maximization of profits but the minimization of loss' (Drucker, 1986). The bottom line in most business decisions is usually one of profit which either results from increasing revenue or reducing expenses. This chapter will focus on the latter and on how to show its importance to management. Safety must compete with every other enterprise within the business to show a return on investment. In a recent conversation with a manager in a large corporation, he stated that any new initiative must show a return of at least 10 percent. To put this in perspective, at the time of writing savings accounts are now returning less than 2 percent; certificates of deposit are paying between 3.3 and 4.5 percent depending on the period of time invested between six months and fifteen years; mortgage rates are in the 6 percent range and treasury bills are returning less than 5 percent. In other words, the company is expecting a rate of return on any project better than would be expected in most other healthy, well-planned investments. That is the management perspective and must be the perspective considered when making the case for any safety initiative. It is up to the safety professional to provide the evidence to management that the investment in safety is worthwhile, and this can be done. Aside from understanding return, management is also required by law to provide a safe environment for its customers and its employees. Lastly, providing safety is good for the company, its customers and its employees; it is the morally correct thing to do.

Safety is Required by Law

Common carriers are required to provide their customers with safe transportation – by law.

Title 49 of the United States Code, subtitle VII, chapter 447, section 44702 states that:

> When issuing a certificate under this chapter, the Administrator shall consider the duty of an air carrier to provide service with the highest possible degree of safety in the public interest and differences between air transportation and other air commerce.

This section of the public law makes management of safety a specific legal responsibility for air carrier management teams and, as such, is a fundamental principle of the FAA oversight doctrine. While this section applies specifically to air carriers, the FAA expects all certificated organizations to make safety a top priority and holds their managements accountable for doing so (FAA, 2006). In other words, the FAA legally holds air carriers and their management team responsible for the safety of the flying public.

In addition, the Williams-Steiger Occupational Safety and Health Act (OSHA) (General Industry Regulations, 2008) of 1970 requires, in part, that every employer covered under the Act furnish employees employment and a place of employment free from recognized hazards causing or likely to cause death or serious physical harm to employees. The Act requires employers to comply with occupational safety and health standards promulgated under the Act, and to comply with standards, rules, regulations and orders issued under the Act which are applicable to their own actions and conduct. This regulation applies to all ground personnel working for US airlines and all employees in the manufacturing sector. At the time of writing, the flight deck crew is still under the authority of the FAA. Not only is the safety of the flying public protected under federal legislation, the employees are protected as well.

OSHA has concluded that effective management of worker safety and health is decisive in reducing the extent and severity of work-related injuries and illnesses. Effective management addresses all work-related hazards, including potential hazards, whether or not they are regulated by government standards. OSHA reports a basic relationship between effective management of worker safety and health protection and a low incidence and severity of employee injuries. Such management also correlates with the elimination or adequate control of employee exposure to toxic substances and other unhealthful conditions. OSHA states that effective management of safety and health protection improves employee morale and productivity, as well as significantly reduces workers' compensation costs and other less obvious costs of work-related injuries and illnesses (OSHA, 2010). OSHA urges all employers to establish and maintain programs in a manner which addresses the specific operations and conditions of their worksites.

Responsibility for Safety

The legal responsibility for safety clearly rests with top management. It is charged with carrying out the major responsibilities of the organization. This function is referred to as *line* management responsibilities. Organizations generally have two types of reporting relationships: line and staff. Safety typically operates from a staff position in the organization; it has a strictly advisory relationship with management. Line authority flows from the very top to the bottom of the organization in direct reporting relationships. In other words, there is a direct line of reports that carefully follow the chain of command. The line is typically charged with carrying out the major function of the organization. In a company that manufactures aircraft, the line does the building of the aircraft. In an airline, the line flies the planes. Line doesn't operate in a vacuum; it depends on advisors for information regarding accounting, marketing and safety. This information is generally given as advice. Line listens and then decides how it will incorporate that advice into daily actions.

If individuals in accounting, marketing, safety, or other advisory areas are given decision-making authority regarding that function, the advisor is no longer in a staff-authority position, but is in a functional-authority position. Most safety professionals and practitioners are in pure staff positions. They monitor the safety function and they advise management on what to do. Management listens to the staff advice and then decides whether or not it will abide by that advice. It is the staff's responsibility to provide a quality input and then to convince management of its value. That value may be based on legal requirements, but it can also be based on cost–benefit. If management deems the advice to be cost effective; that is, it sees the implementation of the decision as one which will provide a return on investment comparable to that to be received in other operations of the organization, it will likely take the advice. If the decision is likely to yield a lower return, then the advice may be ignored. Consideration of safety advice should also account for intangibles such as positive versus adverse publicity, goodwill in both the employees' and the public eyes, and long-term relationships within the organization. It is up to the safety practitioner to convince management of the importance of those items and how they relate to the advice.

The Chief Executive Officer (CEO) of an organization is considered a part of line management. The CEO's executive vice-president and his or her reports are also considered part of the line. It is the line-management responsibility to ensure the adoption, implementation and accomplishment of appropriate safety rules and regulations within the organization. It is the job of line management to carefully assign roles and responsibilities for safety. Management must adopt all appropriate practices, assign responsibilities for those practices and enforce adoption of them on the part of all employees. Management must assure input is received from all employees, including the safety professional, but management assumes all responsibility for carrying out the policies and assuring that all employees adhere to them. Management must ensure that all parts of the safety program are fully integrated into the organization and that each individual knows their role relative to safety. This requires effective communication and promotion of policies; evaluation and rewarding of employees based partially on safety criteria; and an effective, progressive disciplinary system to be applied when employees don't follow the safety rules. It is critical that management assures safety operations are in place and fully integrated into the employee network, and that it always maintains full responsibility for the safety function. When something doesn't work well, the question becomes, what can management do differently to assure this and similar incidents do not happen in the future? Management must look for and correct inappropriate safety practices. Both the FAA and OSHA make this clear: safety is the legal responsibility of management. As mentioned earlier, it is the role of the professional safety department to monitor the overall system and assist management by providing advice as to steps it should take regarding safety. The safety professional must make convincing arguments based on the business case for safety. It is essential that the safety department make management fully aware of the cost benefit of the safety function.

Identification of Loss Producing Exposures

Minimization of loss, known as *loss control*, is largely a function of priorities. The first step is to determine where losses may occur; that is, find the process, activities, devices, or people likely to lead to an accident. This is done through a systematic evaluation of all exposures. *Systematic* is the key to doing this well. Attempt to find a technique that comes close to assuring that all high-risk activities are accounted for and considered. Most companies begin by reviewing accident-investigation records, insurance and workers' compensation claims and any other means they have of accumulating accident and near-miss reports. If internal records are not available, industry records, available on government websites and through industry sources can be useful. OSHA reports inspection results by Standard Industrial Classification (SIC) code, so users can narrow their results to their own specific industry. Typical problems found in companies similar to your own are ones which may create a loss-producing incident for your company if proactive steps are not taken to prevent them. The FAA reports accident and incident data. The reports are available by aircraft type and summaries of specific events are also online. The FAA also promotes the use of the Aviation Safety Information Analysis and Sharing System (ASIAS) on its website. Through ASIAS, the FAA promotes the open exchange of information regarding safety, weather, turbulence and aviation accidents. The information is openly shared with the goal of alerting others as to problems that might eventually adversely affect their own operations.

Another common approach to anticipating problems involves following the chain of events leading to the final delivery of a product or service. In an airline operation, the approach may be to look at how the purchasing or leasing of an aircraft takes place. What safety considerations are there and how are they incorporated? Once the aircraft is delivered, how is it placed into service and what processes are instituted to assure continuous safe operation? Follow each of those processes from beginning to end and look for hazards or potential pitfalls along the way. Consider the expected life of the plane and every interaction that occurs in terms of employees, passengers, maintenance personnel, or any person or object inside or outside the organization. Consider the *potential* hazards associated with each,and inventory them. In this phase of the process the ultimate goal is to identify hazards and potential hazards that may have an adverse effect on the operations and the bottom line of the company. Once this has been done, the next step is to consider the potential costs of those hazards. It is the exposure and the consequence of an event that determines its risk. In other words, *risk* is the likelihood of a loss times its potential cost.

Evaluation of Risk

Given a tax rate of approximately 50 percent on every dollar earned, every dollar saved equates to roughly two dollars in pre-tax earnings. Pre-tax earnings are often a fraction of the total revenue. A company receiving a 10 percent return on investment must generate 10 dollars to earn one dollar, so a one dollar saving is roughly equivalent to 20 dollars in sales after taxes. Many companies operate in the net revenue range of less than 5 percent – some are even less than 2 percent. If a company is only finding a 2 percent return on total investment, after taxes, it must generate fifty dollars in gross revenues for every one dollar of loss control savings.

Management may not necessarily comprehend the importance of loss minimization, but the typical manager does understand the value of dollar-saving initiatives. The question becomes, how much is it worth to fix a given problem? Assume an essential process in the maintenance shop causes one employee to lose a finger each year. How much would the company spend to fix the problem? What if the finger is lost on an average of once every five or ten years? Would the amount be different? What if the loss is someone's leg or someone's life? Would the amount be different? How much would the company spend? Can a finger, an arm, or a life equate to dollars and cents? The immediate answer to that question is, 'Yes'. Companies and societies make decisions on a cost–benefit basis daily. Lives are sacrificed for the sake of convenience, time, or other factors. Many times, monies can be assigned to a problem and the problem can be solved, but no one is willing to spend the amount of money required.

Questions often arise as to how injuries or deaths equate to risk. The first response is often to state that human life is invaluable and any cost is justified if a life will be saved. As altruistic and selfless as that may sound, it isn't doable. There are risks inherent in most activities and that risk can be dramatically reduced or eliminated at some cost. Full vehicular searches of all cars entering the airport area could eliminate the risk of explosions at the terminal, as could full screening of all people entering the terminal from a remote parking area. Risks are higher on international flights and lower on domestic flights, but risks exist on all flights. For some amount of personal inconvenience, time and even more money, all passengers could be screened and searched to assure they have no weapons. All threats of weapons being carried onboard by passengers could be virtually eliminated if the airlines go to full strip searches of everyone wanting to fly. For some additional amount of money, air marshals could be assigned to all flights. Risks of highjacking and related terrorism could be virtually eliminated, but at what cost?

Of course, these measures aren't reasonable due to the high expenditures in terms of time and manpower; yet lives could and would be saved. Speed bumps on the major highways approaching the airport could virtually eliminate high-speed accidents, thus saving dozens or even hundreds of lives per year. Highway patrol could also be stationed every 200 yards or so. In other words, some unreasonable level of investment could save numerous lives. All loss of life in aviation accidents could be eliminated by eliminating all flying. The cost in terms of time and money are too great. Unintended consequences would be the increased loss of life in other forms of transportation. The costs involved in saving lives are sometimes so great as to make them totally unjustifiable. That cost includes the money, time and resources that neither companies nor society are willing to invest. At what cost do we willingly sacrifice a life? For most loss control evaluations, that decision must be made. In line with that thinking, one must also bear in mind that the life at risk is a live person – a husband or wife, son or daughter, father or mother, brother or sister, or good friend. The life saved may be your own. So although the decision may be one of dollars, it is never without emotional cost and drain on the resources of the workplace, the family and society.

We, as a society, as voters, or as consumers, are willing to accept the residual risk of incurring the consequences of accidents, terrorism, or the results of other risks in order to save time, money and inconvenience. Risks are thought to be low so we assume them. They are 'acceptable'.

Note in Figure 3.1 how risks may be categorized as to priority. The priorities are based on the probabilities times the potential costs, that is, the risk. Risk equals probability multiplied by potential costs. Those events with extreme risk are addressed first, followed by events with high, then moderate, and finally low risk. Risks can be quantified as will be seen later in the chapter.

Likelihood	Consequences				
	Insignificant (Minor problem easily handled by normal day to day processes)	**Minor** (Some disruption possible, e.g., damage equal to $500k)	**Moderate** (Significant time/ resources required, e.g., damage equal to $1 million)	**Major** (Operations severely damaged, e.g., damage equal to $10 million)	**Catastrophic** (Business survival is at risk, damage equal to $25 million)
Almost certain (e.g., >90% change)	High	High	Extreme	Extreme	Extreme
Likely (e.g., between 50% and 90% chance)	Moderate	High	High	Extreme	Extreme
Moderate (e.g., between 10% and 50% chance)	Low	Moderate	High	Extreme	Extreme
Unlikely (e.g., between 3% and 10% chance)	Low	Low	Moderate	High	Extreme
Rare (e.g., <3% chance)	Low	Low	Moderate	High	High

Figure 3.1   Risk matrix (Drservices.com, 2010)

Losses

Hazards and their effects on assets result in losses. Once hazards are determined, an estimation of the effect of the hazard on assets must be made and an average cost of a loss is established. This can be a relatively straightforward task in terms of a problem area where multiple losses have occurred. In the case of a fleet of vehicles, the average cost per incident may be readily determined. The same could be true with any group of injuries such as baggage-handling accidents. The goal is to generate a list of possible losses and the costs of each. In a case where the potential losses could cover a wide range, an estimated average will suffice. The goal is to generate an educated estimate of relative values. In areas where exposures are low or infrequent, industry averages may be used to determine an expected loss rate. If losses occur an average of once every other year, with no other mitigating factors, the probability of a loss in a given year will be 50 percent, so 50 percent becomes the multiplier in the determination of risk. If multiple losses occur per year, the probability of a loss will be 100 percent or 1, so instead of using probability of loss to determine level of risk, use a multiplier equal to the number of losses occurring per year. In other words, if a loss occurs two times per year on average, the multiplier is two.

The next step in determining level of risk is to assign an estimated amount to each loss or potential for loss. This can be done through examination of historical records or consideration of industry figures. Sometimes, the loss is capped by law. For example, an examination of records shows that airline attendants working for a given airline suffer a company-wide average of ten injuries per year. Closer examination reveals that company direct medical costs equate to US$5,000 per injury. Other indirect costs associated with the injury include additional administrative time for rescheduling, time spent on investigation of each of the incidents, other related costs reveal an additional US$5,000 that can be accounted for. Total costs spent on injuries of this nature reveal that approximately US$100,000 per year is spent by the airline in direct and indirect costs.

Direct Costs

Direct costs in any accident involve much more than the costs paid out in medical claims. Sehn (2008) provides examples of direct costs in motor fleet accidents to be considered to include those that follow. The applicability to the airline industry is inherently evident:

- repair of cargo damage
- repair of vehicle damage
- medical treatment
- loss of revenue
- administrative
- police report
- possible effect on the cost of insurance
- possible effect on the cost of workers' compensation insurance
- towing
- storage of damaged vehicles.

In addition to Sehn's list, other direct costs could include those associated with site cleanup, damage to company property, damage to hangars or other facilities and more. All need to be captured when calculating the cost of accidents.

Indirect Costs

Once an accident occurs, there are always additional indirect costs that should be included in considerations. These could include such items as necessary training or education for a substitute for the person injured or killed, additional administrative costs to handle hiring of a replacement, or increased insurance premiums. A well-known rule of thumb is that for every dollar spent in direct costs for an accident another two dollars will be expended in indirect costs. Indirect costs are much harder to justify and much harder to sell to management and, in turn, may be suspect by them. Don't base the argument on indirect costs unless you can make a firm case that they are likely to be spent. In any case, your goal is to add up the total costs of dollars expected to be expended. If there are multiple, possible outcomes of the accident with multiple, possible dollar losses, consider what you think will be the average cost of a given accident. Multiply the total cost by the total number of accidents you have per year to get your annual cost. According to Sehn (2008), indirect costs include:

- (lost) revenue from clients or customers
- (lost) revenue from sales
- meetings missed
- salaries and wages paid to employees in the accident
- lost time at work
- cost to hire and train replacement personnel
- supervisor's time
- loss of personal property
- replacement vehicle rental(s)
- damaged equipment downtime
- accident reporting
- medical costs paid by the company
- poor public relations/publicity
- increased public relations costs
- government agency costs.

Review accident and incident reports to help determine where losses are occurring. By first addressing those losses which are most expensive, companies are also taking basic steps toward loss-control. Of course, the problem with this methodology is that it tends to be reactive in nature. In other words, instead of anticipating where losses may occur, the concept is dependent upon losses occurring and then applying the fix. This can be expensive in an industry where mistakes can have high consequences. The approach would be totally impractical without incorporating some proactive tools into the process. Consider the use of proactive tools to anticipate losses. Such tools are mentioned throughout this book.

Putting a Dollar Value on Safety

The solution to nearly any safety problem involves how much it will cost to fix the problem and whether the cost is worth it. Once the costs have been calculated for a single occurrence, determine how many occurrences will take place over the working life of the staff, the aircraft, the company, and the organization. If an accident has occurred and continues to occur on a regular basis, your job is easy. Calculate the cost of each occurrence and multiply the number of occurrences per year. You know the dollar cost of the event. If the accident does not occur regularly, you may have to estimate the probability of the event in one year. Base this on the number of years of exposure and the number of times loss-producing accidents have occurred. As mentioned above, if an accident occurred once per year, the probability of occurrence in one year would be one. If it only occurs once every five years, the estimate of occurrence in a given year is 1/5 or 20 percent or 0.20. This probability will be combined with the estimated cost of the accident as determined above.

If accidents in this category cost US$100,000 per year in each year over a five-year period, the total cost equals US$500,000. You have already accounted for the multiple times per year in the earlier step when you multiplied total cost by total number of accidents per year. If, on the other hand, your total cost of an accident is US$100,000 per year and the probability of an accident occurring in a given year is .20, the estimated value of *annual* dollar savings is US$100,000 × 0.20 = US$20,000. The key word here is *annual*, because you can expect to save this amount of money each and every year from now on, if you have completely eliminated the problem. The question becomes, what is the value of a savings that continues year after year and never ends?

From a practical standpoint, management won't write you a blank check and tell you to spend whatever you want since the savings go on forever. The reality of the matter is that, even a forward-thinking management team isn't likely to look beyond five to eight years of savings. The most they would typically consider would be that this fix would save the company from US$100,000–160,000 over the life of the change. Normally, the life of a fix is never considered to last beyond eight years. The technology, the methodology, or the whole strategy may change in that amount of time anyway. To speculate beyond the eight-year mark is normally not considered viable. Consider also that US$1 received in the future is not as valuable as US$1 received today. The value of that future dollar depends on the return on investment management can get elsewhere. Safety competes with every other entity in the organization for operating dollars. Money that goes to safety can't be spent on market development, new product design, or pension plans. It must provide a solid return on its own and that return should at least match what is available elsewhere. If management expects to get a 10 percent return on money invested in other parts of the operation, then 1 dollar to be received a year from now is only worth about US$0.91 today. You may ask why it isn't worth US$0.90. That's because of the compounding of the interest that occurs throughout the year. One dollar to be received two years from now is only worth about US$0.83 today. A dollar to be received five years from now is worth only about US$0.62 today. In order to calculate the total value of the savings for the five-year period, the value of the dollar to be received each year in the future is added together.

These calculations are easily made through the use of a present value of an annuity table (Figure 3.2) which will tell you the present value factor of US$1 to be received at the end of each year for a given number of years at various percentages.

This brings us to the bottom line. In our case, how much we would be willing to spend to save approximately US$20,000 (or any amount) per year from now on (assuming projected savings the company expects a given return on its investment)? Again, additional assumptions are in order:

1. Management believes that investments must provide their full return within eight years. Too many variables exist beyond the eight-year period; therefore, any savings can only be demonstrated over that period of time. Although we expect to save US$20,000 per year from now on, we'll cap those savings at the eight-year mark and assume they only continue through then.

2. Management expects at least a 10 percent return on investment. This figure is based on what management expects to get with alternative investments. If the normal return expected for money spent on marketing, engineering, product development, or new air routes is typically 10 percent they'll expect at least the same from safety. We use that figure as the basis for our determination. If they expect more or less, we can adjust accordingly by moving to another column in the table.

3. Money to be received today is worth more than money to be received in the future and vice versa. Again, this depends on the interest rate and typical return on investment. At 10 percent interest, a dollar to be received eight years from now is only worth about half of what a dollar in the hand today is worth. This is due to the fact that a dollar will almost double in value over an eight-year period at a 10 percent interest rate.

By looking at the present value table (Figure 3.2), we learn that a savings of US$1 per year for each of the next eight years has a present value today of US$5.33 (assuming a 10 percent interest rate). That means that a savings of US$20,000 per year for each of the next eight years is worth 5.33493 × US$20,000 or US$106,699.

Period	PRESENT VALUE OF ORDINARY ANNUITY (annuity in arrears – end of period payments									
	RATE PER PERIOD									
	0.25%	0.50%	0.75%	1.00%	1.50%	2.00%	2.50%	3.00%	4.00%	5.00%
1	0.99751	0.99502	0.99256	0.99010	0.98522	0.98039	0.97561	0.97087	0.96154	0.95238
2	1.99252	1.98510	1.97772	1.97040	1.95588	1.94156	1.92742	1.91347	1.88609	
3	2.98506	2.97025	2.95556	2.94099	2.91220	2.88388	2.85602	2.82861		
4	3.97512	3.95050	3.92611	3.90197	3.85438	3.80773	3.76197			
5	4.96272	4.92587	4.88944	4.85343	4.78264	4.71346				
6	5.94785	5.89638	5.84560	5.79548	5.69719					
7	6.93052	6.86207	6.79464	6.72819						
8	7.91074	7.82296	7.73661							
9	8.88852	8.77906								
10	7.91074									

(excerpted for brevity)

Figure 3.2   Present value of an annuity (principlesofaccounting.com, 2010)

Example Problem

You have been investigating the problem of baggage handlers and their strains and sprains. During your investigation, you learn that during the last five years incidents depicted in Table 3.1 were recorded.

Table 3.1   Recorded incidents in the last five years

Year	Number of incidents	Average medical claim per incident (US$)	Total cost (US$)
1	3	3126	9378
2	5	3653	18 265
3	5	3801	19 005
4	6	4100	24 600
5	8	4800	38 400

There are numerous ways to approach this problem as suggested by the data. Any approach will require certain givens, or assumptions. First, consider the trends. Generally, the numbers of incidents and the costs per claim are rising. Over a four-year period (the end of year one to the end of year five or 5-1), the number of incidents has increased by 5 or by approximately 1.25 per year. They have also nearly tripled. During the next four years, it would be reasonable to estimate that, given the same growth in business and numbers of employees, the trend will continue and the number of incidents will grow in a similar pattern. At the existing growth rate of 8/3 = 2.67, the next four years could yield a number of incidents approaching 8 × 2.67 or 21.36. If growth in the number of employees is slowing or even remaining stable, the expectation may be for the number of incidents to do the same. A similar approach can be taken when costs are considered. Costs have quadrupled in the last four years, as a result of both the increase in cost per incident and the likely additional number of employees. If costs continue to grow at the present rate, they may be expected to quadruple in the next four years. This number will likely be adjusted by any decline in the expected growth of numbers of employees. Assuming the number of employees is expected to average ten per year over the next four years, the calculation of expected costs would be made by considering the expected average cost per claim and multiplying that times ten to come up with an expected total cost per year. In this case, since rates increased by 52 percent (US$4,800 – 3,126 = US$1,624 and US$1,624/3,126 = 52%). We can expect the same growth in the next four years. At the current cost of US$4,800 and a growth rate of 52 percent over a 4-year period, we expect costs to increase a little each year and reach US$7,296 in four years and continue to do so for the subsequent four years. In other words, over the next eight years, the average cost of incident per year will be US$7,296. If we estimate injuries at a rate of ten per year (ten becomes our multiplier), the average cost per baggage handling incidents per year, over the next eight years, will be US$72,960. Also consider the costs of adverse publicity, lost customers and the resulting lost sales. Indirect costs are a second key consideration. Carefully review the costs, outside the direct medical costs, associated with all accidents. As mentioned above, there are a number of items to be considered. In this case, a review of internal administrative time required to handle the incident including clerical wages and benefits is calculated. A conversation with the workers' compensation provider reveals that if these claims had not been submitted, insurance costs would be reduced by a certain amount and so forth. Indirect costs must be carefully considered and thoroughly documented. Once complete, we learn that our estimated savings in indirect costs will be approximately equal to our direct costs and we can document this number for management. By adding both the direct and indirect costs, we learn that we expect to save approximately US$140,000 per year from this point forward. In a discussion with the finance department we ascertain that the average expected return on investment is approximately 7 percent per year. In looking at the present value of an annuity (Figure 3.2) we determine that a savings of one dollar per year for each of the next eight years is worth US$5.97130 today. By multiplying US$140,000 times that factor, we anticipate our total savings of US$835,982. This is the maximum amount of money we could present to management to fix the problem. If our fix results in additional costs elsewhere, such as additional fuel use due to additional weight on all flights, then this increased cost must also be deducted from the savings.

Non-monetary Costs

Before closing the book on costs, also consider any non-monetary costs. What is the value of saving an injury to the person who may suffer that injury? What if you are the person whose limbs or life may be saved? What if that limb or life belongs to your closest friend or relative? Aircraft and workplace accidents have implications well beyond the monetary costs. There are also intangible human costs that cannot be measured. Those costs should also be built, as an addendum, into the business case. They should be an integral part of the argument presented to management regarding the investment of dollars into safety.

The Correct Approach to Loss Control

The most effective approach to dealing with any hazard is to engineer the hazard away so it presents no threat to anyone (General Industry Regulations, 2008). Ideally, the hazard will be completely eliminated. For example, to prevent back injuries to employees loading luggage on a plane, automate the whole process. This would engineer the problem away – but be very expensive. Closer examination of the back injuries reveals that nearly all of the injuries occur while helping passengers place their luggage into overhead bins. The questions arise as to whether there is any cost-effective method to engineer this problem away? Could bins be designed to make it easier to load the luggage? Are there lightweight mechanical devices that might help? What would be the additional fuel costs to carry those devices? If engineering approaches are not feasible, are there administrative controls that might help?

The institution of administrative controls is normally the second approach to be considered in the possible reduction of hazards. Administrative controls normally include such practices as prohibiting an employee from being within a certain distance of an engine on a plane or only starting an engine during times when employees are occupied elsewhere. Rotating employees so they only work limited times or establishing workplace rules are all examples of administrative controls. In this case, could training on lifting techniques help alleviate the problem? How effective would they be and what is the cost of training? Could passengers be required to check luggage likely to cause injury – that is, luggage of a certain size or weight? What are the marketing implications of that and what are the long-term costs? Could attendants be prohibited from handling luggage? A brainstorming session with affected parties and informed managers may generate engineering or administrative solutions.

If neither engineering nor administrative controls are appropriate, is there personal protection equipment (PPE) that can help alleviate the problem? With luggage loading, PPE can be an especially difficult problem. Some companies have placed alarms on shirt collars so that when an attendant bends at an incorrect angle to lift a load, the alarm goes off, warning against the improper angle. A more controversial approach would be to insist on the use of back belts to help ensure the appropriate lifting posture. With the engine example, PPE includes hearing protection such as ear plugs or ear muffs. Again, the cost must be considered and deducted from the savings.

In summary, although the engineering approach is usually considered the most effective, it is also frequently the most expensive approach – if it even falls within the realm of feasibility. If engineering controls don't work, utilize administrative controls. Administrative controls include such practices as prohibiting an employee from being within a certain distance of an engine on a plane or only starting an engine during times when employees are occupied elsewhere. Rotating employees so they only work limited times or establishing workplace rules are all examples of administrative controls. When administrative controls won't do the job, the last resort is to rely on PPE. With a problem like luggage loading, PPE can be especially difficult. Regardless of the intervention selected, the correct approach to the problem is to first, attempt to engineer it away; second, attempt to install administrative controls that will correct the problem; and finally, use PPE if nothing else works. Always go from the most effective to the least effective method. If none of the above approaches are suitable, then some of the projects or work may be subcontracted in an effort to assign some or all of the risk to a third party. The risk is still there but it can be handled by a specialized company better suited to handle the risk. Due to size or specialization in a particular area, the subcontracted company may be able to minimize or eliminate the risk in a more cost-effective manner. Sometimes the risk is so great and the cost–benefit is so low, that the process is eliminated altogether or postponed until technology permits a cost-effective approach.

When risk cannot be completely eliminated or when a low probability event leads to a high-consequence loss, insurance is often the answer. Insurance is designed to *indemnify* the party suffering the loss. In other words, the party is compensated for the loss so that they are in roughly an equivalent financial position post-loss as pre-loss. Of course, insurance companies attempt to calculate their own loss exposure and charge premiums to cover the costs of the losses as well as overhead and usually, profit. Rather than pay the insurance bill, large companies facing risk may choose to self-insure. Self-insurance is no insurance. A more effective approach might be to cover relatively low dollar losses and purchase insurance over a certain amount, such as for losses exceeding one million or even multi-millions of dollars. The first dollars of loss that an insurance company must recompense are the most expensive to cover when purchasing insurance. Although high dollar deductibles may not seem attractive when shopping for insurance, the long-term cost benefit is in favor of the purchaser.

Regardless of the approach taken – engineering, administrative, PPE, or transfer of the risk through contracting or insuring, the key consideration is cost. The cost of implementing the loss control must be deducted from any dollar savings anticipated.

Strategies for Getting CEO Support for SMS

A strong business case has value only if the right person hears the case and has the authority to act on it to effect change. It is not uncommon for safety and the safety department to be somewhat hidden within a staff function such as personnel or human relations. When this occurs, all communications to top management and the CEO are filtered and the risk is that they may never make it to the top. When they do reach the top, there is also the risk that they have been modified or their effectiveness impaired due to the personal biases of those in the middle.

Strategy 1

Safety should be located in a staff position with direct reporting responsibilities to the CEO. Relationships to management are on an advisory basis only. The safety department will monitor safety throughout the organization and make all recommendations to top management. The full responsibility for implementing and carrying out any and all safety directives will rest with management and the management team. The safety department must be ready and able to present cost-justified strategies for safety to management and effectively argue the rationale for implementation of any and all safety directions. All strategies will be carried out by the line management team only. If the safety staff attempts to assume the role of implementing and enforcing safety directives, that staff will run the risk of conflict with line management. The response from line management, particularly at the lower levels, is often one of resistance. When given the choice between production and safety, production often wins. If all safety directives flow from the top of the line management hierarchy, and no choice is made available to any member of the line management team, safety will prevail.

Strategy 2 (General Industry Regulations, 2008)

Insist on a strong commitment to safety from management and the management team. This commitment should begin with a written document outlining the dedication of line management toward all aspects of safety to include those aimed at both customers and employees. It should contain clearly established safety policies and emphasize the priority that management expects safety to be given relative to other organizational values. Management should also set forth clear goals for the safety programs and assure that all members of the management team are committed to those goals. With input from employees and all members of the management team, clear plans should be set forth for the accomplishment of those goals within a specified timetable. Encourage management to set strong examples regarding safety by always personally practicing safety, incorporating safety performance into the evaluation of all members of the line management team, and incorporating a progressive disciplinary process into company policy for matters dealing with critical safety issues. Management should address safety issues in all public forums as well as those that include employees. Safety should not be considered the highest priority, but it should be incorporated into everything the company does. The company doesn't just *fly*; it flies safely. The company doesn't just *produce airplanes*; it produces them safely. Safety needs to be incorporated into all critical aspects of the overall management function to include those involving planning, organizing, directing, assuring compliance with company policy, staffing and communicating with the public and employees.

Strategy 3 (General Industry Regulations, 2008)

Ensure employee involvement in all aspects of safety. Employees should be clearly involved in the structure and operation of any safety program, particularly in decisions that affect their personal safety and health. They should be given the opportunity for input through committee and task force assignments, a system to incorporate their ideas and suggestion, the ability to provide feedback in terms of anonymous reporting of errors and problems, and direct channels of communication to the safety staff whenever they deem it necessary. Roles and responsibilities for safety should be as clear to all non-supervisory employees as they are to management. This will only occur through the incorporation of evaluation of safety performance into the routine performance evaluations of the company; they too will be accountable for incorporating safety into their performance. Establish committees centered on departments, work processes, or tasks. At the same time, establish committees using a vertical slice of the organization. Have representation from different levels, but choose your members carefully. Vertical-slice committees have a tendency to move in the direction desired by those who have more formal power in the organization. As you organize the committees, think of all the ways each can be involved in the safety processes. They can participate in inspections, audits, investigations, follow-ups on near-misses, job hazard analyses, training and other tasks germane to the safety function. Involve as many employees as possible on the committees and keep the committees involved in the safety processes. They should be especially concerned with establishing policies and procedures regarding safety and working safely. This will require their reviewing existing regulations, policies and procedures available from other, similar companies or industry organizations and the ongoing operations of the company.

The key here is to involve as many employees as possible at the grass roots, non-manager/non-supervisory levels. Groups should be given clear direction and told any constraints under which they will be working. It is critical that the rules don't change midstream. If there are budget, legal, or existing policy restrictions, they should know what they are before they pursue their respective charges. Involve the committees in the writing of new policies and the review of existing policies. Those closest to the work are often in the best position to review it and to provide relevant feedback in how it is reviewed. Ensure that regular meetings and participation by all committees occur. This may be accomplished by integrating the committee mission into the routine workday. For example, a departmental committee is charged with developing a safety checklist and checking the department against standards prior to each workday. Members of work teams observe one another's work and make suggestions relative to safe work practices. Management must give the work teams the authority, responsibility and resources necessary to improve the system. This may come in terms of advisory authority only when it comes to adopting new policy. When it comes to daily practice of safety, these teams may be given budgets and physical resources to assure this happens.

Strategy 4 (American National Standards Institute, 2005)

Create a series of mechanisms to assure regular review and improvement of all safety systems and the overall safety culture. The historical tool used for this exercise is the safety audit. A key consideration is an understanding of how the audit differs from an inspection. An inspection is simply an observation of whether the safety items are in place and according to standard. For example, all machine guards are in place and working during the operation of any machine or all fire extinguishers are fully charged and operational. During an inspection, the inspector typically responds to the point of inspection as a 'yes' or 'no'. It either is or isn't safe. The guards are or are not in place, the extinguishers are or are not operational. Inspections take place on a regular basis. They are scheduled and unscheduled, but critical for an effective safety operation. Inspections provide a great opportunity for employee involvement. Employees can inspect their own work areas and those of others. Employees can use checklists to assure that critical safeguards are in place. They can do the same for any job.

Job Safety Analyses (JSAs) are a variation of inspections wherein employees break down a job into specific steps, identify the hazards associated with each step and select measures to reduce or eliminate the potential hazards. Before beginning a process, the JSA is reviewed to assure that all appropriate steps are followed and that appropriate safety measures are in place. The JSA can provide a checklist for a given job and thereby serve as an inspection. Employees who do the specific job should be actively involved in the development of the JSAs. Audits are set up to determine whether or not the systems are in place and appropriate to assure a safe and healthful working environment. Normally audit items are assembled and audits are conducted by a team of experts and those knowledgeable about the workings of the particular operation. A typical audit team may consist of managers in a given operation, a safety professional, even representatives from other, similar companies. Audits also offer another opportunity for employee involvement. The team determines the key points to be considered. Management ensures that employees and authorized representatives have the necessary resources to participate in the planning of the overall safety management system. The audit team would evaluate this statement, using a Likert Scale of 1–5: 1 would indicate that no assurance is in place, 5 would indicate complete assurance is in place and all tools necessary to do the job are operational. The audit would be conducted periodically and the audit team would indicate opportunities to improve the system based upon areas where the evaluation indicated less than a 5 is appropriate. Once an appropriate audit tool in a given area is developed, this evaluation of the overall safety system is ideally ongoing so that constant improvements can be made in the safety system. For example, during month one of the year, the planning process is evaluated and a report is made. The company responds by attempting to make the improvements suggested. The process is repeated in month two by considering another aspect of the system such as implementation. Periodically, the team repeats the evaluation of planning and the loop continues. Another approach is to consider various departments or operations. Operations might be considered from a safety perspective. For example, in month one, fire operations are evaluated. In month two, the emphasis moves to ergonomics, and so on. The instrument itself is constantly undergoing evaluation and evolution and processes, technologies, and management techniques change.

Strategy 5 (General Industry Regulations, 2006)

Establish programs to correct the hazards, risk and system deficiencies – on a proactive basis. These problems should be anticipated and corrections should be made before a problem manifests itself in terms of an accident. This sometimes occurs in the long-term, somewhat laborious task of identifying program deficiencies, determining corrective actions and aiming resources toward overall, continuous improvement of the system. When hazards creating an immediate threat are identified, they should be corrected at once. In any case, a steady system of continuous and periodic improvement must take place. Not only are inspections and audits critical parts of that process, but other tools can be used to aid in the overall program. Companies may also rely on near-miss reports, unplanned inspections (rotating responsibilities), employee interviews involving incident recall or accident imaging, planned job/task observation, group meetings, records reviews (that is OSHA 300, first-aid, insurance and workers' compensation reports), new process reviews and periodic process reviews. All of these tools have been successfully used to identify hazards and potential hazards on a proactive basis. Integration of these tools into the ongoing, continuous process improvement loop can be an invaluable part of the overall process. When hazards or potential hazards are identified, carefully consider the options available to your company. Of course, the standard approaches include engineering the hazard away. If that is unsuccessful, the implementation of administrative controls is attempted next. The last opportunity for correction is the utilization of PPE. The above approaches are always used in the same sequence – engineering, then administrative, then PPE as a last resort. Sometimes overlooked are approaches such as contracting the work to another firm or attempting to eliminate a step or process altogether.

Conclusion

Safety is proactive and it is part of business. The safety program must be forward-thinking and it must involve all members of the organization – beginning with the CEO and working down through the company. The emphasis on the CEO is critical in that CEO language and commitment is to the business. As we saw in the quote by Milton Friedman, 'The business of business is business'. From a practical standpoint, the business of safety is business. It is critical that objectives of the safety program be carefully aligned with those of the organization. The most appropriate way to accomplish that is through the establishment of metrics for the evaluation of any safety proposal or program based upon the return on investment expected of any other operation within the company. In spite of the old adage, safety is never first in the operation. It is an essential part of each operation. Every employee, from the top to the bottom of the organization, must maintain that emphasis to assure that the objectives of safety are accomplished.

References

American National Standards Institute (2005). *American National Standard – Occupational Health and Safety Management Systems*. Fairfax, Virginia: American Industrial Hygiene Association.

Davis, I. (2005). *What is the Business of Business?* New York: McKinsey & Company, The McKinsey Quarterly, cited 21 December 2005. Available from http://www.mckinseyquarterly.com/article_page.aspx?ar=1638&L2=21&L3=3.

Drservices.com (2010). *Risk Matrix*. Available from http://www.drservices.com.au/images/tramatrix.jpg.

Drucker, P. (1986). *The Practice of Management*. New York: Harper Collins Publishers, Inc.

Federal Aviation Administration (2006). *Introduction to Safety Management Systems for Air Operators*, Advisory Circular No. 120-92 Initiated by AFS-800. Washington, DC: US Government Printing Office.

General Industry Regulations (2008). *29 CFR 1910 OSHA*. Davenport, IA: MANCOMM.

Occupational Safety and Health Administration (2010) Available from http://www.osha.gov.

Principles of Accounting (2010) *Present Value of an Annuity*. Available from http://www.principlesofaccounting.com/ART/fv.pv.tables/pvofordinaryannuity.htm

Sehn, F. (2008). Cost analysis and budgeting. In Joel Haight (ed.), *The Safety Professionals Handbook: Management Applications* (pp. 759–761). Des Plaines, IL: American Society of Safety Engineers.

United States Code Title 49 Transportation, USC Sec. 44702. Available at http://usocde.house.gov/download/pls/49C447.txt.

Williams–Steiger Occupational Safety and Health Act, 84 Stat. 1590 et seq., 29 U.S.C. 651 et seq. (1970).

Chapter 4

Embracing a Safety Culture in Coast Guard Aviation[1]

Roberto H. Torres

Overview

Understanding organizational culture is critical to organizational success. In their seminal work *Built to Last*, Collins and Porras (1994) documented the importance of organizational culture in companies with records of success spanning at least 50 years. Disney, for instance, does not have employees; they have cast members, thereby reinforcing their primary mission of public entertainment. Wal-Mart does not have employees, but associates.

Command climate, a form of organizational culture, is central to military organizations. More specifically, safety culture, a significant component of command climate, is critical to the success of aviation services. This chapter addresses safety culture in aviation from the perspective of an experienced United States (US) Coast Guard Flight Safety Officer.

Beginning with its fundamental importance to the successful implementation of safety management systems (SMS), safety culture is defined and related to Coast Guard command climates. Concepts such as informed culture, reporting culture, just culture, flexible culture and learning culture, are presented along with practical suggestions for developing and implementing a viable safety culture. The chapter concludes with a case study that demonstrates the importance of the Coast Guard's command climate and safety culture to its extraordinary success in responding to the devastation created by hurricanes Katrina and Rita.

Safety Culture Defined

Aviation SMS rest on four critical pillars: safety policy, safety assurance, safety risk management and safety promotion. Even well-designed systems, however, will only succeed and thrive in organizations that foster a safety culture which is inherently embedded within its operational construct.

Recognizing this, influential aviation entities worldwide publish formal definitions of safety culture. The International Civil Aviation Organization's (ICAO) *Safety Management Manual* (SMM, 2009) defines safety culture as:

---

1 The views expressed herein are those of the author and are not to be construed as official or reflecting the views of the Commandant or of the US Coast Guard.

the context in which safety practices are fostered within an organization. These safety practices include a series of organizational processes, procedures, and policies that aim to achieve a specific outcome – the identification of hazards. The processes (effective safety reporting), procedures (hazard reporting system), and policies (safety policy) are complex, specific ideas and behaviors that can be packaged in such a way as to make them easily understandable to a wide audience and therefore easier to apply on a large scale. (p. 2-30)

In the US, the Federal Aviation Administration (FAA, 2006) defines safety culture as:

the product of individual and group values, attitudes, competencies, and patterns of behavior that determine the commitment to, and the style and proficiency of, the organization's management of safety. Organizations with a positive safety culture are characterized by communications founded on mutual trust, by shared perceptions of the importance of safety, and by the confidence in the efficacy of preventive measures. (Appendix 1, p. 4)

Hudson (2001) produced a model using information and trust as its dimensions to show the evolutionary development of safety culture. Interestingly, as the culture evolved through the following stages, personnel became increasingly informed and the level of trust grew within the organization:

1. Pathological: '*Who cares as long as we are not caught?*' This is a particularly dangerous attitude in the dynamic pace of aviation operations and is, in fact, negligent in that violations are not only excused but essentially encouraged. Thankfully, few organizations exist at this point on the safety culture continuum.
2. Reactive: '*Safety is important; we do a lot every time we have an accident.*' Unfortunately, many aviation organizations around the world are stuck in this vicious cycle of only reacting to hazards after the damage is done. Moving beyond this reactive stage is perhaps the single biggest step an organization can take to change its safety culture.
3. Calculative: '*We have a system in place to manage all hazards.*' This safety culture stage lies somewhere between a reactive safety culture and the more desirable proactive safety culture. Organizations found at the calculative stage are forward-thinking in managing existing hazards, but have not yet made a commitment to fully address all hazards identified through their evaluation initiatives.
4. Proactive: '*We work on the problems that we still find.*' A proactive safety culture actively looks for and seeks out the subtle, underlying hazards that plague all aviation organizations. Tools such as safety audits, surveys and voluntary reporting systems are used to shift the organizational mindset

to a more proactive one. Once this happens, the challenge is to keep the culture moving forward.

5. Generative: '*Safety is how we do business around here.*' Also known as predictive, this type of safety culture is the ultimate goal toward which aviation organizations should strive. A generative, or predictive, safety culture uses forecasting tools and modeling techniques to discover potential vulnerabilities before they develop into hazards. Advanced reporting systems and data analysis are used to achieve this level.

In 2004, the Global Aviation Information Network (GAIN): Working Group E published its findings in *A Roadmap to a Just Culture: Enhancing the Safety Environment*. Citing research from Dr. James Reason, the report described a safety culture as consisting of five systematic components:

1. informed culture – people are knowledgeable about the system
2. reporting culture – people share within the system
3. just culture – people are held accountable to the system
4. flexible culture – people adapt to the system, and
5. learning culture – people improve the system.

GAIN: Working Group E summarized the benefits of a viable safety culture: increased safety reporting, trust building and more effective safety and operational management.

While more and more aviation organizations around the world are realizing the benefits of a strong safety culture, the US Coast Guard's aviation sector has recognized these immense benefits for many years and actively uses it to enhance operational effectiveness. They reference it, however, as *command climate*.

Command Climate in the Military

In the Coast Guard, as in other military services, the term command climate is often substituted for safety culture. Although command climate incorporates other tangible and intangible notions, such as promotion and retention rates, overall job satisfaction, human relations and freedom from harassment, safety culture is central to a just and effective command climate. As a result, command climate and safety culture are used interchangeably throughout this chapter when referring to Coast Guard aviation.

Command climate is one of the most observed but least understood concepts in the military. The goal of every commanding officer is a cohesive and effective unit, and this requires nurturing an exemplary command climate. A review of military literature reveals that field commanders must be able to:

1. Articulate a clear vision and establish attainable goals.
2. Allow subordinates freedom to exercise initiative.

3. Establish accountability at appropriate level.
4. Show confidence in subordinates.
5. Encourage and reward prudent risk-taking.
6. Achieve high performance through positive motivation and rewards.
7. Give clear missions within boundaries of autonomy.
8. Listen to subordinates and seek ideas.
9. Demonstrate concern about the welfare of subordinates.
10. Establish and model high ethical standards.

The Coast Guard is well aware of the benefits of a positive safety culture or command climate. Their Air Operations Manual states that 'Effective aviation safety requires continuous command emphasis and leadership example. Experience has shown that a strong command mishap prevention (loss control) policy will reduce aircraft mishap potential and thereby enhance overall mission effectiveness' (US Coast Guard, 2008, p. 1-9). A major indicator of a positive command climate is the credibility of the commander. Where creditability is high, expect to find increased reporting, trust and confidence in the overall system and clear communications up and down the chain of command.

Coast Guard Aviation

The smallest of the five armed forces, the Coast Guard uses aviation assets (both fixed- and rotary-winged aircraft) to protect US maritime interests through multi-mission integration of maritime safety, maritime security and maritime stewardship. As America's maritime guardian, the Coast Guard is 'always ready for all hazards and all threats' (US Coast Guard, 2009, p. ii). Key to the Coast Guard's culture is the Guardian Ethos:

> I am America's Maritime Guardian.
> I serve the citizens of the United States.
> I will protect them.
> I will defend them.
> I will save them.
> I am their Shield.
> For them I am Semper Paratus.
> I live the Coast Guard Core Values.
> I am a Guardian.
> We are the United States Coast Guard.

The Coast Guard is the premier, worldwide aviation search and rescue, law enforcement and counter-drug organization. It operates a variety of aircraft based out of 26 air stations and 5 air facilities throughout the US, including Hawaii, Alaska and Puerto Rico. Aircrews are standing by 24 hours a day, 7 days a week and 365 days a year; ready to launch within 30 minutes in all weather conditions.

The sizes of the aviation units vary from small three-helicopter air stations to large multi-aircraft facilities.

The general chain of command structure is similar across all units and typically includes a commanding officer, an executive officer, an operations officer and an engineering officer. This command cadre leads a team of pilots, aircrew and support personnel who make up the core of the unit. Through this hierarchical structure and a fleet of modern aircraft, Coast Guard aviation runs the gamut of service missions, with its body of personnel perpetually encouraged to model by the core values of honor, respect, and devotion to duty.

Coast Guard aviation boasts a robust safety program, dedicated to fostering a culture that promotes aviation professionalism, supporting successful completion of aviation operations, increasing operational efficiency and maximizing loss control. By Commandant mandate, all air stations have a safety department or division, normally consisting of a department/division head, a flight safety officer (per type of aircraft) and a ground safety officer. The *Safety and Environmental Health Manual* (US Coast Guard, 2006) clearly outlines the responsibilities of an air station's safety department or division. Among a multitude of other responsibilities, the most significant duties of the flight safety officer include:

1. Act as the commanding officer's representative and advisor on all aviation safety matters.
2. Report to the commanding officer at least monthly regarding the unit's safety posture.
3. Distribute aviation safety literature to ensure it receives widest readership possible and that all hands have access to it. Consideration should be given to publishing a unit newsletter.
4. Manage a unit safety incentive/suggestion program stressing individual achievement.
5. Coordinate and present aviation safety training.
6. Submit to the command, at least annually, a written unit aviation safety survey covering all phases of the unit's aviation operations.
7. Ensure completion of aviation mishap reports according to protocol. Monitor and report to the commanding officer progress of corrective actions.
8. Maintain files of unit and other mishap reports. It is recommended that an aviation safety trend analysis be conducted and presented to the commanding officer on a regular basis. An annual compilation and review of mishaps and trends can be included in the unit aviation safety survey.
9. Administer a unit-level anonymous reporting program for identifying unsafe conditions (p. 2-7).

The heads of all Coast Guard aviation components – operations, engineering, and safety – informally form a notional entity called the tripartite or tri-p for short. The *Merriam-Webster Online Dictionary* (2009) defines tripartite as 'having three corresponding parts or copies', which is key to the success of this concept.

In Coast Guard aviation, the tri-p works together and equally at all levels, from headquarters to the area to the district to the unit. In this unique relationship, the three critical areas of aviation work concurrently at all echelons. For example, at the higher levels, the platform manager of a specific type of aircraft concurrently works closely with the systems manager of the same aircraft and the safety program manager. Of course, these informal relationships are intertwined with other important associations including acquisitions, training, and capabilities, to name a few. At the unit level, there is constant and decisive interaction between the operations department which includes scheduling, training, standardization; the maintenance department which includes repairs, production and logistics; and the safety department which includes risk management, hazard identification and mishap investigation. The tri-p concept benefits Coast Guard aviation, as a whole, by increasing operational effectiveness, maintenance availability, and safety oversight.

Coast Guard aviation is currently in a challenging period of dynamic transformation brought on by new homeland security roles in response to the terrorist attacks on the US of September 11, 2001. As the Coast Guard invests large sums of capital in new aircraft acquisitions and diversifies its responsibilities into lesser-known capabilities such as airborne use of force and unmanned aerial vehicles, the positive safety culture prominent throughout the organization is more important than ever.

What You Need to Know

Using the five components briefly described earlier, Coast Guard aviation's safety culture is described below with examples of effective implementation. These sections are replete with practical examples of ways to build a safety culture in a military aviation organization, in this case, America's maritime guardian.

Informed Culture

James Reason (1997, p. 195) describes an informed culture as 'one in which those who manage and operate the system have current knowledge about the human, technical, organizational, and environmental factors that determine the safety of the system as a whole'. Simply put, those operating in an informed culture know operationally where they have been, where they are and where they are going. Transfer of this information throughout an organization is one of the biggest leadership challenges.

While there is no single approach to establishing and maintaining a positive safety culture or command climate, clearly no effort can succeed without top management commitment. In the Coast Guard's chain of command environment, commanding officers lead the way by modeling such a complete commitment. A formal safety policy statement is critical in establishing an informed safety culture. A written safety policy transforms safety – often an abstract concept – into a tangible and attainable goal.

In the case of the US Coast Guard, the Commandant's Safety and Environmental Health Policy Statement is directive in that it charges every member of the organization to honor a commitment to preserve the health and safety of all personnel. It unmistakably establishes responsibility for this mandate at all levels of the organization. Additionally, the formal safety policy statement emphasizes the concept that consistently successful performance requires thorough preparation. The Commandant asserts that while the Coast Guard cannot reasonably remove all levels of risk inherent to its daily operations, the organization can certainly strive to reduce those risks and keep them within acceptable limits. The fundamental SMS principle of safety risk management is applied by identifying potential hazards, assessing the risks associated with those hazards and controlling risks to acceptable levels. Finally, it concludes that reducing risks protects individual members while decreasing the number of fatalities, the incidence of injury or disease, and the loss of property – and this ultimately preserves mission readiness.

I have been privileged to assist many air station commanding officers in development of their official safety policy statements. All of the statements have been directive, clear, concise and explicit in the expectations of the command with regard to the balance of personnel safety, operational requirements and risk mitigation. An abridged version of an effective and powerful safety policy statement follows:

> Our commitment to safety is based on a vision to enhance operational readiness through the preservation of human and material resources. As commanding officer, I am personally committed to maintaining a safe, healthy, and sustainable working environment where we operate. To reach our goal, it is the paramount responsibility of each of us to be safety conscious in every aspect of our lives, both on and off duty. The Safety Division will continually educate, train, and monitor the overall safety program and safety climate for full compliance and full effectiveness. To achieve this commitment, we must:
>
> * Exercise mature and professional judgment in the absence of clear guidance.
> * Eliminate unacceptable risk. Never accept degraded equipment or conditions which exceed the limits of you or your crew.
> * Be proactive in identifying safety hazards, the first step of Operational Risk Management. Once a hazard is identified, we must do something about it.
> * Realize continual preparedness and training is crucial to safety.
> * Use Operational Risk Management principles in every aspect of our lives.
> * Prevent mishap recurrence through comprehensive incident reporting.
> * Use Crew Resource Management and Maintenance Resource Management principles to carry out our roles in the aircraft and in the hangar deck.

- At no time allow the Tempo of Operations to dictate not adhering to established safety procedures in both operations and maintenance.
- Be a team. When one succeeds, we all succeed. When one fails, we all do. We therefore openly discuss our close calls and miscues knowing that all will benefit.
- We tactfully suggest improvements to others while likewise welcoming their input to us.
- Learn from each other. Look out for each other.
- Suggest and make improvements to our equipment, techniques, and procedures to benefit the entire fleet.

Safety training takes place continuously throughout Coast Guard aviation. Promulgated from the Division of Aviation Safety at headquarters, topics such as survival training, egress training, aviation physiological training, operational hazard training and safety training for line personnel are taught both initially and recurrently. While training is clearly an important part of an informed culture, it needs to go beyond that; training must be documented and archived for all members. One difficulty with the military environment is the constant change of personnel. During in-briefs for incoming personnel at one unit, the safety division developed a comprehensive presentation to thoroughly cover topics unique to the unit. By systematically discussing and providing training on a variety of important safety topics, the command assures that all personnel possess standardized, baseline knowledge of all pertinent flight safety, ground safety and personal safety matters.

Reporting Culture

Fostering a reporting culture – one in which people are prepared to report their errors and near-misses – is a tremendously important component of a safety culture (Reason, 1997). An SMS relies heavily on the transfer of information, up, down and across the chain of command.

The rank structure in a typical military organization actually lends itself to cultivating a reporting culture. The Coast Guard, for example, has clearly defined guidelines for reporting mishaps. The Coast Guard *Safety and Environmental Health Manual* delineates clear requirements for (at the very least) notifying the unit safety officer. The challenge lies in getting people to actually do it! As safety officer, it is my job to encourage reporting and to extract data from those reports to produce necessary changes to the respective system.

The final step of the safety assurance framework is determining and acting on any preventions or corrections that need to be applied to the process. The Coast Guard's aviation safety directorate tracks mishap recommendations through an innovative program called Recommended Actions Tracking System (affectionately known as RATS). This ensures that recommendations developed through systematic mishap investigations are not forgotten about and are actually tracked from implementation to closure (US Coast Guard, 2006).

Most safety departments at air stations throughout the Coast Guard have a formal recognition process for excellence in reporting. I have seen it called the 'Eagle Eye Award' or the 'Hawk Eye Award'. Recognition occurs in front of a member's peer group, to reinforce the relentless pursuit of risk mitigation. At one unit where I was safety officer, the 'On the Watch Award' was given to members (often junior in rank) who challenged themselves to seek out and destroy hidden and potentially catastrophic hazards.

Recently, an 'On the Watch Award' was given to a junior enlisted member who incredibly found a sheared piston on a yaw control servo from an MH-60J 'Jayhawk' helicopter. A notable find in its own right, the member found this in the early hours of the morning, fighting circadian disruption, while working the dreaded mid-shift. Without hesitation, the young aircraft mechanic declared the aircraft hard down for maintenance and proceeded to inform the supervisor on duty. It is so important to recognize and validate these 'timely reporting' success stories.

Throughout the Service, my counterparts approach mishap reporting in different ways, often reflecting the variability in unit size. One of the commonalities in methods that have proven effective is ease of use, since complex reporting methods often stop the reporting process. Operators must be taught and re-taught the procedures for reporting mishaps, especially for the relatively minor incidents. This issue comes up over and over again in aviation safety surveys (which will be discussed in detail below).

Surprisingly, a still-effective method to collect data and encourage reporting is through easy-to-use anonymous reports. Often called 'any mouse' reports in the Coast Guard vernacular, some people just feel more comfortable reporting hazards on their own time and in their own way. We have places throughout our unit for workers to drop in a report. To make this type of reporting work, two best practices are essential. Most important is to routinely check those suggestion-type boxes and to check them while lots of people see you doing it – this immediately lends credibility to the program. Secondly, I note when personnel do write their contact information on the report, since this is an important indicator of a positive safety environment.

Once a year, every Coast Guard aviation unit is required to conduct a safety survey. Although there is some latitude in its content, most unit aviation safety surveys are very similar. There is even a standardized version available from the Division of Aviation Safety at headquarters.

There is no doubt that a safety survey is essential to a safety culture, however, it must be well-crafted. Ideally, the questions asked must be relevant and appropriate to the unit. These types of surveys usually result in: nominal data (mainly from demographical information); ordinal data from a variety of Likert-type questions (with responses such as: strongly agree, agree, neutral, disagree and strongly disagree); and qualitative data from open-ended questions. Below are sample questions grouped in broad categories typically covered in a Coast Guard aviation safety surveys:

1. Overall:
   a) The command genuinely encourages and holds personnel safety among their highest priorities.
   b) The command has incorporated operational risk management processes in decision-making at all levels.
   c) My supervisors understand and are supportive of personal life stressors (e.g., family, financial, or other) that can affect my job performance.
   d) The unit's morale and esprit de corps is very high.
   e) Do you believe the mishap potential at the unit has increased or decreased and why?
   f) On a scale of 1 to 10, rate the unit's overall safety posture.

2. Operations:
   a) Flight scheduling allows for proper crew rest.
   b) I am adequately trained for all my required duties.
   c) I am proficient in all flight-related activities.
   d) I am confident all of the people whom I stand duty with are proficient in their assigned duty.
   e) Aircrews use appropriate operational risk management.
   f) Job-related fatigue is a factor that affects my ability to do my job safely.

3. Maintenance:
   a) I always follow maintenance procedure cards.
   b) I am given enough time to complete assigned tasks.
   c) I am able to perform my job without distractions.
   d) It is sometimes appropriate to deviate from established rules or regulations when flying or performing aircraft maintenance.
   e) I feel unreasonable pressure to return an aircraft to flying status even though other aircraft are available to meet mission demands.
   f) Unit personnel follow tool control and support equipment issuance programs.

4. Safety:
   a) My supervisor clearly explains the safety hazards, precautions and proper use of protective equipment associated with my job.
   b) My supervisors routinely adhere to and enforce safety practices.
   c) I know how to use the unit's formal hazard reporting program.
   d) I am able to report hazards or unsafe practices without fear of adverse consequences.

e) Unit drills (e.g., aircraft mishap, fire, or adverse weather) occur often enough that I am confident I would know what to do in the event of an actual emergency.

f) When a safety hazard is identified, the safety department responds quickly to eliminate the hazard.

The data generated by aviation safety surveys constitute an incredible tool for infusing a safety culture throughout an organization. Being a military organization, where accountability is expected, it is not uncommon for a survey to generate an 80–90 percent response rate. First, the annual survey provides an opportunity to report hazards; it is the sole reporting venue for some. Although, as a safety officer, I want users to report hazards immediately, some people will only report them through a survey – and I will take what I can get!

Secondly, the survey gives the leadership cadre an annual snapshot of the command climate at their unit. Properly analyzed, this data will help command leadership formulate their strategic agenda, assess progress toward fulfillment of their unit's vision, and guide their selection of areas to immediately emphasize.

Thirdly, a safety survey provides a to-do list of sorts or an action agenda for the safety division. For example, the data from one survey showed that a majority of the people felt that reporting a hazard did not bring unwanted repercussions – obviously, a good thing. However, the results also showed that a majority of people were not clear on the process for reporting minor mishaps or incidents – not a good thing. Clearly, the safety department needed to better advertise the local mishap reporting process.

Finally, data can be collected at regular intervals and analyzed for trends. Soon after aviation safety surveys are conducted at Coast Guard air stations, the results are promptly presented to all personnel. This important last step not only informs the system users of the safety posture at their unit, but the overall safety process gains credibility.

Just Culture

Particularly in aviation, a delicate balance must exist between a no-blame atmosphere and one inhibited by fear of punishment. A just culture is 'an atmosphere of trust in which people are encouraged, even rewarded, for providing essential safety-related information – but in which they are also clear about where the line must be drawn between acceptable and unacceptable behavior' (Reason, 1997, p. 195).

Coast Guard protocol mandates completion of both a safety and an administrative investigation (one conducted mainly for line-of-duty determination) following any serious aviation accident or mishap. According to the *Safety and Environmental Health Manual* (US Coast Guard, 2006), the Commandant has determined that certain investigative reports contain privileged information and should only be used for safety purposes and only reviewed by personnel directly involved with mishap prevention. Those conducting a safety investigation adhere to this concept of privilege, a notion unique to the military. By definition, this concept of privilege is intended to prevent the unnecessary disclosure of privileged safety information outside the safety program. It ensures those participating that any information they disclose to a safety investigator can only be used for the purposes of the safety investigation – a type of forward-looking accountability. According to the *Coast Guard Air Operations Manual* (2008):

> The concept of privilege is intended to provide full disclosure of mishap information (which otherwise may not be disclosed) essential to determining the true causal factors during mishap investigation. When necessary, a written assurance of confidentiality shall be given to a witness in order to obtain complete and candid information about the circumstances surrounding a mishap. (p. 1-13)

Dekker (2007, p. 119) asserts that one of the critical questions for a just culture is, 'How protected against judicial interference are safety data (either safety data from incidents inside of the organization or the safety data that come from formal accident investigations)?' In response to that important question, the Coast Guard goes to great lengths to protect safety data and will not release it – under any circumstances – for anything other than mishap prevention.

In discussing a just culture, the Coast Guard details the importance of aviation mishap communication. Policy states, 'Mishap review is not meant to punish, criticize, or embarrass personnel involved. Essential to this part of the aviation safety program is the free exchange of information or matters relating to the safety of aviation operations' (US Coast Guard, 2008, p. 1-11).

Much has been written recently regarding a just culture. In the end, I believe it is a simple equilibrium of safety reporting tempered with accountability. Interestingly, one of my previous commanding officers would go out of his way to publicly praise a mechanic who, although they may have made a maintenance error, had come forth and admitted their mistake. This type of accountability breaks down the barriers that can easily proliferate in a military organization.

Throughout my experience as safety officer of various aviation units, I have always explained to the people I have been privileged to work with that in the conduct of my job, I am like Switzerland or, in other words, neutral. This simple analogy instills a sense of assurance that I will use any information given only for the purposes of avoiding future mishaps. A safety culture is focused solely upon improving procedures to avoid similar undesirable outcomes.

I strongly believe that by providing safety information and knowledge, people will have a vested interest in assessing the risks in their operating environment. This heightened awareness gained by learning from mistakes clearly outweighs the effects of punishment. So, ask yourself, can an employee openly come forward if they make a mistake, so that your organization can learn from the event?

Flexible Culture

Based on Dr. Reason's research, GAIN: Working Group E (2004, p. 4) defined a flexible culture as, 'a culture in which an organization is able to reconfigure themselves in the face of high tempo operations or certain kinds of danger, often shifting from the conventional hierarchical mode to a flatter mode'. With a reputation for being first on-scene for many recent disasters (for example hurricane Katrina and the Haiti earthquake) and an innate bias for action, not only does the Coast Guard respond and adjust well to surge-type operational demands, but they are constantly evolving and flexing institutionally. The Commandant of the Coast Guard recently called the Service he is charged with a 'change-centric' organization.

In a dynamic organization such as the Coast Guard, flexibility is not only expected but actually needed in order to adequately meet all competing demands. According to the Coast Guard's Publication One – a seminal document reflecting capstone doctrine – the principles of Coast Guard operations include: clear objective, effective presence, unity of effort, on-scene initiative, flexibility, managed risk and restraint. Flexibility is in the DNA of Coast Guard operations; our missions demand it.

The principles of Coast Guard operations also touch on another aspect of a flexible culture – it must be a trusting culture. Specifically, the principles of on-scene initiative and managed risk demonstrate how the Coast Guard empowers all personnel, placing full trust in those at even the lowest levels. In aviation, aircraft commanders are given full autonomy to prosecute all missions. While the *Coast Guard Air Operations Manual* is packed with specific rules and regulations, it includes specific allowance for deviation. Nowhere is a commitment to flexible culture better exemplified than in this (2008) policy statement (2008):

> The probability of saving human life warrants maximum effort. When no suitable alternative exists and the mission has a reasonable chance of success, the risk of damage to or abuse of the aircraft is acceptable, even though such damage or abuse may render the aircraft unrecoverable. (p. 1-8)

Flexibility requires appropriate safety risk management, which is of course a pillar of an SMS. A tool used for continual assessment and management of risk in the Coast Guard is known as *operational risk management* or simply ORM (US Coast Guard, 1999). The ORM process:

1. Is a decision-making tool people at all levels use to increase operational effectiveness by anticipating hazards and reducing the potential for loss, thereby increasing the probability of a successful mission.
2. Advocates harnessing feedback and input from all organizational levels to make the most informed decisions possible.

3. Exists on three levels: time critical, deliberate and strategic. Risk decisions must be made at levels of responsibility that correspond to the degree of risk, considering the mission significance and the timeliness of the required decision. (p. 5)

The Coast Guard ORM process consists of seven steps: (1) identify mission tasks, (2) identify hazards, (3) assess risk, (4) identify options, (5) evaluate risk versus gain, (6) execute decision and (7) monitor situation. When identifying hazards, the ORM process uses the acronym PEACE, which stands for:

1. **P**lanning.
2. **E**vent complexity.
3. **A**sset selection.
4. **C**ommunications and supervision.
5. **E**nvironmental conditions.

When identifying options, the ORM process uses the acronym STAAR, which stands for:

1. **S**pread out: risk is commonly spread out by increasing either the exposure distance or the time between exposures.
2. **T**ransfer: transferring risk does not change probability or severity but rather shifts possible losses or costs to another entity.
3. **A**void: avoiding risk altogether requires canceling or delaying the job, mission, or operation, but this option is rarely exercised due to mission importance. However, it may be possible to avoid specific risks (for example, avoid risks associated with a night operation by planning a day operation instead).
4. **A**ccept: accept risk when the benefits clearly outweigh the costs, but only as much as necessary to accomplish the mission or task.
5. **R**educe: risk can be reduced. Using protective devices, engineering controls and personal protective equipment usually helps control severity. Training, situation awareness, attitude change and stress/fatigue reduction usually helps control probability. Reducing the number of people involved or the number of events, cycles, or evolutions usually helps control exposure. (p. 8)

This ORM process defines risk (R) as a function (f) of severity (S) times probability (P) times exposure (E), or $R = f(S \times P \times E)$. When evaluating risk versus gain, the Coast Guard uses a matrix called the GAR Model, which stands for Green-Amber-Red. The GAR Model plots the calculated risk against the probable gain from a certain operation. This brings a level of objectivity and consistency to the continual and ever-important ORM process.

Learning Culture

Finally, for an organization to reap the benefits of a safety culture, it must possess a learning culture with 'the willingness and the competence to draw the right conclusions from its safety information system, and the will to implement major reforms when their need is indicated' (Reason, 1997, p. 196). Implementing a learning culture may seem easy to leadership until they reach its most important stage: acting on what is learned. There are a variety of tools available to observe, analyze and identify necessary changes. Many organizations, however, stop there and do not proceed with implementing needed changes in policy and procedures.

An organization's desire to achieve a safety culture is commendable, but not sufficient. Leadership must have a process to get there. A simple quality management tool – at the core of an SMS – is the Plan-Do-Check-Act (PDCA) cycle. The PDCA cycle can be used formally or informally to ensure the learning process never stops. This simple tool encourages action to realize the desired fundamental shifts an effective safety culture requires. In this uncomplicated cycle, the last two components (Check-Act) embody the notion of a learning culture.

Coast Guard written policy delineates a critical step in safety assurance. The *Safety and Environmental Health Manual* (US Coast Guard, 2006) requires the Division of Aviation Safety to conduct recurring Aviation Safety Standardization visits at all air stations to garner and share best practices, and to evaluate unit safety posture. Standardization visits also provide feedback on the Division of Aviation Safety's performance from the field unit perspective. These visits occur every two to three years and are purposefully scheduled within the first year of a new commanding officer's tenure. The rationale is to make recommendations to air station leadership early in their term to help ensure an appropriate command climate and a positive safety culture.

The *Coast Guard Air Operations Manual* requires that all aviation commands have a formal pre-mishap plan (similar to an emergency response plan). According to the *Safety and Environmental Health Manual*, it must be exercised and subsequently reviewed annually. Continuing from the earlier description of flight safety officer duties, the following items specifically address the responsibilities regarding pre-mishap plans and preparations:

1. Update and annually exercise the unit's pre-mishap plan. Consider conducting alternating tabletop and field exercises of the pre-mishap plan.
2. Conduct annual training for the Unit Permanent Mishap Analysis Board members. Placing particular emphasis on protection of the crash site and wreckage, photographic documentation of crash site hazards and collection of all pertinent logs and records.
3. Maintain and periodically inventory the unit aircraft crash investigation kit.
4. The unit's Pre-Mishap Plan shall provide guidance to ensure the effective completion of the numerous time-critical tasks required as a result of a major mishap. Permanent Mishap Board members and their alternates must

be clearly identified in the Pre-Mishap Plan. Their respective duties must be delineated prior to the mishap. (p. 2-7)

In order to validate the effectiveness of the pre-mishap plan, I try to bring in experts from outside the unit to observe or proctor the annual drill. This technique has worked extremely well and has certainly been beneficial. These unbiased safety professionals have discovered weaknesses that would be difficult to overcome in the event of a real mishap. Often fixes are easy, but might have gone undetected without engaging persons who bring a fresh perspective.

Of utmost importance is the emphasis placed on continually updating the pre-mishap plan. Mishap drills are a rich source of data for making needed changes. Following a mishap drill, users provide input in their respective areas of expertise to identify and implement needed modifications.

As discussed by Stolzer, Halford and Goglia (2008), a critical component of safety management – specifically safety assurance – is often administered through a safety council. When properly implemented, leadership from various departments come together to set certain priorities and determine the overall risk that an organization is willing to accept. Per Coast Guard regulations, all units, no matter how small, must establish and use a safety council. At an air station, a safety council typically meets quarterly and includes representatives from various aviation engineering shops (engine, metal and avionics), operations, safety and support. Safety councils systematically review mishaps and near-misses to recommend necessary changes to policies, procedures and regulations. Command cadre, though often in attendance, always see the formal minutes of these meetings and determines final action on various items or concerns.

A popular method by which safety information is passed is through safety pauses or stand-downs, in which units halt their operations to focus on relevant safety topics. During these events, which are required at least annually by Coast Guard policy, the unit takes a break to discuss current issues relevant to their type operation. Positive feedback garnered from personnel on our internal safety surveys validates the effectiveness of these events.

Common Errors

Organizations with sustained performance excellence operate in climates that infuse organizational values at all institutional levels. The Coast Guard is no exception. Coast Guard aviation clearly reaps the benefits of a command climate that embraces safety culture. Organizations that undertake the development and implementation of a viable safety culture for the first time, however, will face difficulties or challenges. Becoming aware of common errors and obstacles faced by those undertaking this process helps avoid or overcome potential pitfalls.

First, as noted earlier, a top-down buy-in must exist from the command. Without this, persuading senior management to accept and support change will be futile. Second, comprehensive planning is a must. Organizations must determine roles and responsibilities, review relevant polices and regulations and make sure needed resources are identified and available. Third, the new or revised reporting process must be easy to use, clear and unambiguous. A well-designed system will target the right audience and capture relevant data. A poorly designed system will not. At best, personnel will do only what command requires, paying only lip service to a system they do not find relevant and important. Finally, to maintain the right safety culture requires dissemination of relevant information. If users are properly trained and the new system appropriately implemented, personnel will grow to value the system, report data in a timely fashion and engage in a culture of continuous improvement.

To instill a safety culture, change must be well managed. Thousand and Villa (1995) contended that to manage complex change, the following factors were necessary: vision, mission, values and goals; skills, knowledge and attitudes; incentives and rewards; human, material and other resources; a strategic action plan; and data to measure progress. If just one of these factors is missing, the impact is substantial. Table 4.1 describes the theoretical outcomes; pay close attention to the right-hand column, as it shows the potential effect when a respective factor is neglected or missing entirely.

Table 4.1   Managing complex change

Vision	+	Skills	+	Incentives	+	Resources	+	Plan	+	Data	=	Change
__?__	+	Skills	+	Incentives	+	Resources	+	Plan	+	Data	=	Confusion
Vision	+	__?__	+	Incentives	+	Resources	+	Plan	+	Data	=	Anxiety
Vision	+	Skills	+	__?__	+	Resources	+	Plan	+	Data	=	Resistance
Vision	+	Skills	+	Incentives	+	__?__	+	Plan	+	Data	=	Frustration
Vision	+	Skills	+	Incentives	+	Resources	+	__?__	+	Data	=	Treadmill
Vision	+	Skills	+	Incentives	+	Resources	+	Plan	+	__?__	=	Concerns

(Adapted from Thousand and Villa (1995).

Remember, the process to make cultural changes within an organization is a marathon, not a sprint!

Change Within

Commanding officers of Coast Guard air stations change every two to three years, making it difficult to nurture and sustain a safety culture or command climate. As commanding officers come and go, leadership styles change – and change is difficult for any organization. Stolzer et al. (2008) allude to the importance of change management in SMS.

The people of any organization predictably go through various stages of concern when experiencing change. When taking over as safety officer or when witnessing various changes of command over the years, I have kept the following stages (from Hiam, 1999) in perspective:

1. Information: 'What is going on?'
2. Personal: 'How will it affect me?'
3. Management: 'What do I need to do?'
4. Consequences: 'How will it affect the organization?'
5. Collaboration: 'What more can I do to help implement change?'
6. Refocus/refinement: 'What else can we change to get more benefits?

When addressing safety culture, it is essential to consider how making necessary changes will affect your greatest resource: the people within your organization.

In the words of the Commandant of the Coast Guard, 'Change is hard, but not as hard as recovering from a missed opportunity or the loss of confidence that comes when leaders fail to act.' Understanding that change to a positive safety culture or command climate is necessary, another good reference for affecting organizational transformation is Blanchard's (1992) seven dynamics of change (described below *in italics*, along with practical examples and possible resolutions):

1. *People will feel awkward, ill-at-ease, and self-conscious.* An organizational change to a reporting culture will be uncomfortable for most, as habits and routines will be disrupted. Supervisors must genuinely encourage and show people that safety reporting truly benefits the unit as a whole. In the military, maintaining the promise that safety investigations will only be used for the purpose of avoiding a future mishap is absolutely critical.
2. *People initially focus on what they have to give up.* While implementing a safety culture, people may erroneously feel like they are giving up certain freedoms of anonymity and non-compliance. In reality, the knowledge they will gain and the level of safety that will be accomplished will far outweigh anything they may feel like they are giving up. In either case, a highly respected command cadre and supportive supervision will ease their concerns.
3. *People will feel alone even if everyone else is going through the same change.* Change tends to isolate some individual personalities or small groups of people. In a military climate, supervisors should ensure that new safety policies are enforced equally and equitably. When conducting safety investigations, as a safety professional I strive to treat people – regardless of rank – with the same emphasis on finding out 'why', not just 'who' and 'how'.
4. *People can handle only so much change.* In the typical dynamic environment of an aviation organization, there are many uncontrollable factors that come into play. Similar to the concept of threat and error management used

in today's crew resource management training, a unit must strive to control the changes that are controllable. Piling on change upon change will not be effective.

5. *People are at different levels of readiness for change.* Both the ICAO and the FAA advocate a gradual implementation of an SMS. Specifically with safety culture, a true transformation cannot and will not happen overnight. In some cases, changes will take time. Open communications up and down the chain of command enhance people's ability to accept change.

6. *People will be concerned that they do not have enough resources.* Although limited resources are a reality in today's aviation environment, leaders should do everything possible to ensure appropriate resources are available for necessary safety processes. Results from safety surveys, for example, can assist in providing the data often needed to drive appropriate funding for the acquisition of resources.

7. *If you take pressure off, people will revert to their old behavior.* If positive safety changes are not perceived as being taken seriously (especially by top management and leadership), it is likely that new procedures will go away. Setting the example and coaching by supervisors will stop these reversions to old methods. SMS safety assurance protocol is a key part of this concept (for example audits and internal evaluations).

The Sign

Upon reporting to Coast Guard Air Station New Orleans as the new flight safety officer, there were two signs at the unit that immediately caught my attention. The first sign, just inside the main gate, was a continually updated sign tracking the lives saved by aircrews of that unit since its establishment in the 1950s. The impressive count was just short of 3,500 lives saved – people that would no longer exist had it not been for the brave men and women of Air Station New Orleans. The second sign, painted in bold letters on the glass doors to the main building, stated 'Safety First'. I remember discussing with my new commanding officer the irony in that statement. As the busiest all-helicopter search and rescue unit in the Coast Guard, how could 'safety' truly be 'first'? If that were true, the five assigned aircraft would never leave the ground! After promulgating the command's new safety policy, the sign was changed to 'Safe Mission Execution . . . Always'. The point is that all too often organizations approach safety culture in an unrealistic manner. As insignificant or inconsequential as the sign may seem, it was important to set the proper tone at the unit for the command's true expectations. An antiquated slogan became an attainable goal.

So, How Safe is Safe Enough?

A significant difficulty with traditional Coast Guard safety programs is the lack of quantifiable data. In the broad scope of operations, an indicator of success is the amount of programmed flight hours actually flown in support of various missions. In the realm of maintenance, there is aircraft availability and a myriad of other sources of reliability data. In safety, however, the lack of documented mishaps is not really an appropriate measure of safety success or failure.

One of the main issues discussed in ICAO's 2009 SMM is the notion of acceptable level of safety (ALoS), which is established by the State. In order to fully understand ALoS, the SMM discusses two distinct concepts: safety measurement and safety performance measurement. Safety measurement is reactive, as it attempts to quantify high-level or high-consequence events, such as accidents or serious incidents. This type of measurement can include scheduled inspections, audits, or status of safety regulation implementation. Conversely, safety performance measurement is more proactive, as it encompasses continuous monitoring of various safety processes. This type of performance measurement provides a current evaluation of the operational performance of a management system, such as an SMS.

ALoS can be expressed by two measures: safety performance indicators and safety performance targets. Safety performance indicators are short-term, tactical and measurable objectives. For example, the amounts of foreign object debris (FOD) ingestions causing significant engine damage at a specified unit. Safety performance targets, on the contrary, are long-term, strategic and measurable objectives. Using the example above, a performance target would be reduction in the amounts of FOD ingestions causing significant engine damage by 50 percent during a specified period at that same unit.

The SMM discusses ALoS as a three-pronged model: (1) the high-level safety management objectives of an organization (Coast Guard), of which achievement is confirmed by safety measurement; (2) the minimum safety performance the organization should deliver through its safety program, of which achievement is confirmed through safety performance measurement; and (3) the aggregate safety performance measurement of specific units' SMS, which is an indirect indication of the overall organization's safety performance.

In managing an organizational ALoS, the techniques discussed throughout this chapter – when appropriately used – will facilitate decision-making. A positive safety culture or command climate will enable this progress and ensure it occurs at a proper level.

Summary

Establishing a positive safety culture or command climate will undoubtedly transform any organization. Of prime importance to the success of any aviation organization, a positive safety culture forms the foundation that supports the four symbolic pillars of an SMS. A viable safety culture is infused within all organizational levels, is adaptable, and continually seeks out and eliminates hazards. A safety culture allows for calculated risk by being professional and systematic about risk. Of critical importance is the understanding that a safety culture is not just about systems, but it is about individuals in the system ... and their trust is central to its success.

According to the FAA (2008), the many attributes of a positive safety culture include:

1. Competent personnel who understand hazards and associated safety risk, are properly trained, have the skill and experience to work safely and ensure safe products/services are produced.
2. Individual opinion is valued within the organization and personnel are encouraged to identify threats to safety and to seek the changes necessary to overcome them.
3. An environment where people are encouraged to develop and apply their skill and knowledge to enhance safety.
4. Processes to analyze information from employees' reports, assess their content, develop actions as necessary and communicate results to the workforce and the public.
5. Effective communications, including a non-punitive environment for reporting safety concerns.
6. There are clear standards of behavior where there is a commonly understood difference between acceptable and unacceptable actions.
7. Adequate resources to support the commitment to safety.
8. A process for sharing safety information to develop and apply lessons learned with regard to hazard identification, safety risk analysis and assessment, safety risk controls and other safety risk management responses. Sharing of information related to corrective actions, and results of management reviews is encouraged.
9. Safety is a core value of the organization that endures over time, even in the face of significant personnel changes at any level.
10. Willingness to recognize when basic assumptions should be challenged and changes are warranted – an adaptive and agile organization.
11. Decisions are made based on knowing the risk involved in the consequences of the decision (p. 10).

In order to effect permanent and long-lasting change such as positive safety culture, an aviation organization must address the various factors systematically with dedication and resolve. Transformational change requires hard work and commitment. Developed by Thousand and Villa (1995), Table 4.2 is both a check list and a good starting point for assessing organizational readiness to engage in the required systemic change.

Table 4.2   Consequence of component omission checklist

Component	Consequence of omission
Vision	Confusion: *Why are we doing this?*
Skill	Anxiety: *I want to do this, but I don't know how!*
Incentives	Resistance: *Something not worth doing is not worth doing well!*
Resources	Frustration: *How can we be expected to do this without the right tools?*
Action plan	Treadmill: *I don't know what's expected of me, so I'll just try this.*
Data	Concerns: *Is this really working?*

Adapted from Thousand and Villa (1995).

Finally, the question is not if an organization can afford to expend resources to embody a safety culture, but can it afford not to?

Case Study

When hurricane Katrina struck the Gulf Coast in August 2005, Coast Guard Air Station New Orleans became the hub or focal point for one of the largest air rescue operations in modern history. Within hours of the epic storm passing Louisiana, the five helicopters from Coast Guard Air Station New Orleans were in the air assessing damage and picking up stranded survivors. A robust safety culture, deeply embedded in all aspects of operations and maintenance, paid huge dividends as seemingly overnight the unit effectively quintupled in the number of aircraft assigned and ultimately supported another 85 aircraft throughout the entire operation. Coast Guard Air *Station* New Orleans informally became known as Coast Guard Air *Group* New Orleans. Service men and women from all 26 aviation units – from Alaska to Hawaii to Puerto Rico to all points in between – came together to save men, women and children from rising waters. Relying on exceptional leadership, standardization, training and with the consummate support of Coast Guard Aviation Training Center Mobile, Alabama, these aircrews flew 4,423 hours on 1,856 sorties, and saved over 7,100 lives, twice the number of rescues previously conducted at Air Station New Orleans – all with no casualties or serious aircraft mishaps!

A pervasive safety culture and a positive command climate played a significant role in the success of the Coast Guard's surge operation following hurricane Katrina. Each of the components of safety culture, including informed, reporting, just, flexible and learning, functioned in unison and provided the basis for a clear and consistent attitude of safe mission execution – even in the face of such a large-scale operation.

An informed culture: as the unit made preparations for the impending storm, personnel were constantly updated on command actions and expectations. Using the Coast Guard Principles of Operations, plans were developed and executed while ensuring the line operators and maintainers were appropriately informed. I vividly remember our commanding officer giving us clear instructions on his expectation that we use complete autonomy when operating our helicopters. He was directing us to use appropriate risk-assessment techniques, but was ordering us to mentally prepare to make difficult life-changing decisions.

Once aircrews returned to the area after having evacuated to safe havens, keeping the information flowing was certainly a challenge. Prior relationships and credibility established by the safety department before the storm played a significant role in nurturing trust throughout the operation.

A reporting culture: amid the controlled chaos of a surge operation of this sort, we found innovative ways to remain, more than ever, a reporting culture. Since we were operating in unknown territory, it was critically important to share pertinent information. As aircrews arrived from units all over the Coast Guard, we provided thorough on-the-spot briefs regarding items such as course rules, known hazards, communication protocol, ramp procedures and safety practices. With the safety office damaged, we had to make our presence known by purposefully increasing our physical contact with the operators and maintainers to collect and share relevant information.

A just culture: despite our best efforts, we were forced to extend and even bend certain rules and procedures due to the severity of the situation. We used operational risk-management techniques to make complex decisions, especially during the first few days of the rescue effort. With the full support of the command, we flew longer than authorized hours with little to no rest. When support arrived from other units, we set in place many safeguards to keep the operational pace running smoothly and effectively. Immediately, maximum flight time limitations were reduced, hot refueling operations (where the aircraft are fueled with rotor blades turning) were prohibited, and precise ramp traffic procedures were put in place; even in the face of adversity, personnel were focused on safe mission execution.

A flexible culture: as a sea-going service, the Coast Guard is charged will all US maritime search and rescue. We do not train for extensive urban rescue missions. During the surge operations that took place in the Gulf Coast in the aftermath of hurricane Katrina, on-scene commanders relied heavily on the flexible and adaptable Principles of Coast Guard Operations to safely carry out countless rescues in harrowing conditions. As front-line operators facing insurmountable odds in the face of one of the largest natural disasters in American history, aircrew kept their focus by remaining flexible in their actions. There are countless stories of thinking outside the proverbial box. Rescue swimmers armed themselves with axes to extricate stranded New Orleanians, flight mechanics improvised hoisting techniques to safely remove people from the tightest of balconies, and pilots skillfully hovered within feet of power lines in power-limited helicopters.

A learning culture: less than one month later, Air Station New Orleans was again challenged operationally when hurricane Rita, another record-breaking storm, slammed into the Texas and Louisiana border. The unit put into effect the many lessons learned from hurricane Katrina. In the end, the heroic actions of the men and women of Coast Guard Air Station New Orleans saved another 71 lives.

There is no doubt that a positive safety culture and a strong command climate were key to the overwhelming success of Coast Guard Air Group New Orleans' rescue efforts following the unprecedented destruction of hurricanes Katrina and Rita. During those memorable days in August and September of 2005, the safety culture ingrained in the mindset of all Coastguardsmen and women was the difference between triumph and failure as they fulfilled the Coast Guard's mission under the most extraordinary circumstances. Along with the honor, respect and devotion to duty displayed by these extraordinary guardians, the Coast Guard's command climate and exemplary safety culture – both centered on trust – fostered top-down and bottom-up communications and empowered rescuers to do whatever they deemed necessary to effectively carry out their duties, earning the respect and admiration of a grateful nation.

References

Blanchard, K. H. (1992). The seven dynamics of change. *Quality Digest, 12,* 18 and 78.

Collins, J. and Porras, J. I. (1994). *Built to Last: Successful Habits of Visionary Companies*. New York: HarperCollins.

Dekker, S. (2007). *Just Culture: Balancing Safety and Accountability*. Farnham: Ashgate.

Federal Aviation Administration (2006). *Introduction to Safety Management Systems for Air Operators*. Advisory Circular 120-92. Washington, DC: Federal Aviation Administration.

Federal Aviation Administration (2008). *Safety Management System Guidance*. Order 8000.369. Washington, DC: Federal Aviation Administration.

Global Aviation Information Network Working Group E. (2004). *A Roadmap to a Just Culture: Enhancing the Safety Environment*. Retrieved December 22, 2009 from http://www.coloradofirecamp.com/just-culture/definitions-principles.htm.

Hiam, A. (1999). *Motivating and Rewarding Employees: New and Better Ways to Inspire Your People*. Holbrook, MA: Adams Media Corporation.

Hudson, P. T. W. (2001). Safety management and safety culture: The long, hard and winding road. In W. Pearse, C. Gallagher and L. Bluff (eds), *Occupational Health and Safety Management Systems* (pp. 3–32). Melbourne, Australia: Crown Content.

International Civil Aviation Organization (2009). *Safety Management Manual (SMM)*, 2nd edn, Doc 9859. Montreal, Canada: International Civil Aviation Organization.

Merriam-Webster Online Dictionary. (2009). Retrieved December 22, 2009 from http://www.merriam-webster.com/dictionary/tripartate.

Reason, J. (1997). *Managing the Risks of Organizational Accidents*. Aldershot: Ashgate.

Stolzer, A. J., Halford, C. D. and Goglia, J. J. (2008). *Safety Management Systems in Aviation*. Aldershot: Ashgate.

Thousand, J. S. and Villa, R. A. (1995). Managing complex change towards inclusive schooling. In R. A. Villa and J. S. Thousand (eds), *Creating an Inclusive School*. Alexandria, VT: Association for Supervision and Curriculum Development (ASCD).

US Coast Guard. (1999). *Coast Guard Operational Risk Management*. Commandant Instruction M3500.3. Washington, DC: US Coast Guard.

US Coast Guard. (2006). *Safety and Environmental Health Manual*. Commandant Instruction M5100.47. Washington, DC: US Coast Guard.

US Coast Guard (2008). *Coast Guard Air Operations Manual*. Commandant Instruction M3710.1F. Washington, DC: US Coast Guard .

US Coast Guard. (2009). *Coast Guard Publication 1*. Washington, DC: US Coast Guard.

Chapter 5

Safety Culture in Your Safety Management System

Dan McCune, Curt Lewis and Don Arendt

> No matter how interested individual employees might be or what assistance a manufacturer offers, or how insistent a certificating authority might be – none of these factors will have a significant effect on safety without support from top management.
>
> John O'Brien, ALPA's Engineering and Air Safety Department

Introduction: Why a Safety Culture?

Of all of the components of the Safety Management System (SMS), safety culture is often the most difficult to foster. Interestingly, there may be a reason for this difficulty, particularly within the field of aviation. The concept of safety culture was discussed in FAA advisory circular 120-92a (FAA, 2010) where elements of the SMS form a framework for a healthy safety culture.

Experts agree that safety culture is a fundamental element to the foundation of accident prevention and SMSs. In fact, a strong safety culture is one of the most effective and systemic ways to reduce accident and incident levels within an organization. However, in order to make a safety culture truly effective, safety promotion activities must take place in such a way that they instill and reinforce that culture throughout the organization. Unfortunately, one of the major challenges to implementing and sustaining an effective safety culture lies in the age-old battle between protection and production. For many industry leaders, operational activities are seen as direct contributors to the bottom line, while safety is sometimes looked upon as a necessary evil that continually draws upon the resources of the company. Therefore, it is very hard for decision-makers to invest in a program that does not have direct contributions to production or ultimate fiduciary value. For instance, a CEO immediately sees the value of investing in new aircraft, but does not readily see the return on investment when the organization loses production time because pilots need to attend training to enhance their flying skills.

A strong safety culture will help prevent accidents, but it is difficult to prove to senior management that a zero accident rate is attributed to the safety culture. Safety professionals cannot show senior leadership how many accidents have been prevented by a strong safety culture, but when an accident does occur, it usually uncovers a flaw in the organization's safety culture.

SMS and the Development of Effective Safety Culture

Whenever you turn on the nightly news, chances are you will hear some mention of the aviation industry. It may be an account of an aviation-related incident. It may be a report detailing changing airline policies or the rising cost of air transportation. Or, more recently, it may be a discussion of air safety and new FAA regulations. As we all know, SMSs have been of primary importance to the FAA for a number of years. SMS was recognized as a vital component to the continued growth and success of the aviation industry. It was also acknowledged that a set of uniform safety standards should be created, which would conform to international SMS aviation protocols while being flexible enough to accommodate the needs of individual American airports. To this end, the FAA implemented a pilot program to study and compare the current Airport Certification Programs and developing SMS principles. The results of this pilot program are already being put to use by the FAA, resulting in the implementation of new safety protocols for the aviation industry, and the establishment of SMS as a U.S. aviation regulatory standard. Though we know why SMS should be adopted and what it should accomplish, SMSs are about more than just regulations and enforcement.

In order for a SMS to not only work but to remain effective, the aviation industry needs to create a culture of safety. Safety culture can be very simply defined as an organizational commitment to safety at all levels of operation. Establishing an effective safety culture, however, is anything but simple. Effective safety cultures distinguish themselves through having clearly defined procedures, a well understood hierarchy of responsibilities at all levels and clear lines of reporting to facilitate effective and useful communications regarding safety issues. A more detailed list of the attributes of an effective safety culture was presented by the International Civil Aviation Organization (ICAO), which placed a strong emphasis on the role of senior management and the importance of communication. All levels of aviation management must make it clear that safety culture is concerned with the safety of not only airline passengers, but also of airport and airline employees. Safety management should not be viewed as simply a means to an end or a blind adherence to industry standards, but rather as a company and industry-wide commitment to best practices and continuous improvement of everything safety related. In an effective safety culture under SMS human error is seen as inevitable, and the focus is shifted from a reactive to proactive method of managing risk. The prevailing view of risk should be professional and realistic, focusing on eliminating or maintaining optimum levels of acceptable risk using past incidents, perspective and insight.

The aviation industry has, in the past, been comfortable maintaining a reactive position to safety regarding occurrences as isolated incidents, and consistently only taking action when something happens. This attitude gradually became more calculated, growing into a regulatory system and developing a bureaucracy to enforce it. The introduction of SMS is shifting the focus from enforcement-centered to a more proactive approach, and hopefully will give rise to a culture of safety so firmly established that the perception will be that safety is simply the best, most effective and most profitable way to do business.

In a good safety culture, senior management does the following:

- places a strong emphasis on safety
- has an understanding of hazards within the workplace
- accepts criticism and is open to opposing views
- fosters a climate that encourages feedback
- emphasizes the importance of communicating relevant safety information
- promotes realistic and workable safety rules
- ensures staff are well educated and trained so that the consequences of unsafe acts are understood (Flannery, 2001).

The Evolution of Safety Culture

How organizations react to safety information displays their level of safety maturity. Listed below are a few examples, from the basic pathological to the sophisticated generative level.

- *Pathological*: we don't care as long as we don't get caught
- *Reactive*: we take action only in response to incidents
- *Calculative*: our approach to safety is systematic, through an established bureaucracy
- *Proactive*: we take steps to deal with issues before incidents occur
- *Generative*: safety is how we do business (Hudson, 2001).

Effective safety management is a learned skill and, as with any skill, continues to grow and develop over time with practice. Therefore, an effective culture of safety is one that has practiced safety management until that skill set has become second nature – safety is simply the way business is done, and improvements to the system are considered improvements to the company as a whole. Of course, this procedure for creating and maintaining a safety culture sounds much easier than it actually is; road blocks must be expected throughout the process at all levels. Management, initially on board with the implementation of SMS, may become less enthusiastic as they realize that some changes will not be cheap or simple to implement. Managers may be uncomfortable soliciting and responding to negative feedback and lower-echelon staff members may be difficult to convince that reporting honestly on current or potential problems is in their best interest. In some groups, such as pilots or physicians, where perception of infallibility can be closely linked to professional reputation, the idea of admitting personal error may be akin to admitting personal and professional failure – or possibly to committing professional suicide. These are all hurdles which must be overcome systematically at an organizational level, with a major top-down emphasis on building trust and establishing non-punitive reporting systems. Without these two factors in place, SMS cannot be successful and a culture of safety will not develop successfully. Similarly, the basic conditions which must exist in order for safety culture to flourish are:

- trust
- a non-punitive policy toward error
- commitment to taking action to reduce risk-inducing conditions
- diagnostic data that show the nature of threats and the types of errors occurring
- training in threat recognition and error avoidance and management strategies for crews – crew resource management (CRM)
- training in evaluating and reinforcing threat recognition and error management for instructors and evaluators (Helmreich, 1999).

The concept of crew resource management, or CRM, is based on the idea that organizations must recognize that human error is unavoidable and that it is the responsibility of a mature organization to effectively manage that error (Hayward, 1997).

CRM seeks to:

- reduce the likelihood of error
- isolate errors before they have an operational effect
- reduce the consequences of errors when they do occur.

CRM, as it is known today, is an outgrowth of Cockpit Resource Management training, which was developed in the early 1980s and gradually expanded into other aspects of aviation and outward from there into other industries. Properly implemented according to the specific needs and culture of a particular organization, this approach to the handling of incidents and reporting can be highly effective for combating and correcting issues with reporting, feedback and admission of fallibility. Establishing and maintaining such systems requires a firm commitment from management to 'stay the course' even when, from a purely financial perspective, it would be more advantageous not to.

Data-gathering, for example, can be a costly and time-consuming process, as can the creation and implementation of new training programs. Management must not only be cognizant of the long-term benefits of those costly, inconvenient actions, they must also be aware that employee and public perception of their willingness to pursue safety ahead of, or at least on a visibly equal basis with, profit will greatly contribute to the trust-building which is such a vital element of effective safety culture.

Finally, the concept of safety culture cannot be discussed without also touching on the related concept of safety climate. These terms are sometimes used interchangeably, but they actually define different dimensions of the issue of safety. Safety culture, so closely tied to SMS, speaks to the development of safety regulations and related organizational safety systems, which work to create a stable and long lasting environment. Safety climate, on the other hand, more often refers to the psychological perception of the state of safety at a particular time (Zhang et al., 2002), which, of course, can be expected to change frequently under the influence of any number of social and environmental factors. Monitoring the safety climate within an organization, therefore, should provide valuable insights into the state of that organization's developing culture of safety, especially during the implementation phase of new systems and procedures.

All in all, safety culture should be seen as a natural outgrowth of the application of well thought-out SMSs, the commitment of senior management to safety as the best way to do business and the growth and development of safety oriented organizational norms. Like SMS, the evolution of safety culture is a continuous process, not a means to an end or a static goal to be reached; a healthy culture of safety should maintain its stability while constantly reaching toward new heights, never stopping in place and saying, 'That's good enough, we don't need to do any more.' Though this is a continuous process, the aviation industry, and other industries as well, can proactively expect to reach a goal where safety truly will become just the way we do business.

Models of Organizational Culture

What is an Organizational (Safety) Culture?

Before we look into models, it's beneficial to describe what we mean by an organizational culture, of which safety culture could be said to be a subset. It will also be important to see how the aspects of safety culture, and factors affecting it, are related to tangible management programs, activities, and decisions. Hofstede and Hofstede (2005) wrote that organizational (or corporate) cultures are acquired when we enter a work organization ... with our values firmly in place and [that] they consist mainly of the organization's practices – they are more superficial [than national cultures]. Thus they offer that the defining characteristics of organizational cultures are made up of the things that the organization does or in how it does them.

It's not entirely correct to say that we want an organization to have a safety culture or that they lack having one. Culture isn't something with an on/off switch, nor is it something that regulators can require an organization to posses. Like attitudes and personalities of individuals, cultures tell us something about the corporate 'personality' as well as the collective attitudes, beliefs, values and behaviors of the organization's members and their leaders.

Why do we Care about an Organization's Culture?

The question of why we even care about culture may come to many practitioners' minds. After all, aren't we mostly concerned about 'bottom line' performance both in terms of business and safety? The important thing about studying organizational culture is that it allows us to predict the practices, the behaviors or performance, of the organization. We want to know how individuals in an organization and the organization as a whole typically perform under given circumstances. Understanding certain aspects of an organization's culture can help us to gain that knowledge, and also allows the organizations' leadership to shape the environment in which their organizations work to foster the behaviors that support desirable outcomes. Whether we call it 'culture' or just 'organizational behavior', the bottom line for safety management is in understanding and enhancing organizational performance related to safety.

Three Aspects of Organizational Systems

Psychologist Albert Bandura (1985) describes a system of three interacting elements – behavior, psychology and environment – that make up any social system, including organizations. These elements also interact in sometimes complex ways.

Behavioral

- Individual
  a) Management
  b) Employees/members

- Group
  a) Collective behavior
  b) Organizational norms

Psychological

- Affective domain – 'feeling', 'believing', attitude
- Cognitive domain – 'thinking': for example, decision-making

System/Environmental

- Practices, programs, facilities, processes, procedures

Figure 5.1 gives us a picture of how these three factors interact and how factors outside of the organization can affect the various aspects of the organization as a social system.

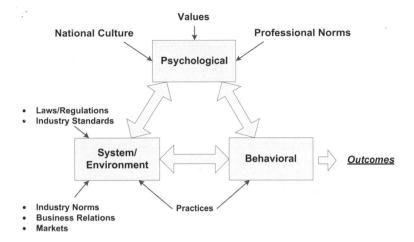

Figure 5.1   Interaction of social system elements

Next, we'll consider the three elements individually, as well as some of the implications for managing the organization. Then we'll return to the discussion of how the aspects interact and how they impact organizational performance – the bottom line in both safety and productivity.

Behavioral Aspect: Individual and Collective Performance

Hofstede and Hofstede (2005) go on to note that practices can be 'labeled conventions, customer, habits, mores, traditions, usages'. All of these could be categorized as behaviors. Thus, one of the significant aspects of culture is the patterns of *behaviors* of an organization and its members.

Behavior concerns how people act. Policies and procedures can affect how people in the system behave, but their behavior is also affected by the way they think and feel (psychology) and the resources they're given to do the job (system/ environment). Thus, there may be a difference between the way the system is designed and how employees function in actual operations.

While management can observe employee behavior, they can't do this all the time. Also, management behavior and the behavior of other employees can affect the way people think and feel, and subsequent behavior. Both the psychological and behavioral elements can also affect the future of the system/environmental element in ways such as changes in procedures, procurement of equipment, design of training, policy changes, etc.

Psychological Aspect: Thinking, Feeling and Believing

The psychological element concerns how the people in the organization think and feel about various aspects of organizational performance, including safety. Some of the factors that affect the psychological element are brought to the organization by its members. The national cultures, professional cultures and industry norms are among these factors. These factors, and the tone set by management, combine to affect the organization's values. However, some of the factors in this element, such as those related to national and professional cultures, will be highly resistant to change and may be shared across other organizations of the same type, location, etc.

The psychological element is one of the most powerful in making up the unique culture of the organization, but is the one that is the least tangible as well as being the one least under direct control of management. You can't control how people think and feel about a matter of policy, although policies can affect how people think and feel.

The psychological aspect of culture also has a cognitive[1] side (thinking). Management consists of both psychological and behavioral elements of decision-making and action. Decision-making itself has both psychological and behavioral elements. The thought processes involved in decision-making (cognitive domain) and value system of the decision-maker (affective domain) are closely related to the behavioral aspect of the actions taken by the decision-maker. These actions may, in turn, interact with the attitudes and perceptions (psychology) of the organization's members, as well as the changes in the working environment brought about by the decision-maker's actions.

Environmental Aspect: Surroundings and Situation

Individuals in a system work within the *environment* provided by those systems – the tools, facilities, technology, processes, procedures, rules and social and management structures that exist within the organization. Management decisions can directly affect the structure and content of the working environment, as can factors in the external environment such as economic, political, legal and mission or market factors.

This is where the processes of the SMS live. This is also the most tangible of the elements and the one that can be most directly affected by management actions. The organization's policy, organizational structure, accountability frameworks, procedures, controls, facilities, equipment and software that make up the workplace conditions under which employees work, all reside in this element. Elements of the operational environment such as markets, industry standards, legal and regulatory frameworks and business relations such as contracts and alliances

---

1   For further discussion of the affective and cognitive domains of human behavior with respect training and learning, see the *FAA Aviation Instructor's Handbook* (FAA 2009).

also affect the make-up part of the system's environment. These elements together form the vital underpinnings of this thing we call culture.

Hollenbeck and colleagues (2002) point out two aspects of the external environment that are important to system functioning – environmental complexity and environmental change. In situations where an organizations business or mission is stable, and relations both within the organization and between the organization and other organizations are simple, little is needed to adapt to the environment to survive and thrive. However, few aviation enterprises enjoy this situation, so the components of flexibility and learning, pointed out by Reason, become important. Flexibility is needed for the organization to adapt expeditiously, and learning is needed for them to accomplish change effectively and efficiently.

Interactions between Aspects

It should be easy to see how these three aspects interact. For example, if management decides (psychological–cognitive aspect) to obtain information through an employee reporting system, they must provide the infrastructure to do so and motivate employees to use it (create an environment). Non-reprisal policies are important to provide employees a belief (psychological–affective aspect) that they will be safe to use the system to submit reports (behavior).

At this point, further action on the part of management is pivotal. If they renege on their promise of non-attribution for reporters, employees will quickly change their beliefs regarding the reporting system, with behavioral change (ceasing to report) quickly following. If, on the other hand, management exercises proactive decision-making behaviors and improves the system (environmental change), based upon the reports received, positive beliefs and behaviors are reinforced, potentially increasing the strength of both the psychological and behavioral aspects of the culture.

Effects of Reward Systems: Motivators and Potential Pitfalls

Another situation where the aspects of culture interact is in the way in which behavior is rewarded in the organization. While it is a common and certainly legitimate practice to reward employees for accomplishment of productive or mission goals, such rewards can easily lead to goal conflicts for employees. Where management's stated goal to 'be the safest company in the industry' appears to conflict with the goals of on-time performance, maximum load factors, cost control and other production objectives, employees are left to sort things out on their own, often under intense time pressure.

In these cases, management may not recognize the ambiguity that they have caused in their employees. Management perception of safety and the impact of their decisions can often differ from that of line employees (Von Thaden and Gibbons, 2008). In this case, management behaviors (resulting from management's decision-making process) are observed, processed and perceived by employees, whose resulting behaviors may not be what were understood by managers at the time that they made the decisions and set the rewards. In this case, employees may feel that they are being tacitly punished if they take the extra time, expense, and actions to ensure safety. Moreover, if reward systems are dependent upon group performance (either in production of safety), social pressures can further confuse the issue for line employees.

Managers should also look carefully at the latter factor – reward systems for safe behaviors. Systems that reward employees of employee groups for lowering injury and damage rates can easily backfire and further exacerbate the gap between management's awareness of system performance and 'what's really going on'. Employees may be more inclined not to report minor injuries or occurrences, not only giving a false impression of the real situation to management but also depriving them of information on potentially hazardous conditions. Most people are familiar with the saying 'perception is reality'. In the case of safety reporting, this can easily translate to 'reporting is reality'. Unfortunately, anything that distorts the content of safety reporting also distorts our perception of reality.

The decisions, perceptions and behaviors of managers and line employees form the individual and collective behaviors that make up what we call norms. They can also result in a significant divergence of organizational practices that are expected by managers and regulators. The next two sections explore the development of norms and practices.

Norms: Everybody's Doing it But What is Everybody Doing?

Characteristic patterns of behaviors are also called *norms*. Hollenbeck and colleagues (2002) write that norms exist when 'the members of a group or organization share a set of beliefs about the acceptability of particular types of behavior, leading them to behave in ways that are generally approved'. It is these patterns of beliefs and behaviors that most often characterize what we call culture.

Beliefs come from the affective domain of psychology,[2] the feeling, believing and attitudinal side of human mental and emotional functioning. From this we see the effect of psychology in the form of beliefs and the social nature of acceptability of behavior. Also, we can see the interactions between group and individual psychology and behaviors.

Norms can come from either formal or informal means. Organizations create norms through formal means in the form of policies, rules, SOPs, checklists and the like. Formal training programs are also a means of creating norms. By these means, organizations deliver expectations regarding performance and behavior to employees. An important aspect of formal norms is having clear standards of acceptable and unacceptable behavior and performance, which are then reinforced

---

2   Ibid., FAA H-8083-9A.

through the training, counseling, communication, supervisory direction and discipline programs and practices within the organization.

Organizations and their members also create norms by informal means. In many cases, this is through what is reinforced by the organization's management, regardless of what has been delivered through normal means. Most of us can recount experiences from early in our careers where, having gone though whatever initial training, briefings and familiarization with process documentation, we were shown 'how it's really done here' by the old hands in the organization.

A less tangible, but often more powerful means of informal norm shaping is through observational learning. Here, the behavioral and psychological aspects interact. Much of what we learn is through observation of others. This can be both in learning technical skills as well as in non-technical, such as crew resource management, adherence to SOPs, risk perception, etc. If novices observe the behaviors and attitudes of their peers and mentors to differ from the expectations that are officially presented, these behaviors can be learned by the novices without a word being said. It is thus especially important for instructors, supervisors and managers to 'walk the talk', and for training programs that the instructors and supervisors emphasize this point.

The final form of informal creation of norms is through behavioral reinforcement. Here, programs for discipline and rewards are crucial. It has long been known that reinforcement of behaviors is a powerful antecedent to subsequent behavior. That is, if people are rewarded for particular behaviors, they will probably repeat them in future similar circumstances. In this area, care must be established to avoid unintended consequences. Reward systems that offer rewards for production goals may lead to an inadvertent, and often unconscious, relaxation or subordination of safety objectives. Likewise, rewards such as those given for time without accidents, incidents, lost time, or equipment damage may inadvertently discourage open reporting. Thus, while reward systems for both production and safety objectives can be both effective and appropriate, care needs to be taken to foresee the full range of potential interpretations and behavioral outcomes.

Practices: What's Really Going On?

It's important to recognize that an organization's behavioral norms, its underlying culture, may differ significantly from management's policies and formally established procedures. Degani and Wiener (1994) outline four levels of information and performance in organizations:

1. philosophy (general high level operating principles)
2. policy (expectations as to what the management wants)
3. procedures (specified methods and ways to achieve goals), and
4. practices (how activities are actually accomplished).

It's also important for managers to be sensitive to the fact that while the first three of these elements are more or less directly under their control, the last element, practice, is only partially under their control. Moreover, this element is the one that most directly reflects the organization's norms and, therefore, its culture. Where there is a difference between expectations and actual practice, it's also important for managers to discern why the difference exists before taking action. It may be easy to assume that more training or enforcement of rules is the solution to bringing practices into conformance with prescribed policies and procedures. However, upon closer analysis, managers may find that employees are actually using 'work arounds' to compensate for unworkable procedures, confused over competing goals, or coping with environmental difficulties such as poor facilities, unworkable schedules, or understaffing.

Reason's Components of Safety Culture

James Reason (1997) defined several components of a safety culture. These could also be viewed as traits similar to personality traits of individuals. The first three traits relate to how the organization fosters an environment conducive to learning, successful change, and management of risk. The latter two are related to how an organization adapts itself to its external environment.

Informed

> creating a safety information system that collects, analyzes, and disseminates information from near misses as well as from regular proactive checks on the system's vital signs … those who operate the system have current knowledge about the human, technological, organizational, and environmental factors that determine the safety of the system as a whole. (Reason, 1997, p. 195)

Safety management, as any management, is largely a practice driven by decision-making. Consistent decision-making is, in turn, supported by acquisition and use of sound information. Thus an organization that constantly informs itself is more likely to succeed both in business and safety performance.

Reporting – 'Any safety information system depends on the willing participation of the workforce, the people in direct contact with the hazards' (Reason, 1997, pp. 195–6). A key source of safety information comes from reports from all levels of the organization – from top management through all levels of employees. An organization must then make sense of the acquired data by turning the reports into useful information through analysis.

There are two proven ways to kill a safety reporting system – burn the reporter or burn the data. If safety reports are used as a source of information for disciplinary action, the reporting system will likely suffer an almost immediate demise. Employees will quickly lose trust in the organization's motives. The second way is slower but just as deadly to the system. If employees are not convinced that the organization is serious about acting on the situations that they report, they will ultimately lose faith in the system and discontinue using it.

Just – an environment that fosters two main ideals. The first is the reinforcement of open reporting; 'an atmosphere of trust in which people are encouraged, even rewarded, for providing essential safety related information but' (Reason, 1997, p. 195). The second are certain standards of behavior; '[a culture] in which they are also clear about where the line must be drawn between acceptable and unacceptable behavior' (Reason, 1997, p. 195).

It is important to understand that 'just' or 'non-punitive reporting' implies a discipline-free organization. To the contrary, it is important for the organization to have clearly stated standards that are consistently and fairly enforced. An organization's members must know what is expected of them in terms of behaviors and performance, but they also need to be assured that they will not be sanctioned for reporting safety problems even when they result from inadvertent errors.

Another point that is often misunderstood is that of accountability. This often takes on the connotation of being synonymous with blame. Being responsible for one's actions and performance includes being up front with one's mistakes and taking responsibility for correcting them. This includes the actions and possible mistakes on the part of management, as well as line employees. Sidney Dekker (2007) describes *forward-looking accountability* versus *rearward-looking accountability*.

Forward-looking accountability concerns taking responsibility for seeking and using information to avoid errors and, when they occur, being forthright about mistakes, reporting them, and taking action to continuously improve. Concentration on rearward looking accountability, blame and sanctions for undesired outcomes, can promote hiding information, shifting blame and avoiding responsibility for improvement.

Flexible – high-reliability organizations (HROs, discussed further in the next section) 'possess the ability to reconfigure themselves in the face of high tempo operations or certain kinds of danger' (Reason, 1997, p. 196). These organizations use information effectively, and couple it with constant improvements to their systems. Effective change management depends on consistent reporting, data collection, analysis and system adjustment. Policies, procedures and practices should be under continual review and adjustment to meet changing demands. This makes this attribute an essential element of an effective safety management system.

Learning – learning can be described as 'the willingness and the competence to draw the right conclusions from its safety information system, and the will to implement major reforms when their need is indicated' (Reason, 1997, p. 196). Reporting and other data collection are of little value unless accompanied by sound analysis. This doesn't always have to take the form of sophisticated analytical or statistical methods. Simple review and discussion of reports, audit findings and other data is often all that is required. Whatever methods are used for analysis, however, managers of business functions, as well as line employees, must be directly involved. It would be a mistake to assume that the safety or quality assurance departments or officers can perform these functions in isolation.

Components of High Reliability Organizations

Managing change is closely allied with managing risk. Organizations that are consistently able to successfully manage change in high risk environments are termed *high-reliability organizations* (HROs). Researchers Karlene Roberts and Carolyn Libuser defined the following components of HROs (Ciavarelli, 2008, p. 21). Note that these components are closely aligned with the processes of an effective SMS.

- Safety process auditing – 'a system of checks and reviews to monitor and improve processes' (part of the safety assurance component of the SMS).
- Safety culture and reward system – 'social recognition that reinforces desired behavior or corrects undesired behavior' (part of the organizational policy component of the SMS).
- Quality assurance (QA) – 'policies and procedures that promote high quality performance and work performance' (part of the policy and safety assurance components of the SMS).
- Risk management – 'whether or not leaders perceive operational risks and take corrective action' (safety risk management component of the SMS).
- Leadership and supervision – 'policies, procedures, and communication processes used to improve people's skills and to proactively manage work activities and operational risk' (parts of the policy and safety promotion components of the SMS).

It should be straightforward to see that the four components of a SMS, policy, safety risk management, safety assurance and safety promotion, are closely allied with the components of Roberts and Libuser's description of HROs. It should also be clear that Reason's components of a safety culture are highly related to the ability of an organization to become an HRO. Finally, the interacting, behavioral, psychological and environmental elements of organizational cultures form the basis upon which the HRO and safety culture factors interact.

Traits of High Reliability Organizations

The Roberts and Libuser work provides insight into components of HROs that are highly suggestive of the elements of an SMS. Another pair of HRO researchers, Karl Weick and Kathleen Sutcliffe (2007), have defined a set of characteristics of HROs that describe them in terms reminiscent of personality traits, which are commonly used to describe the psychological and behavioral characteristics of individual people. These characteristics also leverage highly the components of Reason's informed culture. Moreover, they are likely to thrive in an organizational environment shaped by an SMS.

Preoccupation with Failure

HROs are constantly aware of small failures that may indicate system deficiencies that could lead to more serious future consequences. In this area, multiple approaches need to be used in hazard identification and safety assurance. The importance of Reason's reporting culture is emphasized but it must also be accompanied by sufficient analysis to make sense of the information received

Reluctance to Simplify

Albert Einstein has often been paraphrased with words to the effect of, 'make things as simple as possible but no simpler.' In a world of 'keep it simple, stupid,' we tend to emphasize the first part of Einstein's advice, to the neglect of the latter. However, Einstein was really cautioning about the dangers of over-simplification. Certainly we must make our systems, processes, procedures, and documents as straightforward and understandable as feasible but our safety management efforts need to avoid glossing over aspects of our operations that may be important. Safety analyses, while needing to avoid 'analysis paralysis,' also need to be sensitive enough to pick up the nuances and details that may make the difference between the routine and the hazardous.

Sensitivity to Operations

Weick and Sutcliffe (2011) remind us that our operating environments are not static. This is especially true in complex industries. They also point out that many organizational relationships are not linear but that these interfaces are important to effective organizational functioning. We also need to remember that organizations do not exist to 'be safe.' The commonly used phrase 'safety is our highest priority' is seldom correct. Organizations exist to provide products and services to their customers or other recipients, such as the citizenry, in the case of public service organizations. However, HROs need to 'keep their eye on the ball.' They have to make sure that they remain informed, aware of the small failures and potential flaws in operational systems that could lead to injurious breakdowns. This is the purpose of the safety risk management and safety assurance functions of the SMS. Deference to Expertise. HROs recognize that real knowledge and expertise doesn't only reside in the higher echelons of the organization. They recognize that they have to seek out and respect this knowledge wherever it resides. The reporting culture is part of this. Managers need to 'keep their ears (and eyes) open.' The orientation that Reason describes as the just culture is also important. Employees often have knowledge that management won't gain access to if the employees are either afraid to tell them or if they feel that their input won't be respected.

Commitment to Resilience

The first four characteristics of HROs are primarily associated with knowledge and information; being sensitive to the sights and sounds of potential hazards, listening to the organization's members and paying attention to the organization's operations. However, the organization also needs to be action oriented. All the knowledge about hazards isn't valuable unless the organization is willing to act on it. Reason's flexible culture suggests that the organization has to be willing to change their operations to correct problems that they become aware of. This includes having a strategy to recover from small problems and major ones as well. The orientation to all of levels of failures needs to reflect the learning culture. Organizations need to be more interested in finding out what went wrong than whom to blame. Resilience also takes us to another point. Most people and organizations subscribe to the philosophy that 'we want to be proactive, not reactive.' This is good as far as it goes but we can't allow ourselves to believe that we've been smart enough and insightful enough to perceive all threats and develop effective countermeasures all the time. We have to recognize that sometimes we will fail. That means that we really have to have effective reactive tools in our kit. We have to be able to quickly respond to failure, learn from it, and move forward.

Measurement

Measurements of the three elements differ. The system/environmental area is the one for which we can most easily set standards which can subsequently be audited. We can set requirements for policy, processes, procedures and resources and we can determine if these standards have been met in a rather straightforward manner. The behavioral element can also be measured, although mostly through sampling, similar to what is done in traditional surveillance activities. However, while we can see what employees are doing while being observed, behavioral audits only provide a snapshot and there is no guarantee that the behavior is the same when people aren't being observed. This is one reason that employee reporting systems are very important to SMS – they provide a means of finding out 'what's really going on'.

Even though the psychological element is neither auditable nor directly observable, various tools exist for sampling it, such as interviews and surveys. If survey tools are constructed properly and the data are analyzed carefully, we can obtain insight into underlying behaviors that wouldn't be readily available otherwise.

Any measurement and assessment of organizational cultures should be less concerned with classifying the patterns of attitudes, norms and behaviors that define the organization's culture as good or positive, than in returning tangible, actionable information that managers can use to affect changes in investments, organizational structures, policies and infrastructure investments, that have the potential to improve safety performance.

Effects of National Cultures

Hofstede (2009) cautions that it is a mistake to assume that all people, in all cultures, are basically alike and will, therefore, react similarly in similar situations. This, says Hofstede, can be a serious mistake in international business relationships. Differences that can influence individual and group behaviors and performance include differences in perceptions of individual vs group collective values. There are also significant differences in perceptions of how status within the culture is regarded both between individuals as well as subgroups within the society, differences in perception of risk and differences in perceived acceptability of individual behavior such as assertiveness vs deference to authority. These aspects can have a strong influence on shaping organizational cultures in a given region or nation.

These differences can have important implications for safety as well as business. The problems with assumptions about the effects of national cultures that Hofstede points out need to be factored into implementation of safety management systems, and especially in efforts to shape and assess organizational cultures. This is certainly an area where one size does not fit all. Attempts to impose assumptions regarding safety culture in ways that are contrary to national and ethnic cultures may be met with difficulty, if not resistance. Better to study the values, traditions and patterns of relationships in a nation's culture and find ways to leverage these aspects to foster the behaviors that produce desired safety outcomes.

As we've mentioned earlier, the objective is not to influence an organization's culture in the abstract, but to stimulate the patterns of thinking, behavior and decision-making that result in desirable safety performance. Detailed comparative study of national culture is beyond the scope of this book, but managers, consultants and other safety professionals engaged in international business would be advised to study the works of Hofstede and others to gain a deeper understanding of this aspect of safety management.

Implications for Practicing Managers

The first thing that we should mention is that safety management is not in the domains of the 'safety manager'. Safety must be integrated into the fabric of the organization's business functions. In the words of retired UAL captain and current FAA Inspector Rick Clarke, the focus is not on *being safe* but on *operating safely*. The business leaders are the ones who control the assets and personnel activities of the business enterprise in which risk is incurred and they must be the ones to shape the organization's culture. The key difference between a traditional safety program and a safety management system is not in the particular practices of risk analysis, auditing, investigation, safety training and the like. Instead, it is in the use of the system by operational managers as a decision-making tool.

Desirable attributes of organizational culture also cannot be achieved by mandate or declaration; they must be fostered through the demonstrated commitment and practices of the company's management. The most important and tangible things that management can do is to build systems that give employees a safe, productive environment in which to work and to visibly demonstrate, through their own behavior, their commitment to safety. Participation, attendance at safety meetings, receiving briefings, etc. are not enough. Business leaders must also be safety leaders.

The sections below summarize some of the actions that can be taken by management to foster such an environment. For the most part, these processes follow the four components of SMS that are by now familiar to most practitioners. However, we've added the aspect of management behaviors that is part of the culture model outlined earlier. This element is so pivotal to the success of an SMS that it merits separate treatment.

Safety Management Policies

- clear organizational goals and objectives
- minimization of goal conflicts (production vs protection)
- clear delineation of management and employee accountability
- clear standards and expectations of behavior and performance.

Demonstrated Management Behaviors

- personal, visible demonstration of management involvement (beyond the statements and rhetoric – active, personal, visible participation in safety activities)
- inclusion of employee groups in safety management activities
- allocation of resources to safety management activities.

Safety Risk Management as a Core Value and Process

- risk management practiced as a way of doing business rather than an add-on.

Safety Assurance

- implementation and use of a non-punitive employee reporting system
- data-driven decision-making

- regular management review of safety data and analysis
- continuous improvement based on organizational learning.

Safety Promotion

- employee standards of competence and adequate training to attain and maintain job competencies
- clear and regular communication to and from employees.

Tools to Measure Your Culture

The FAA's guidance from advisory circular 120-92a understates the complexity and difficulty of effectively instituting a safety culture; however, safety professionals from the trenches to the corporate office realize the complexity and importance of building an environment from which a healthly safety culture can emerge. We have to foster an environment that will infuse positive safety beliefs, thought processes, and behaviors into the collective value system of the organization. Establishing and maintaining a safety culture is undoubtedly the hardest part of an SMS. The works of Dr. James Reason, Dr. Robert Helmreich and Dr. Sidney Dekker all agreed on the axiom: one can see a broken safety culture, but can someone see a strong safety culture?

I have studied safety culture for years and found that safety culture is the seed from which all safety programs stem. In essence, safety culture is the foundation for accident prevention within an organization. That being said, if a safety culture does not have a well established root system, it can also be fragile. Dr. Tony Kern (1999) said it best in his book *Darker Shades of Blue: The Rogue Pilot*, 'A safety culture must be inspired and constantly nurtured to prevent that downward spiral into disaster.'

This section will discuss the senior leader's role in developing and maintaining an effective safety culture and will present a variety of tools to help in this process which have shown success in other industries. The section will also provide ideas on how to gauge the effectiveness of a safety culture.

When I was researching material for this, I looked back on my 25 years in the field. I reflected on assignments that ranged from managing safety for small organizations with five aircraft, to a large, multi-state organization with hundreds of aircraft in fleet. However, the role was always the same: prevent all accidents. The solution was always the same: build and maintain a strong safety culture. The crusade was always the same: manage the risk and maintain a balance between production and protection.

The Human Factors Analysis and Classification System Framework

One approach to instituting an effective safety culture draws upon the use of a common safety language, or taxonomy. An organizational taxonomy can help serve as a model or blueprint of what a proper organization should look like from a safety perspective.

There are several organizational taxonomies presently available, but the one I found most successful was the Human Factors Analysis and Classification System (HFACS), developed by Drs Doug Wiegmann and Scott Shappell (2000). The HFACS framework is based on the 'Swiss cheese' model of human error developed by Dr. James Reason (1997) and is a comprehensive tool that can be used to identify both the latent and active causal factors associated with unsafe events. HFACS had its genesis in the US Navy and was successfully used to help reduce aviation accidents due to human error. Since then, it has been used in a variety of industries to help improve safety. Akin to the 'Swiss cheese' model, the HFACS framework is based on the premise that unsafe acts can be prevented by reducing deficiencies among preconditions for unsafe acts, unsafe supervision and ultimately organizational influences (see Figure 5.2).

While the entire HFACS framework is certainly necessary to ensure systemic improvements within one's safety environment, for the purposes of this chapter I would like to focus the discussion primarily on the top two tiers: organizational influences and unsafe supervision (for more information on these and the other tiers, please refer to Weigmann and Shappell (2003).

The organizational tier aims to focus users on high-level factors that ultimately influence the occurrence of unsafe acts by an individual. These factors are often far removed from the event itself, and can sometimes be difficult to identify during accident investigation. Nonetheless, organizational influences are undoubtedly critical to understanding why people commit unsafe acts, particularly since they often have the most far-reaching impacts. According to the framework, the organizational influences tier contains three specific causal factors: operational process, organizational climate and resource management.

Operational Process

This category of factors refers to the decisions and rules that govern an organization's day-to-day activities. It focuses on the policies and procedures that guide operations. When evaluating the organizational processes of your organization, you must consider whether the operating procedures are fairly and justly designed. Ideally, if the organizational processes are designed to facilitate a balance between production and protection, then there will be no hidden pressures to sacrifice safety in favor of operations. Unfortunately, many organizations suffer a disconnect between the policies that are written and the policies that are unspoken. For example, one organization may have a 'no fault go-around' policy clearly stated in the operating manual. At the same time, they may also have a definitive pay structure for their crew members, which is based purely on time to delivery. Therefore, the reality is that if a crew does miss an approach, then they will not be paid for that time, thus creating a plain conflict between safety policies and performance policies.

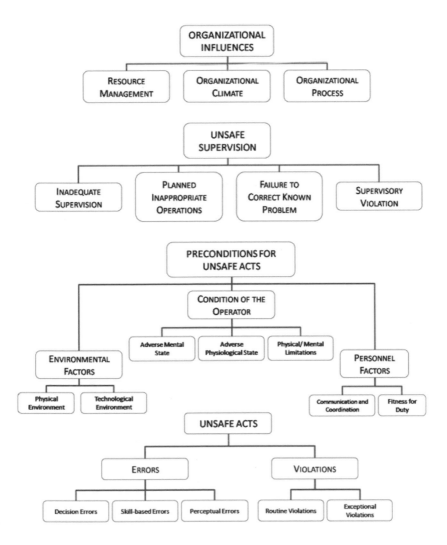

Figure 5.2   The HFACS framework

Adapted from Weigmann and Shappell (2003)

Organizational Climate

The next area within organizational influences is known as the organizational climate. This category refers to the corporate and ground-level cultures that dictate the actual performance of a workforce. For instance, some airlines still maintain an extremely hierarchical ranking structure wherein the captain dominates the crew, and questions by subordinates are seen as a disrespectful infraction upon the captain's authority. This understanding may not be written in any manual, but failure to abide by these 'rules' will undoubtedly result in debasement. The downstream results of this type of culture are ominous and often result in decision errors and violations due to poor communication.

It is within this causal category that the concept of a true safety culture should be explored. One way to evaluate your safety culture is to see if the CEO has a safety culture statement readily in place, and if he truly believes in that statement. Many times, safety is simply seen as a poster on the wall and people in the organization know that mission comes before safety. The president or CEO has to make everyone understand that safety should be ingrained in every action, every operation, and that is not just 'okay' to be safe, but rather safe behavior is expected.

One feature of an organization that has direct consequence on the quality of the organizational climate is the structure of the organization itself. Many organizations have a safety director, but oftentimes the structure is confusing and non-aligned with best practices. For example, consider a director of safety who actually reports to the director of operations; that structure cannot work because production is the main goal of an operation and when safety falls under the governance of operations, safety will always fall in second.

Resource Management

The final causal factor under organizational influences refers to resource management. This last category focuses on how an organization manages its resources with regard to safety management and considers the allocation and maintenance of organizational assets such as human resources (personnel), fiduciary assets and equipment/facilities. Earlier in the chapter, we discussed the importance of executive support in fostering an effective safety culture. When we look at the resource management category, we need to consider how and if executive-level management intend to support safe behavior. If the organization indicates that safety is paramount, but is not willing to devote the necessary resources to ensuring its success, then the consequences of this condition are twofold. First, adequate resources may not be available to support safe operations and second, this failure sends an express message to the workforce, 'We [the organization] want you to value safety, but we don't really value safety ourselves.' It is this type of perception that is the biggest threat to the success of any attempt at establishing an effective safety culture.

Gap Analysis

One particularly useful way of systematically evaluating the weak points or 'holes in the Swiss cheese' of an organization is to use the HFACS framework to conduct a gap analysis. A gap analysis will highlight those areas within your SMS that are deficient and are in need of improvement. To conduct a gap analysis, complete the following steps:

1. Lay the complete HFACS framework out in front of you.
2. Starting at the top (organizational influences), visit each causal factor and list the various defenses (safety programs, organizational policies, etc.) that are in place within your organization to prevent causal factors in each category from occurring. At this point, do not qualify the relative effectiveness of each defense, but simply indicate whether or not a defense exists.
3. Once all of the causal factors have been evaluated, review the framework and identify particular areas where little or no defenses exist. These gaps, or 'holes in the Swiss cheese', illustrate immediate areas in need of improvement.
4. Finally, work with your safety team to generate specific 'fixes' or interventions to close those gaps and opportunities for failure within your organization.

Keep in mind that defenses placed at higher levels in the HFACS framework (e.g. organizational influences and unsafe supervision) will have a greater downstream impact on preventing preconditions for unsafe acts and the unsafe acts themselves.

Tools to Enhance your Safety Culture

Any student of safety culture knows that to be effective, a safety culture must originate at the top-levels of the organization. In Helmreich's book, *A Safety Culture in Aviation and Medicine* (1998), he said that the norms and values of the profession must be passed on to its recruits. One example of this is a strong safety orientation program. In this program the CEO or the president, or whoever holds the highest position in the organization, can address new hires and help them understand his or her vision of the safety culture. This may sound impractical to an organization employing 45,000 people; however, it has been accomplished with the simplest of technological means. While the CEO cannot attend every safety orientation, their safety culture philosophy can be recorded in a high-quality video, which can then be played at the beginning of every employee's orientation via the Internet or DVD. Establish the culture early on, and a strong safety program will follow.

Establishing a safety culture early is important, but it must be constantly nurtured to prevent the 'downward spiral into disaster'. One president of a large aviation related university has demonstrated this tenant by instituting safety culture enhancement days to keep faculty and staff mindful of the university's value of safety. For example, on a hot August day in Florida, the president will hop in his golf cart and take drinks to all of the employees working outside. He will go to the maintenance hangars, pass out water and cold drinks, and even pat employees on the back in appreciation for their efforts all the while reminding them to be safe. Safety culture is enhanced when employees know that senior leadership believes in them and cares about their well-being; it makes it easier to adhere to standards and helps prevent the organization from falling prey to normalization of deviance.

The *safety culture coin* is another positive tool that can be used to remind individuals of their role in the safety culture. The coin can be designed by any organization with symbols that represent its business (see Figure 5.3). One example of a coin that has had a very positive effect on safety culture says, 'Live the Safety Culture'. This coin is earned by any individual doing something to enhance the safety culture of the organization. The purpose of the coin is to serve as a reminder that every individual has a role in the fostering of safe practices within the company. The coin lets the individual know that the organization expects them to do the right thing, the safe thing, every day, every time, on every critical task! The coin serves as a reminder that the organization is in their pocket to back them up when they make the tough, but safe decision. The coin is just one more tool to remind the individual of their role in an organization's safety culture.

Figure 5.3   Safety coin

The challenge coin is a tradition that has origins in the United States military, and as such the concept of the coin carries with it many of the convivial associations inherent to military comradeship. Essentially, there is a responsibility when you are a member of the coin brotherhood: you *must* take the safe course of action every time. You must believe in safety all the time, even when no one is looking. For instance, in the US Army a coin-holder must carry a coin at all times, and if that coin-holder is challenged by a fellow coin-holder, they must present their own coin in return. If the challenged individual produces a coin in return, then the challenger must buy the challenged a beverage. On the other hand, if the challenged coin holder does not produce a coin, then they must buy the challenger a beverage. The coin is another reminder to 'LIVE THE CULTURE'.

References

Bandura, A. (1985). *Social Foundations of Thought and Action: A Social Cognitive Theory*. Upper Saddle River, NJ: Prentice-Hall.

Ciavarelli, A. (2008). Culture counts. *Aerosafety World*, 3(2), 18–23.

Degani, A. and Wiener, E. L. (1994). Philosophy, policies, procedures, and practices: The four 'P's' of flight deck operations. In N. Johnston, N. McDonald and R. Fuller (eds), *Aviation Psychology in Practice* (pp. 44–67). Aldershot: Ashgate Publishing.

Dekker, S. (2007). *Just Culture*. Aldershot: Ashgate Publishing.

FAA (Federal Aviation Administration) (2006). *Introduction to Safety Management Systems for Air Operators. Advisory circular 120-92a*. Retrieved July 18, 2009 from http://www.faa.gov/regulations_policies/advisory_circulars/index.cfm/go/document.information/documentID/319228.

FAA (Federal Aviation Administration) (2009). *Aviation Instructor's Handbook: FAA-H-8083-9A*. Washington, DC: United States Government Printing Office.

Flannery, J. A. (2001). Safety culture and its measurement in aviation. Masters thesis, University of Newcastle, Australia. Retrieved from http://asasi.org/papers/other/safety_culture_measurement_aviation.pdf.

Hayward, B. (1997). Culture, CRM and aviation safety. Paper presented at the ANZASI 1997 Asia Pacific Air Safety Seminar. Retrieved from http://asasi.dynamite.com.au/brent1.htm.

Helmreich, R. (1998). *Culture at Work in Aviation and Medicine*. Aldershot: Ashgate Publishing.

Helmreich, R. L. (1999). Building safety on the three cultures of aviation. In *Proceedings of the IATA Human Factors Seminar* (pp. 39–43). Bangkok, Thailand, August 12, 1998.

Hofstede, G. and Hofstede, G. J. (2005). *Cultures and Organizations: Software of the Mind*. New York: McGraw-Hill.

Hofstede, G. (2009). *Cultural Dimensions*. Retrieved March 21 2010 from: http://www.geert-hofstede.com.

Hollenbeck, J.R., Moon, H., Ellis, A., West, B., Ilgen, D.R., Sheppard, L., Porter, C.O.L.H. and Wagner, J.A. III (2002). Structural contingency theory and individual differences: Examination of external and internal person–team fit. *Journal of Applied Psychology, 87*, 599–606.

Hudson, P. (2001). Aviation safety culture. Presented at Safe skies Conference, Canberra, Australia.

Kern, A. (1999). *Darker Shades of Blue: The Rogue Pilot*. New York, McGraw-Hill Companies.

Reason, J. T. (1997). *Managing the Risks of the Organizational Accident*. Aldershot: Ashgate.

Von Thaden, T. L. and Gibbons, A. M. (2008). *The Safety Culture Indicator Measurement System (SCISMS)*. Technical Report *HFD-08-03/FAA-08-02*. Urbana-Champaign, IL: University of Illinois.

Weick K. E. and Sutcliffe, K. M. (2007). *Managing the Unexpected: Resilient Performance in an Age of Uncertainty*. San Francisco, CA: Jossey-Bass.

Weick K.E. and Sutcliffe, K. M. (2011). *Managing the Unexpected through Mindfulness*. Retrieved April 2, 2011 from: http://high-reliability.org/HRO_Weick_Sutcliffe.html.

Wiegmann, D. and Shappell, S. (2003). *A Human Error Approach to Aviation Accident Analysis*. Aldershot: Ashgate Publishing.

Zhang, H., Wiegmann, D. A., von Thaden, T. L., Sharma, G. and Mitchell, A. A. (2002). Safety culture: A concept in chaos? Presented at the 46th Annual Meeting of the Human Factors and Ergonomics Society, Santa Monica, California.

Chapter 6
SMS Implementation

Bill Yantiss

Introduction

The topic of safety management systems (SMS) has reached center stage for both aviation service organizations and civil aviation authorities worldwide. A number of publications have portrayed safety management as a new and emerging science based on recent industry revelations that provide new insight on how to drastically improve aviation safety. Many safety practitioners find this portrayal of safety management somewhat bewildering because the aviation industry has methodically focused on improving safety performance since the first accident of the Wright brothers. Air carriers have sponsored development of crew training enhancements such as, crew resource management (CRM) training, wind shear training, Advanced Qualification Program (AQP) concepts, and the extensive use of simulation as the primary means of providing realistic duplication of emergency situations that are impractical to accomplish in an actual aircraft. Air carriers also aggressively retrofitted their fleets with the latest technology, including, ground proximity warning systems (GPWS), predictive wind shear warning systems, weather radar and flight data-recording hardware. As the aviation industry expanded globally, there was a recognized need for global safety standards, resulting in the creation of the International Civil Aviation Organization (ICAO) in 1944, to provide guidance and oversight of State civil aviation authorities.

Prior to the 1990s, many Civil Aviation Authorities (CAA) operated primarily with a regional mentality, that is, focused on developing civil regulations based on familiar ICAO standards, but written to meet the needs of their own State, with little regard to international harmonization. Operators recognized that compliance with these standards represented the minimum level of safety to be granted an Air Operators Certificate (AOC); however, they also learned the difficult lesson that merely attaining regulatory standards did not always achieve the desired level of safety performance, as evidenced by incident and accident rates. Simply stated, a regulatory 'compliant' airline is not necessarily a safe airline. Regulations provide a good foundation for operational safety, but organizational leaders must establish effective management systems and appropriate documentation of processes and procedures. Over several decades, airline leaders have reduced the accident rate by embracing best practices, such as structured operational management systems, operational procedures based on human factor considerations and implementation of vigorous accident prevention programs.

Code-share relationships became popular in the early 1990s and several regulatory authorities and individual operators established a new requirement for an on-site operational safety audit prior to approving a code-share marketing agreement. A code-share agreement is much like that of a supplier, that is, the supplier provides products or services based on service level or quality agreements. As these marketing agreements proliferated among large international operators, several realities became apparent with respect to common safety standards.

First, operators embraced different sets of operational and safety standards to approve a code-share partner as a supplier. Although the intent was to ensure that the supplier provided an 'equivalent level of safety' to the customer, there were significant differences in the safety standards. For operators having 30-plus code share agreements, there was the potential of receiving a reciprocal safety audit using 30 different standards (checklists), each developed by the individual operator. Needless to say, line managers found this situation inefficient and frustrating. They asked, 'Is it possible to use only one standard and one checklist for these audits?'

Second, the proliferation of international safety audits was becoming very expensive. Several operators were investing over US$2 million annually to host safety audits of their own carrier and to simultaneously conduct audits of other code-share partners. If all 30 code-share partners performed a reciprocal audit of an operator on a biennial schedule, this would result in hosting an audit every month. The cost to prepare an organization for an audit and the time spent to manage an on-site five-member audit team can easily approach US$100,000 per audit. Operators recognized the need to gain control of this 'audit frenzy' menace that was both cumbersome and inefficient.

Third, the proliferation of safety audits revealed that there were significant differences in the expectations of State regulatory authorities, regarding the level of safety performance of an operator flying into their jurisdiction, and of those participating in code-share arrangements. They tended to operate independently with little appreciation of the value of harmonized global standards. Through the Universal Safety Oversight Audit Program (USOAP), ICAO verified significant variations in the health of State CAAs and their ability to provide adequate regulatory oversight; however, the challenge was how to raise the bar of safety for both the CAA and the operator.

Fourth, airlines understood that a well-managed airline is a safe airline. Essentially, if management systems are well-designed, operational processes and procedures are well-documented and employees are receiving appropriate training and supervision, then the fundamental elements are in place to produce consistent and predictable results. This infrastructure serves as the foundation to achieve excellent results in terms of safety and quality.

Was it possible to develop a balanced operational standard that focuses not only on regulatory compliance but fundamental management principles, resulting in a high level of public safety?

Operators Raise the Bar

In response to the operational and fiscal realities described above, a number of carriers approached the International Air Transport Association (IATA) to sponsor development of an international safety standard. Over 125 international air carriers and CAA representatives participated in the development of the IATA Operational Safety Audit (IOSA) program, a concept based on ISO 9001:2008 principles, ICAO international safety standards and a host of industry best practices. This revolutionary standard was intended to be accepted by all international scheduled air carriers and regulatory authorities worldwide.

When the IOSA standards were first being crafted in 2001, IATA recognized that SMS would eventually be required by ICAO for all air operators. With this in mind, fundamental safety management concepts were incorporated into the first edition of the *IOSA Standards Manual* (ISM), published in 2002. Many of these best practices were considered so important to safety management that they were identified as requirements in Section 1, Organization and Management System (ORG), of the ISM.

For example, the ISM has always reflected a requirement for a documented organizational management system that was based on fundamental safety and quality principles. Key features of this management system included:

- the identification of the accountable executive
- defined management roles and responsibilities
- a safety policy
- the provision of resources
- a management review process
- a planning process
- communication processes
- a flight accident prevention program
- a quality assurance program
- an emergency response plan (ERP).

IOSA represented the first global scheduled air carrier standard that addressed not only existing ICAO requirements, but also fundamental management principles that would eventually be accepted by the aviation industry as essential elements of a SMS. IOSA registration became a significant milestone for many operators, as it required months of hard work to ensure IOSA standards were met, particularly in documenting the management system and implementing associated systems, programs, policies, processes and procedures. As SMS becomes mandatory across the industry, registered IOSA operators are finding that conformity with the IOSA standards provides a beneficial intermediate step toward full SMS implementation.

Similarly, the International Business Aircraft Council (IBAC), based in Montreal, Canada, has produced a comparable safety standard for the business aviation and charter operations communities. This concept is labeled the International Standard – Business Aircraft Operations (IS-BAO), and was developed to define a SMS and capture best practices for an important segment of the aviation industry.

These two safety performance standards were developed and published by industry representatives long before ICAO released the first edition of the *Safety Management Manual* in 2006. The safety and quality management fundamental elements laced throughout IOSA and IS-BAO standards are clearly evident in the new Appendix 7 to ICAO Annex 6. The entire aviation community has benefited from the collaborative partnership of IATA and IBAC with ICAO to move forward with harmonizing SMS standards in all segments of commercial and private aviation. Perhaps some of the most significant changes are occurring within the regulatory community itself.

Regulatory Landscape

There are two issues within the regulatory landscape. First, the source document for SMS standards is the ICAO Annex 6, Appendix 7. These international standards are then reflected in the individual State (country) civil aviation regulations. Second, each State's Civil Aviation Authority, such as the US Federal Aviation Administration (FAA) or European Aviation Safety Agency (EASA), has the option to augment the international standard with additional requirements. These additional requirements are being developed during the current SMS rule-making process within each State.

What is Safety Management?

As defined on ICAO's safety management website, safety management is a managerial process for establishing lines of safety accountability throughout the organization, including the senior managers. The concept of safety management is realized through the development and implementation of a State Safety Program (SSP) and a SMS. The SSP is an integrated set of regulations and activities aimed at improving safety. An SMS is a systematic approach to managing safety, including the necessary organizational structures, accountabilities and policies developed by and for operators.

The second edition of the *ICAO Safety Management Manual* (International Civil Aviation Organization, 2009) (SMM – Doc 9859) describes the functional framework for global implementation of SMSs. ICAO guidance is directed at two audience groups: State CAAs and individual service providers. CAAs, the FAA, EASA and Transport Canada, as examples, are required to establish a SSP that will enable a transition from regulatory compliance and oversight to a performance-based approach of risk management, utilizing safety indicators (categories) and safety targets (metrics).

Implementation of an SSP may require a significant change throughout the regulatory authority organization in order to transition from a reactive environment to a proactive methodology of integrating inspections, with comprehensive evaluations of a management system for safety that can differ significantly from operator to operator. This transition may be turbulent for many CAAs. ICAO is currently delivering a training program for State CAA representatives to provide additional knowledge of safety management concepts, and to understand the spirit and intent of the ICAO Standards and Recommended Practices (SARPs).

States are in various stages of SMS adoption and implementation, an obvious source of frustration for international operators, because they have to deal with these regional variations. For commercial airline (FAR Part 121) international operators, many State CAAs have accepted the IOSA standard for code-share operations because it reflects all applicable ICAO standards. Additionally, the ISM will be reviewed annually to ensure evolving SMS requirements are reflected in the standard. Although the global business aircraft community has begun to rally around the IS-BAO, IBAC is currently revising the program to instill additional rigor in the auditor qualification and on-site audit process. This author believes that the IOSA and IS-BAO standards will continue to evolve and reflect the latest operational best practices; thus, becoming the two baseline standards that reflect accepted SMS requirements for the three major operator segments – scheduled commercial operations, charter operations and business aviation.

Regulatory Challenges

If it's any consolation to the reader, the regulatory authorities are also struggling to get their arms around their roles and responsibilities to accept and oversee the development, implementation and operational performance of the operator's SMS. This is a huge challenge given that safety inspectors must expand their skill set to effectively transition from performing benign cockpit, cabin, or ramp inspections. The transition from *inspector* to *system auditor* represents a significant leap in complexity.

A View from Transport Canada

As a regulator with a great deal of SMS experience, Transport Canada has cleverly captured the transition from traditional surveillance to monitoring the operator's internal management systems and safety performance metrics as five separate phases. The concept recognizes that SMS implementation occurs over time, particularly instituting the changes required in the organization's safety culture. Of particular interest is the gradual shift in regulatory philosophy from one of active compliance inspections to that of closely monitoring the performance of the operator's SMS 12 elements. For example, the regulator will take a renewed interest in incident investigations, follow-up on hazard reports, internal evaluation audit reports, functional manager audit results and the entire suite of performance metrics. This change in philosophy will be evidenced by the CAA spending a lot more time assessing the operator's management systems, processes and safety metrics.

The reaction to SMS concepts by the accountable executive has ranged from 'SMS makes absolute sense – let's get moving' to 'Wow, SMS seems very complex and expensive!' Perhaps these reactions are reflective of their corporate culture and operating philosophy. Perhaps it's only an inadequate understanding of the value of an SMS in relation to the myriad of business activities that occurs in every airline, service organization, or manufacturer.

The reality is that the organization's commitment to safe operations varies from operator to operator. At one extreme, an operator may view safety as meeting the minimum regulatory standards and occasional non-compliance is considered acceptable. Safety and compliance issues are aggressively addressed only after they have been caught by the regulatory authority. This poor safety culture is characterized by a 'blame someone else' reaction, typically evidenced by taking action against the employees who got caught, perhaps using dismissal to make the issue go away. Unfortunately, this punitive environment holds the individual accountable for system, process, procedure and training deficiencies that are endemic in the organization. Executive team members are resistant to change, and react only to accidents or incidents that clearly dictate an operational change. An occasional civil penalty issued by a regulatory authority is considered a cost of doing business. This organization is not in a position to proactively identify emerging safety issues, but selectively reacts only when forced to do so. The FAA may view this type of operator as 'high-risk' or 'borderline', thus requiring additional oversight.

At the opposite end of the spectrum, we find an organization that has integrated safety management into the enterprise management system (an integrated management system) by making safety management an integral part of the business model, that is, the 'way we do business around here', and reflective of the operator's core values. The operator aggressively embraces industry best practices to far exceed minimum safety regulations, and actively solicits input from all employees to identify and resolve operational hazards. The leadership team understands that meeting minimum safety standards does not ensure a safe operation, and operational risk is managed by involving all employee groups and partner organizations as stakeholders in an integrated management system.

The SMS concept represents the transition from a reactive culture to a proactive culture. Globally, the accident rate will never be zero because human error will always be a component of nearly every operational activity. However, the accident rate can be reduced by implementing proactive and predictive methods into the organization's management system. Safety performance is a management responsibility and may be viewed as the 'outcome' of the effective management of day-to-day operational activities. The SMS concept represents the transition from a reactive culture, driven by investigating the most recent smoking hole, to a proactive environment of identifying, predicting and resolving operational hazards associated with significant changes in the operation, before they are implemented. One of the most frequent questions that invariably surfaces from the accountable executive is, 'After I make all of these changes to our management system, what do I really get for this time and effort?', 'Am I really safer?', 'How do I implement SMS without costs getting totally out of control?' All of these are great questions!

Production vs. Protection

ICAO and several CAAs have framed the safety discussion in terms of production goals (delivery of services) and protection goals (safety). This balance may be captured in terms of risk management, that is, what are the hazards in the operation, and what do I do to prevent an accident or major incident from occurring? This demands a continuous process of hazard identification, data collection and analysis, safety risk estimation and implementation of mitigation strategies. There are always at least two relevant discussions for managers and safety practitioners regarding operational safety:

1. What are the existing hazards (for you golfers – sand traps) in the operation and are they acceptable in terms of risk (defined by severity and probability)?
2. What are the 'predicted' hazards or safety issues introduced into the operation as a result of changing operational procedures, operating to a new location, changing suppliers, etc.?

A fully developed SMS provides the necessary tools for the leadership team to effectively manage safety, that is, understand and manage the risk inherent in each facet of its operation. It should be obvious that this shift in emphasis from strict regulatory compliance to safety management will dramatically change the conversations between the aviation operator and the regulatory authority. This discussion will be more data-driven to identify hazards inherent in the operation, and eliminate or mitigate those hazards before an actual incident occurs.

Implementation of an SMS is an evolutionary process that will yield significant benefits in terms of operational efficiency, effectiveness and cost reduction. Let's look at these issues more closely.

Reduced Costs

Implementation of an SMS can have huge results in terms of cost savings. For example, implementation of several elements of an SMS for a major operator's ground-handling program reduced lost time injury (LTI) costs by 13 percent and ground damage costs by 41 percent (slightly less than US$10 million).

Let's briefly walk through the process. Soft spots (hazards) in the operation are identified by analyzing a steady data stream by safety and quality practitioners. Assessment of multiple data streams enables a holistic (system-wide) perspective of the risks associated with operational activities. Representative data sources include the following:

- hazard and incident reports (flight, maintenance, ground handling, etc.)
- incident investigation (reactive)
- internal evaluation program (IEP) reports (corporate level)
- internal audit reports (functional level)
- flight data analysis
- external sources (CAA, safety organizations, manufacturers, etc.)

Following data acquisition and analysis, operational hazards were prioritized in terms of frequency and severity. Corrective actions were developed and approved through the management review process. The results were impressive because the accountable executive, employee unions and line management were included in the entire process.

Reduced Insurance Premiums

Insurance underwriters are becoming increasingly aware of the intrinsic value of safety management systems and the potential impact on their bottom line. Conversations between the insurance representative and the operator are shifting from a reactive environment of incident investigation theme to one of 'show me your proactive safety management system'. For example, a Marsh representative recently addressed a group of international operators and stated that registration as an IOSA operator was expected as a baseline operational standard. Why the interest? Marsh risk managers understand that IOSA requires many best practices in the management system that reduces risk and the probability of a hull loss. The entire insurance community is recognizing that a solid enterprise management system is absolutely essential to identifying and managing operational risk.

Although insurance premiums are heavily influenced by the global accident rate, the premiums for your organization may be negotiated downward by demonstrating implementation of a comprehensive SMS, supported with safety and operational metrics that exhibit success.

Continued Authorization to Operate

At some point in the future, each State will require proof of SMS implementation as a prerequisite for operations at airports within their jurisdiction. Please keep in mind that many States will merely adopt the ICAO standards without additional requirements. For sake of discussion, let's refer to the ICAO standard as a baseline SMS. However, some States will invariably add a number of regional requirements to the baseline ICAO standard. These variations in State standards present the potential for confusion, frustration and noncompliance.

As mentioned above, one solution for dealing with these so-called differences is to develop a global standard that all regulatory authorities can accept. There are two existing standards that may serve this purpose. For commercial airline international operations (FAR Part 121), the IOSA standard will continue revision to reflect ICAO baseline SMS standards and IOSA certification can be used as proof to a State CAA of SMS compliance. For charter and business aircraft operations, the IS-BAO may also be used for proof of compliance with the ICAO baseline SMS standard.

Continuous Improvement and Management of Change

Any commercial enterprise must embrace the concept of continuous improvement in order to remain viable in today's competitive world. Management systems and processes are never static. Continuous improvement is a characteristic of a learning culture that enables proactive risk management through process improvement. Stated differently, the objective is to identify 'holes in the cheese' as described by Dr. James Reason. Each hole in the cheese represents a deficiency in one of the layers of protection that prevents a serious incident or accident.

An enterprise management system is structured with multiple systems, such as a SMS, quality management system (QMS), security management system (SeMS), financial management system (FMS), etc. Within an enterprise, there may be ten to thirty management systems depending upon the complexity of the operation.

Exceed Customer Expectations – Reliability

A robust SMS requires an IEP to review the health of operational processes and procedures. Another term for this management system assessment is *quality assurance*. Regardless of the term used in your organization, the purpose of these programs is to ensure regulatory compliance, verify that employees are conforming to organizational procedures and to identify opportunities for improvement. An on-site audit is the typical tool used in quality assurance or internal evaluation, and is an excellent means of identifying deficiencies that may negatively affect operational reliability, aircraft condition, interface breakdowns, etc. A quality system using auditors who are trained in system design will provide immense value, and improve the integrity of the operation by identifying and recommending correction action to operational deficiencies. Most of these deficiencies result in customer inconvenience.

Proactive – Not Reactive – Management Team

A well-designed management system enables managers at all levels to be proactive rather than continually putting out fires. How many times have managers concluded that they were doing a great job because they single-handedly held the operation together through personal intervention, endless phone calls, and numerous face-to-face discussions? Going home tired at the end of the day doesn't necessarily mean that we have done a good job as a manager. The most obvious symptom of a system design problem is that it requires someone to continually intervene in an operation where the employees should have the tools, training and procedures to get the job done without frequent manipulation.

In a proactive environment, the manager has implemented a well-designed system, thus freeing up time to identify opportunities for improvement, and intervenes only when confronted by unusual or unanticipated circumstances. Effective processes and procedures are clearly defined in organizational manuals. Employees are supported with initial and recurrent training to reinforce proper application and knowledge of procedures. Safety communication identifies potential hazards and how to manage operational risks while maintaining efficiency and reliability.

Implementation

Senior leadership should expect positive results as incremental implementation of SMS principles, concepts and programs take root. Experts tell us that SMS implementation may vary from one to four years, depending upon the complexity of an organization. However, the cultural change associated with non-punitive, proactive safety initiatives may take up to five to eight years to be embraced as 'the way we do business here' and fully accepted by all employees.

There is no silver bullet yielding immediate success and exceptional safety performance. It is important to recognize that a safety system is a collage of activities, programs and processes that work together in an integrated, proactive and predictive fashion to identify operational hazards and manage the associated risk effectively. Each of the 4 SMS components and associated 12 elements must receive equal emphasis during design and implementation of your SMS. It is through active interface between these elements that operational hazards will be proactively identified and properly mitigated.

Shared Lessons Learned

The opportunity to conduct IOSA and IS-BAO audits of 70+ operators has provided a unique opportunity to observe how these organizations grapple with the balance of *production* versus *protection*. Some have been very successful in creating a healthy and positive safety culture, primarily as a result of senior executive leadership engagement and behavior modeling. Others have not been so successful, for reasons that vary greatly between organizations. In the following pages, we will attempt to capture the essence of the 12 SMS elements and techniques that have been effective for many operators. We will attempt to provide practical examples of 'how' to implement key features of an SMS, techniques that will work for all operators, regardless of size and complexity.

An Enterprise Perspective – A 'Macro' View

ICAO (2009, 6.2) Safety Management System guidance is focused on two audience groups: States and service providers. ICAO safety management SARPs are contained in Annexes 1: 6, Parts I and II; 8; 11; 13 and 14. These annexes address the activities of service providers that include the following organizations:

- approved training organizations
- international aircraft operators
- approved maintenance organizations
- organizations responsible for type design and/or manufacture of aircraft
- air traffic service providers
- certified aerodromes.

The regulatory community is primarily concerned with operational risk management and safety outcomes, typically reflected by safety performance metrics. As guardians of the flying public, State regulatory authorities focus nearly all of their activities to developing regulations that define minimum operational standards for aircraft design and their operation. Recognizing that these standards are inadequate to ensure public safety, ICAO is raising the bar by establishing minimum requirements for safety management systems. Although acknowledging the existence of other essential organizational management systems, ICAO has focused primarily on safety management, as you would expect. However, for the organizational leadership team, safety management is only one important component that must be defined and integrated into the organization as a whole.

Airline, manufacturing and business aviation leadership teams have long recognized that an organization is composed of a *system of systems* that are integrated and intra-supportive. Service provider[1] management teams must continuously adjust to dynamic financial, competitive market, and operational pressures that characterize the complex aviation environment. They are held responsible by the State CAA to provide products and services with the highest possible degree of safety. Legislation, such as Sarbanes–Oxley and Basel II, has reinforced corporate officer accountability for both financial and operational results along with accurate reporting to investors (Yantiss, 2006).

Although the bulk of this book addresses the various components of a safety management system, the chief executive's challenge is far more complex than simply developing an SMS that implements the 12 elements as described in the ICAO SMM. Each of the service provider's complex management systems requires multiple cross-functional activities. Although each is designed and monitored by an 'accountable leader', the activities of each system involved are integrated with functional activities across the organization. For example, the turnaround of an aircraft at an airport requires carefully orchestrated work flows of ground handling, customer service, flight operations, cabin operations, aircraft maintenance, etc. What ties these activities together? More broadly, what mechanism organizes and integrates thousands of airline activities into a harmonious work flow? What are the risks to the enterprise that must be clearly understood and managed? These are concepts that safety and quality practitioners must understand in order to effectively communicate operational risk issues to organizational leaders.

A discussion of safety management will only make sense if it is described within the framework of the organization as a whole. An organization or enterprise operates efficiently through harmonious integration of organizational management systems designed to achieve organizational goals, that is, provide customers with a product or service. (In our discussion, the terms organization and enterprise are synonymous.) If these products and services do not meet customer requirements, the organization will not survive the competitive pressures of today's aviation environment. A number of air carriers often use the slogan 'Safety is our Number One Priority' to convey their concern for passenger safety. Although this slogan may resonate positively with customers and the media, it clearly is not the number one priority for the enterprise.

The number one priority of the enterprise is to deliver the service for which the organization was created in the first place, to achieve production objectives, and eventually deliver dividends to stakeholders (ICAO, 2009, 3.2.2). Management has the responsibility to establish the balance of production goals (delivery of services) versus protection goals (safety). For example, operations must continue in inclement conditions (physical and economic) and safeguards (procedures, training, etc.) must be developed to ensure a consistent and competitive operation. Stated differently, the management challenge is to define the organization's 'acceptable level of safety' that may be measured by the number of operational

---

1   The terms service provider, enterprise, organization and operator are used interchangeably

events, ground damage, or employee injury. These issues represent significant costs to the organization, and efforts to reduce them to acceptable levels requires an investment in process change such as policies, procedures, training and leadership engagement.

Safety is increasingly viewed as the consequence of the management of organizational processes, with the final objective of keeping the safety risks of the consequences of hazards in operational contexts under organizational control. Organizational control of operational activities occurs through the integration of the systems that are designed to deliver products and services. Although large organizations may have 30+ systems, the following list will highlight a few that are fundamental in an airline:

- an organizational management system (corporate governance)
- a SMS
- a QMS
- a SeM
- an environmental management system (EMS)
- a documentation management system (DMS)
- a flight operations management system
- a cabin operations management system
- an aircraft maintenance management system
- a dispatch management system (DiMS)
- a cargo operations management system
- a ground handling management system
- a supplier management system (SuMS)
- a FMS
- a marketing management system (MMS)
- a personnel management system.

Quality Management – A Foundation for Safety

The International Standards Organization (ISO) publication 9001:2008 defines a QMS as the organizational structure, accountabilities, corporate resources, processes and procedures necessary to establish and promote a system of continual improvement while delivering a product or service. *The QMS is the organization, its people, and all operational processes dedicated to meeting customer requirements*. In an airline, this means providing on-time departures and arrivals, clean and dependable aircraft, courteous and helpful customer service to include check-in, baggage handling, aircraft servicing, cabin service, etc., while conducting operations in compliance with thousands of regulatory requirements. Aircraft emergencies resulting in delays or diverts, inoperative cabin equipment, lost baggage, diverts due to toilet servicing errors and cancellations due to ground damage are examples of not meeting customer requirements. Customers buy a ticket with several expectations: arriving alive (the most important criteria), arriving on time and arriving with a satisfactory travel experience.

The United Kingdom Civil Aviation Authority has defined the role of the QMS, that is, to monitor compliance with and the adequacy of procedures required to ensure safe operational practices and airworthy airplanes. The QMS and SMS have complementary but independent functions with the QMS monitoring the SMS.

A close look at the ICAO SMM reveals that there is a recognition that multiple management systems exist in each organization. There is a recognition that safety, quality, security and environmental management systems exist in an integrated arrangement; however, there is little information provided as to how to accomplish this recommendation. Therefore, it is up to each leadership team to integrate these systems for strength and efficiency.

The QMS makes the organization operate! ISO 9001:2008 highlights the importance of organizational leadership, a systems approach to management, people, process development, continual improvement, data analysis and supplier oversight. These fundamental quality principles, as described in more detail below, form the foundation of a safety management system:

- Principle 1 – Customer focus. Organizations depend on their customers and therefore should understand current and future customer needs, meet customer requirements and strive to exceed customer expectations.
- Principle 2 – Leadership. Leaders establish unity of purpose and direction of the organization. They should create and maintain the internal environment in which people can become fully involved in achieving the organization's objectives.
- Principle 3 – Involvement of people. People at all levels are the essence of an organization and their full involvement enables their abilities to be used for the organization's benefit.
- Principle 4 – Process approach. A desired result is achieved more efficiently when activities and related resources are managed as a process.
- Principle 5 – Systems approach to management. Identifying, understanding and managing interrelated processes as a system contributes to the organization's effectiveness and efficiency in achieving its objectives.
- Principle 6 – Continual improvement. Continual improvement of the organization's overall performance should be a permanent objective of the organization.
- Principle 7 – Factual approach to decision making. Effective decisions are based on analysis of data and information.
- Principle 8 – Mutually beneficial supplier relations. An organization and its suppliers are interdependent and a mutually beneficial relationship enhances the ability of both to create value.

If you look carefully at each of the 4 ICAO SMS components and associated 12 elements, these eight quality principles form much of the new SMS philosophy, structure, strategy and methodology. This evolution represents a clear signal to both safety and quality practitioners that they must focus on system management and think in these terms. This by no means diminishes the importance of safety and quality programs such as flight data analysis, employee reporting programs and the like; these are essential programs that enable the SMS to achieve the ultimate objective, that is, identify hazards, weaknesses in the daily operation, and to mitigate these issues through a continual improvement process.

Quality Principles at Work

Responsibilities for managing safety can be grouped into four basic categories (ICAO, 2009, 3.8):

- Definition of policies and procedures regarding safety: documented policies and procedures define how senior management desires operations to be conducted.
- Allocation of resources for management activities: senior management balances resources dedicated to production goals (delivery of services) and protection goals (safety).
- Adoption of industry best practices: in the spirit of continual improvement, the sharing of strategies, techniques, processes, procedures, methods, or tools enable commercial operators and service providers to more effectively manage their business and reduce operational risk.
- Incorporation of regulations governing civil aviation safety: regulatory compliance serves as the foundation for SMS. It is a misconception that development of safety, quality and security management systems will make the regulatory framework redundant or unnecessary. Further, ISO 9001:2008 has been revised to reinforce that the management system must perform at a minimum level to meet statutory and regulatory requirements.[2]

Safety Management – An Equal Partner with Quality

ICAO (2009, 2.2.4) defines *safety* as a state in which the risk of harm to persons or property damage is reduced to, and maintained at or below, an acceptable level through a continuous process of hazard identification and risk management. The ICAO definition implies continuing measurement and evaluation of an organization's safety performance and feedback into the management system. Measuring operational activity occurs on the corporate level as safety and quality services departments analyze data received from multiple sources, but perhaps even more importantly, each functional area is assessing its operational activities

---

2   ISO 9001:2008 Clause 0.1.

to identify opportunities for improvement. This culture and passion for continual improvement is an essential change for safety practitioners in the way they approach safety management. More on this later as we review implementation techniques.

Although we all get weary of these definitions, we must remind ourselves that a complete understanding of these concepts and vocabulary is absolutely essential to earn credibility within your own organization and also to understand SMS concepts so that you may craft a logical and understandable explanation to your colleagues. It is important to remind operational managers that realization of an acceptable level of safety is the result of successful management techniques. These techniques are rooted in the QMS foundation and clearly evident in well-designed operating procedures, comprehensive employee training programs and a supportive senior leadership team. In other words, a well-managed operation is a safe operation.

An SMS may be defined as a systematic approach to managing safety, to include the necessary organization structures, accountabilities, policies and procedures. These organizational structures are those that support safety management throughout the organization, that is, woven into the fabric of the organization and becoming an essential element of its culture. These structures include establishing an effective corporate safety services department staffed with trained safety practitioners who manage robust processes that ensure the collection of safety data, the analysis of available data and the mitigation of hazards in the operation. Of equal importance is the safety system structure within each operating division and at the functional manager level. Safety performance and risk management is their responsibility. These managers are responsible for these operational activities, they design procedures and train the employees and they monitor performance of these systems daily. The corporate safety or quality services group function in a supporting role, that is, collecting and analyzing safety data to identify operational hazards and opportunities for improvement from a holistic or big picture organizational perspective.

An Integrated Management System

The integrated management system (ims) is simply a term applied to the integration of all management systems within an organization. An organization is a group of people and facilities with an arrangement of responsibilities, authorities and relationships, examples include a company, firm, enterprise, institution, charity, or association. The foundation is the executive leadership and corporate governance of the organization that is supported by documenting roles and responsibilities of corporate officers, the functions of each operating division or group, management review/decision processes and the integration of multiple work activities. It describes the relationship and operational responsibility of each supporting management system within the overall enterprise. How is this done? The delivery of products and services is through a QMS that organizes company resources, systems, policies and procedures to meet the organizational purpose, that is, to satisfy the customer. Organizational survival is dependent upon how well and efficiently the leadership team aligns work streams and functional activities.

Most organizational management systems are three-dimensional, that is, organizational, functional and cultural. Since many accidents have been caused by what has been labeled as 'organizational factors', it is important that safety and quality practitioners understand organizational dynamics and relationships. Let's briefly review each of these three dimensions:

- *Organizational*: the accountable executive, typically the CEO or COO, establishes the organizational structure to support enterprise activities and ensures selection of division leaders and managers to provide oversight of operational activities. In the technical world of aviation, many functions require careful sequencing of complex activities by highly skilled or licensed technicians. In contrast, safety services or quality services organizations are independent of line organization and provide a number of services to senior leadership and line managers.
- *Functional*: all systems are functional in that a system such as the maintenance management system is structured, staffed, organized, documented and managed to accomplish a specific purpose. The maintenance division leadership team is responsible for designing a system dedicated to complete the multiple work streams associated with all aircraft maintenance activities. Although some of this work may be outsourced to suppliers, it remains a regulatory responsibility of the operator to ensure this work is completed according to specifications. Large, decentralized global organizations present unique challenges to ensure that required interfaces between management teams are efficiently designed.
- *Cultural*: the cultural characteristic of an organization receives special emphasis in all SMS discussions. Culture may be viewed holistically as the organizational or corporate culture and characterized as 'the way things are done around here'. The culture is a mix of several cultural elements under the broad categories of safety, quality, security and just elements. During any operational activity, employees transition from one cultural element or subculture to another. Examples of subcultures include national, safety, learning, reporting, just, informed and professional cultures. A safety culture is defined as the enduring value and prioritization of employee and public safety by each member of each group and in every level of an organization (von Thaden and Gibbons, 2008). The issue of culture will be discussed in greater detail as we review the SMS component of safety promotion.

ISO 9001:2008 clearly identifies a need for organizations to develop integrated management systems that address aspects such as product/service quality, employee safety, environmental performance and financial control. Each of these operational aspects function effectively based on a dedicated management system; for example the QMS, SMS, etc.

The i-MS Visualized

Developing a clear understanding of the IMS across the entire management team may seem easy on the surface, but may present significant hurdles. The daily challenge facing every safety, quality and security manager is convincing the management team that responsibility of these functions and results is the responsibility of the line or operational manager. Safety and quality is the result of well-designed processes, thorough employee training and continual improvement efforts. The functions of safety, quality, security and environmental safety are organizationally independent and provide services to the organization. These four functions also represent cultural elements such as a positive environmental culture or green culture. Figure 6.1 may be of some assistance in your communication efforts (Yantiss, 2006):

Integrated management systems are evident in both government and industry. For example, the European Aviation Safety Agency (EASA) organizes its resources using an Integrated Management System (IMS) concept.[3] EASA defines an IMS

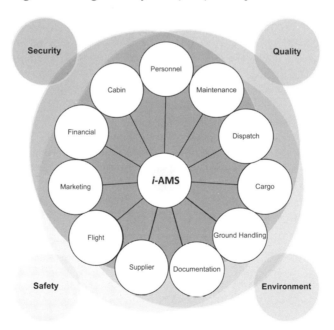

Figure 6.1   Integrated Airline Management System (IAMS)

3   EASA ED Decision 2009/089/E. 3 August 2009.

as a single integrated system used by an organization to manage the totality of its processes in order to meet the organization's objectives and equitably satisfy the stakeholders. Within this system, the manager of the quality section is nominated as the management representative for coordinating the implementation of the IMS. The message here is that safety and quality managers have distinct but related roles that must be harmonized, coordinated, and integrated.

As IOSA audit teams have conducted thorough evaluations of airline management systems, one observation is common – management teams have missed an opportunity to strengthen safety, quality, and security programs by keeping them organized and operating in silos. Operational risk-management processes must focus on identifying deficiencies throughout the organization as revealed through safety, quality, and security data analysis. A proactive safety and quality organization will address weaknesses in processes, procedures, or training prior to them surfacing as a major incident or accident. Each of the four services departments will have an invaluable view of the operational risks that face the enterprise, and bringing these perspectives together into an *integrated* Risk Management System will provide a more insightful summary of operational risks to senior leadership.

The Integration of Safety and Quality – A Progressive Step

Many operators have taken an intermediate step to establish an *IMS* by integrating safety and quality functions into a Safety and Quality Management System (SQMS). Rather than operating in independent silos, these two organizations and respective functions may be merged into an intra-supportive group by integrating data collection, data analysis, risk management, corrective action, and reporting. This may not be an easy transition for some organizations – in some cases, a forced marriage will need to be arranged by senior leadership, but the results will be remarkable. A greater challenge may be the cultural integration of the security function with safety and quality. In the opinion of this author, security issues are also safety issues in that either terrorist attacks or disruptive passengers present a safety risk to both passengers and employees. Again, the value of integration is to enable a holistic perspective of the risk spectrum that confronts every service provider.

Integrated Risk Management – The Next Level

Integration may be perceived as a huge challenge. To simplify this process, a number of operators have structured safety, quality, security, and environmental management systems using a common template. In general, we may structure each system using the ICAO SMS template. Figure 6.2 demonstrates how this alignment may be accomplished.

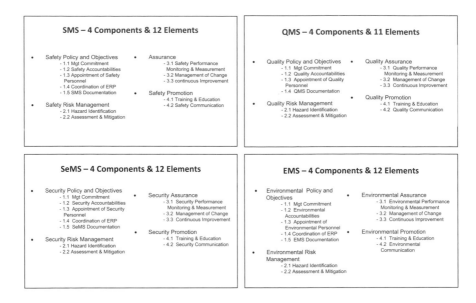

Figure 6.2   IAMS alignment using the ICAO template

With the exception of the QMS, the security and environmental management system may be structured using the ICAO SMS template of 4 components and 12 elements. The QMS is organized without an emergency response plan – all other elements are identical.

Although ICAO has not formally published a framework for quality, security, or environmental management systems, individual organizations have the opportunity to design these systems using the common template. A common template has many advantages over four dissimilar system structures:

- A common design and strategy will make sense to the accountable executive, the management team and individual employees – primarily because they don't have to expend the effort to understand and explain the differences. Insight into the design and function of a single system may be applied directly to the other three systems.
- A common template will tear down departmental barriers because the four groups will begin to understand the standardized application of data collection and analysis tools to manage operational risk, ensure regulatory compliance and adherence to operational procedures. The major difference is that each group is focused on a different body of State law and therefore responds to a different regulatory authority.
- A common risk-management process will align the technical vocabulary throughout the organization. The net result is that security, environmental and quality risks are assessed and communicated using the same tools and

vocabulary used by safety practitioners.

- A common set of management processes will enable consolidation of operational risks as viewed through the eyes of the four departments, thus enabling a holistic organizational risk assessment.
- A common set of management processes can reduce staff expenses as the quality assurance team, which typically conducts on-site operational audits, can be trained to perform on-site safety, security and environmental assurance activities.
- A common risk assessment and reporting process will help ensure the accountable executive understands all of the operational risks facing the organization. Operational risk may be categorized into one of four components: safety, quality, security, or environmental risk.

Value of Integration

The term *integration* always resonates positively in the eyes of the management team. The term suggests teamwork, harmony, efficiency, leverage and alignment. The following are but a few of the advantages of integration efforts:

- reduction in duplication of effort (cost)
- reduction in overall organizational risk (holistic assessment)
- balanced integration of enterprise and operational risk management
- alignment of potentially conflicting operational objectives
- alignment of risk management and assurance processes
- elimination of organizational barriers (silos) to
  a) teamwork
  b) communication
  c) community (trust and transparency)
- diffusion of organizational power systems.

Potential Roadblocks to Integration

Accountable executives will encounter a number of factors that will challenge the integration process and jeopardize its success. Many of these issues are cultural or turf issues that must be understood and effectively countered:

- Operators typically organize safety, quality, security and environmental staff into separate departments. Since each group works with a different body of law, regulatory agencies and technical vocabulary, significant barriers to teamwork, trust and communication frequently develop. Addressing these issues may require strong leadership involvement, organizational changes and a long-term commitment to integrate the four separate subcultures. Full cultural integration may require three to four years to complete.

- The directors (managers) of safety, quality, security and environment may report to the same senior leader or be consolidated into a single division. Regardless of the organizational model, successful integrated risk management may be jeopardized by an individual department head that undermines the integration (team) effort at every turn. This individual may prefer status quo rather than stepping out of their comfort zone and working with other recognized experts in their area of expertise. For example, a safety director may find it uncomfortable to work directly with the security director – or vice versa.
- There may be resistance to integration and alignment based on the excuse that 'the regulatory authority doesn't require it'. Integration, as a best practice, requires a higher level of individual effort than operating in comfortable and ineffective silos.

Integration of these four systems enables the accountable executive to understand the holistic operational risk that must be proactively managed. Ultimately, the accountable executive must understand and accept operational risk as a byproduct of operational activities. These four systems include data collection and analytical processes that produce health assessments of all operational activities. Opportunities for improvement are summarized for senior leadership review and prioritization. Resources may be committed to those initiatives that present the most value in terms of risk management, cost management and service/product quality. A well-managed organization will understand the concept of operational risk management, have established performance measures that reflect acceptable levels of risk, and will schedule frequent management reviews to ensure action is taken when performance deteriorates. The next question is how to organize the four services departments into a harmonious and efficient team?

Organize for Success

For a number of reasons, a number of large international airlines have evolved their organizational structures to enable and accelerate integration of safety, quality, security and environmental teams to capture the advantages of integration. One of the most rewarding and challenging management positions in a large enterprise is the individual who is responsible for leading these individual services departments. Typically, this position is an officer position that serves as the corporate conscience, that is, providing unbiased services for the accountable executive to include data collection and analysis, coaching and verifying regulatory compliance. Each of these departments has many common functions but perhaps the most obvious is to ensure regulatory compliance. Figure 6.3 depicts an organizational model that can be easily modified to suit the needs of any organization.

Figure 6.3   Organizational model

Successful integration of departmental processes may require three to five years. Examples of harmonized processes include the following:

- each director identifies operational risks from his or her unique perspective
- each department monitors compliance with a different body of State law
- each director uses the same risk model to ensure continuity and alignment
- integration of risk-management processes enables a holistic risk assessment
- incidents and accidents are investigated using the same investigative processes and reporting format
- assurance and auditing activities are aligned in processes, checklists, corrective action requests and reports
- each department identifies and prioritizes operational risks that may be communicated to the accountable executive as desired.

Responsibility for Integration

Although organizational decisions rest with senior management, it is important for safety, quality, security and environmental practitioners to understand the concept of an *integrated* management system and to be able to communicate organizational options to decision makers. The biggest challenge for middle managers is to think at an organizational level and fully understand the needs of senior leadership in terms of operational and enterprise risk management. It is important to take advantage of every opportunity to enhance your knowledge about these management systems as they enable the organization to deliver products and services while protecting people, equipment and other resources. The following sections will highlight key techniques and considerations as you implement these systems.

Risk Management – a Holistic Perspective

Introduction

Risk management is the process used by every organization to identify, measure and manage the various types of risk inherent within their operations. It is a broad term for the business discipline that focuses on the protection of assets (equipment and people), the continuing ability of the organization to produce products and services for which it was intended, the ability to produce value for shareholders in terms of a financial return and to anticipate threats or hazards that would threaten the organization's existence. It is important to recognize that the objective of risk management is not to always eliminate risk, but to manage it effectively. Risk management permits the organization to set risk tolerances based on their overall corporate objectives. Risk tolerance may also be described as the organization's 'risk appetite' and will vary greatly between organizations.

Risk management in the aviation business occurs on two levels: Enterprise Risk Management (ERM) and ORM. ERM represents a revolutionary change in the organization's approach to managing the threats to organizational business processes. It includes capturing and understanding every business risk of the organization, comprehensively and systematically (Risk and Insurance Management Society, 2009). Operational risk, in contrast, is limited to those activities that involve production of products and services. Let's review each of these risk-management processes at a very high level.

Enterprise Risk Management

Typically, the organization's chief financial officer (CFO) takes the lead in organizing corporate-level risk management activities. At the corporate level, there are seven categories of risk that confront the organization:

* strategic risk (market dynamics, resource allocation, etc.)
* financial risk (capital structure, liquidity, underwriting, credit, etc.)
* operational risk (assets, people, technology, etc.)
* compliance risk (legal, regulatory, best practices, etc.)
* environmental risk (petroleum products, greenhouse gases, hazardous materials, etc.)
* corporate citizen/image/reputation risk
* project risk (new systems, operations, equipment, etc.)

ERM and ORM have been one of the most significant additions to the insurance industry's vocabulary. Although insurance companies have been using a number of risk modeling techniques as a matter of fundamental practice, ERM and ORM provide new tools with which to assess an organization's ability to manage its business, the threats to its business and the organization's specific efforts to manage these threats. As an example, aviation insurance underwriters strongly desire an annual review of the airline's risk-management processes. During these annual reviews of an airline's safety performance, underwriters are digging deeper into the risk-management processes, the results of risk management efforts and the reduction in significant safety issues – incidents, accidents, employee injuries, damage to equipment and near misses where the organization nearly recorded a hull loss. Risk discussions typically center on the probability of a hull loss and associated number of fatalities. While this may appear to be an insensitive type of discussion, it does represent today's aviation reality, that is, a portion of the enterprise risk is composed of operational risk. The questions for management are, 'What will be the impact of an accident on the enterprise?' and 'How do we effectively manage this risk?'

When the organization makes strategic decisions, operational risk, among the other ERM categories, must be carefully balanced prior to engaging in new venture, for example the addition of A380 aircraft into the inventory. The addition of the A380 or any new aircraft into an airline's inventory represents significant risk in terms of strategic, financial, operational, compliance and project risk. The decision to add such a large aircraft into an airline's existing route structure is dependent upon the organization's risk appetite as a summation of the seven risk categories.

Over a decade ago, the Committee of Sponsoring Organizations of the Treadway Commission (COSO) developed a framework that is readily useable by any manufacturing or service enterprise (Committee of Sponsoring Organizations, 2004). ERM is not strictly a serial process, but is multidirectional in which any component can and does influence another. The COSO model, as an example of ERM, is illustrated in Figure 6.4 and portrays the ability to focus on the entirety of the organization's risk or by subset.

Although there are important benefits to this type of enterprise risk management, there are challenges that must be recognized, particularly in the aviation industry. These challenges result from the realities that human judgment in decision making can be faulty, decisions on responding to risks must consider both costs and benefits, breakdowns in processes occur because of human factor-related errors, controls may be rendered useless through collusion of two or more people, and the obvious ability of management to override risk management decisions. Perhaps the most significant challenge to ERM is the absence of multi-variant mathematical models that are needed to support both operational and enterprise risk management efforts.

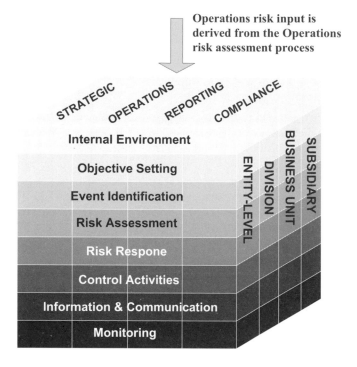

Operations risk input is derived from the Operations risk assessment process

STRATEGIC OPERATIONS REPORTING COMPLIANCE

Internal Environment

Objective Setting

Event Identification

Risk Assessment

Risk Respone

Control Activities

Information & Communication

Monitoring

ENTITY-LEVEL DIVISION BUSINESS UNIT SUBSIDIARY

Figure 6.4   Committee of Sponsoring Organizations framework

Aviation organizations talk frequently about culture and dissecting every component of the organization's corporate culture. At the enterprise level, there is a cultural component that is often overlooked by safety and quality practitioners, that is, the subject of a risk-management culture. As with safety management, ERM starts with the accountable executive. Ultimately, it is the importance that the board of directors and senior leadership team places on risk management that will establish the extent to which the management of risk is integrated across the entire organization. A risk-aware culture is based on a common risk vocabulary and a full understanding of risk among corporate officers that enables collaboration on risk-management issues. Again, insurance companies are ever-increasingly interested in the ability of an organization to understand and manage risk.

Organizations in some industries have concluded that ERM is so crucial to the long-term success of the organization that they have established a new officer position, that is, a corporate risk officer (CRO). Risk management is a discipline, much like safety management, that requires solid technical skills and strong leadership to assist the accountable executive in managing risk holistically. The CRO provides the organizational leadership, coaching, consulting and advocating of risk management concepts throughout the organization.

The most obvious question among safety and quality practitioners is 'who' is primarily responsible for the ERM process (see Figure 6.5). The following group of senior leaders is typically immersed in the ERM process:

- Chief Executive Officer (CEO)
- Chief Finance Officer (CFO)
- Chief Operations Officer (COO)
- Chief Administrative Officer (CAO)
- Chief Information Technology (CIO)
- Chief Marketing Officer (CMO)
- SVP Human Resources
- Corporate and Government Affairs
- Corporate General Counsel

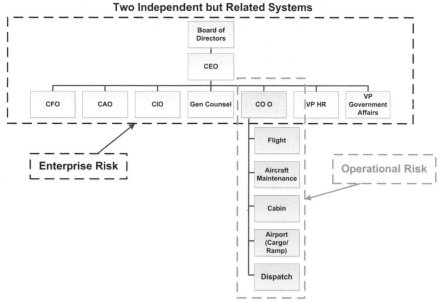

Enterprise and Operational Risk Management
Two Independent but Related Systems

Figure 6.5   Enterprise vs. operational risk

Risks May Be a Surprise

Most organizational leaders are seasoned in their particular industry and recognize the traditional hazards that confront their industry. Over time, precautions and mitigations have been implemented to manage these risks. However, threats or hazards that are not recognized by either the management team or line employees may present considerable risk to the organization. As described by Kaplan et al. (2009) in the *Harvard Business Review*:

> The risks that get you are the ones you're not expecting, and your experience may be what's making you not expect them. For some reason, people do not always just say: Time out. How does this work? Why do we do it this way? What's the problem here? They may assume they know these things from experience. Also, one of the hardest things for a decision maker to admit is his or her ignorance, and the more complicated we get with our metrics and models, the more hesitant people are to admit that they don't understand. (Kaplan, Mikes, Simons, Tufano and Hofmann, 2009)

Low-frequency, high-severity events that are almost impossible to forecast may present the most significant risk to our fragile socioeconomic environment in which we compete. For example, the 2008 global economic and financial crisis was not clearly predicated by many financial risk models and those red flags that surfaced were either ignored or misunderstood by financial experts. The 2008 to 2009 economic crisis 'has been compounded by the banks' so-called risk-management models, which increased their exposure to risk instead of limiting it and rendered the global economic system more fragile than ever (Taleb, Goldstein, and Spitznagel, 2009, pp. 78–79).[4]

Common Mistakes

All management teams, regardless of industry sector, must understand the six most common mistakes that the executive team will make in regard to risk management. These mistakes have some invaluable lessons for the aviation community. The following conclusions are discussed in more detail in the October 2009 *Harvard Business Review* (Taleb, Goldstein, and Spitznagel, 2009, p. 80):

Mistake #1 – We Think We can Manage Risk by Predicting Extreme Events

As a global community, we have been unable to predict or forecast global catastrophes – whether economic or environmental. Recognizing this, executives are shifting from attempting to predict when accidents will occur to crafting well-designed response plans to manage these events. This strategy embraces the idea of lessening our vulnerability to them. Additionally, there is now a huge emphasis on understanding why financial models have failed. What parameters have been

---

4   Mr Taleb is a Professor of Risk Engineering at New York University's Polytechnic Institute. Mr Goldstein is an assistant professor of marketing at London Business School. Mr Spitznagel is a principal at Universa Investments.

omitted from existing models? What are the synergistic relationships of the parameters and how do their interactions generate unpredictable outcomes?

In similar fashion, safety practitioners have evolved from reactive to proactive safety techniques through improved data collection, analysis and reporting. As an example, many safety experts knew that eventually a commercial airplane would fly through a flock of birds and damage both engines to a point that sustained flight was impossible. The red flags were evident in global safety data bases. It was not a matter of what; it was a matter of when. We knew it would occur, but we simply could not predict the airline to which it would occur, and when. We were indeed fortunate that the USAirways A320 engine failure and subsequent landing in the Hudson River did not result in tragic loss of life. Thanks to solid flight crew training in anticipation of such an event, the crew managed to fly the aircraft to a successful ditching in the Hudson River. The A320 ditching procedure is a great example of an emergency response plan.

Mistake #2 – We are Convinced that Studying the Past Will Help us Manage Risk

Risk managers frequently use hindsight (looking in the rearview mirror) rather than using foresight or predictive thinking when assessing risk. In the financial world, the idea of a 'typical' failure is a thing of the past. Each failure must be anticipated in an ever-changing economic environment.

In the world of aviation safety, we must transition from reactive thinking, that is, investigating the most recent 'smoking hole in the ground,' to assessing leading indicators of significant events and implementing mitigations to prevent their occurrence.

Mistake #3 – We Don't Listen to Advice About What We Shouldn't Do

We are bombarded with recommendations of the *do* and *don't* variety. Do this – don't do that. Don'ts are usually more useful in life than do's. For example, telling someone not to smoke outweighs other advice on general health care, for example, do eat a balanced diet. The following excerpt talks in terms of risk management versus acts of commissions and omission:

> Psychologists distinguish between acts of commission and those of omission. Although their impact is the same in economic terms – a dollar not lost is a dollar earned – risk managers don't treat them equally. They place greater emphasis on earning profits than they do on avoided losses (costs). However, a company can be successful by preventing losses while its rivals go bust – and it can then take market share from them. In chess, grand masters focus on avoiding errors; rookies try to win. Similarly, risk managers don't like not to invest and thereby conserve value. But consider where you would be today if your investment portfolio had remained intact over the past two years, when everyone else's fell

by 40 percent. Not losing almost half your retirement is undoubted a victory. (Taleb et al., 2009)

This insightful observation has direct applicability to the aviation industry. Standard operating procedures (SOPs) associated with cockpit procedures, maintenance instructions and ground handling are loaded with *do's* and *don'ts*, intended primarily to enable employees to do their work with minimum of risk. *Selective compliance* with SOPs usually follows a brief, mental risk analysis of the potential ramifications of deviating from procedures. In other words, 'what will happen if I do this?' As mentioned earlier in this book, for every major accident, there are at least 15,000 procedural deviations. The probability of having a major accident as a result of an SOP deviation may be minimal; however, the long term result is a significant deterioration in organizational discipline, evidenced by a culture that accepts SOP deviations as normal and tolerated by functional supervisors as acceptable.

Mistake #4 – We Assume that Risk can be Measured by Standard Deviation

Many enterprise risk management experts suggest that the use of 'standard deviation' be avoided in ERM. As a review, the standard deviation of a statistical population is the square root of its variance. Standard deviation is a widely used measure of the variability or dispersion. It shows how much variation there is from the average (mean). It may be thought of as the average difference of the scores from the mean of distribution, that is, how far they are away from the mean. A low standard deviation indicates that the data points tend to be very close to the mean, whereas a high standard deviation indicates that the data are spread out over a large range of values.

For safety practitioners, the use of data analytical tools is absolutely essential to our craft. For example, aggregate data obtained through flight data analysis (digital data obtained from airplane storage devices) may be easily analyzed for any significant deviation (trends) from operating standards. For example, the data may suggest that flight crews are delaying the stowing of engine reversers during landing, thus increasing the risk of ingesting pebbles and debris into the engines. In response, flight managers can revise training programs and issue alert bulletins advising flight crews of the trends and associated hazards. Similarly, data analysis will permit the identification of outliers, that is, isolated events that represent significant deviation from SOPs and require direct attention with the flight crew. A number of States have developed flight data analysis regulations that enable direct contact with the crew without punitive action. Obviously, the personal contact is intended to gain information on the exact circumstances of the event and develop action plans to minimize reoccurrence.

Mistake #5 – We Don't Appreciate that What's Mathematically Equivalent isn't Psychologically so

In 1965, physicist Richard Feynman wrote that 'Two mathematically equivalent formulations can be unequal in the sense that they present themselves to the human mind in different ways.' Stated differently, the way that risk or probability is stated will greatly influence people's understanding of it.

The message to safety practitioners is to use the same risk matrix, risk-related vocabulary, and risk-management processes throughout the organization. It is important that all operational risk (to include safety, security, quality, and environmental) be evaluated using identical processes and tools.

Mistake #6 – We are Taught that Efficiency and Maximizing Shareholder Value Don't Tolerate Redundancy

Many executives do not fully understand that optimization and elimination of *redundancy* make an organization more vulnerable to change. How do we define redundancy? In the human body, we have spare parts – two lungs, two kidneys, etc. In the aviation industry, redundancy may take the form of excess inventory of spare parts, idle capacities (airplanes, airport gates, facilities) and cash that isn't put to work. In contrast, traditional business thought suggests that *leverage* is good. Leverage is evident in nearly every major corporation. Excess leverage, or debt, creates financial risk when a company misses a sales forecast, crude oil jumps by US$50, or interest rates jump unexpectedly.

We have built the aviation industry by designing airplanes with redundant systems, avionics, navigation systems and even pilots to manage risk. This industry is an excellent example of anticipating nearly every conceivable operational failure and designing redundancy into these systems. The reduction of the global accident rate in the last 40 years has been nothing short of remarkable. As a 1965 snapshot, the global commercial jet aircraft fleet experienced 30 accidents per million departures. Improvements in flight crew training, aircraft design and airport infrastructure have contributed to a global 2009 accident rate of 0.71, or one accident per 1.4 million departures. Certain regions report even better performance. For example, the US reported a 2008 accident rate for scheduled commercial operators of 0.189 with no fatalities. The record is evidence that safety strategies implemented by the Commercial Aviation Safety Team as an industry partnership are clearly turning the dial. Perhaps the following statement from a seasoned statistician says it best:

> According to a Boeing statistic, you would have to live 3,500 years and fly every single day of your life to be involved in a major aircraft accident. Even then, you

would have a 50% chance of surviving the accident. On the other hand, you only have to live to be 700 years old to be shot by your spouse...[5]

In comparison, the US National Transportation Safety Board reports that approximately 40,000 Americans are involved in fatal highway accidents annually. Interestingly, 33 percent of the fatalities involved a single SOP deviation – not wearing a seatbelt.

Leader Overconfidence

Safety practitioners should keep in mind that the Achilles' heel of many corporations is that the leadership team may overestimate its own abilities and underestimate what can go wrong. The leadership team may be overconfident in the ability of line managers to manage risk, and oblivious to the actual operational hazards with which the functional manager grapples with daily. Risk-management processes are designed to identify operational hazards and provide the accountable executive with a real world assessment.

Operational Risk Management

Responsibility for ORM typically rests with the accountable executive. For large commercial airlines, this may be the chief operating officer (COO). Operational risk is a component of the organization's overall risk, that is, enterprise risk. Civil Aviation Authorities and the majority of an airline's employee groups are primarily concerned with operational risk. Further, most of the activities under the SMS, QMS, SeMS and EMS umbrellas are focused on operational risk identification, assessment and mitigation. Ideally, risk is best understood when these four related databases are integrated and information gleaned is collectively analyzed. The strength of the i-MS concept is that the activities and functions of these managers are fully integrated to provide a comprehensive analysis. Specifically, these managers formally compare their independent assessments of operational risk, identify corporate priorities and suggest recommendations for mitigation to senior management. This snapshot of operational risk is one of the inputs into the ERM model.

Risk-management processes involve a series of actions, falling into five general categories:

1. risk identification
2. measurement and evaluation of exposures
3. exposure reduction, transfer, or elimination

---

5   Source unknown.

4. risk reporting
5. risk acceptance.

The objective is simple: tee-up issues, identify responsible managers to develop corrective action, track the effectiveness of these actions, and then report results appropriately. The process is continuous: evaluate the health of operational functions, identify opportunities for change or improvement, make adjustments to operational processes and then re-evaluate process improvements through auditing, employee reporting and operational analyses. Risk-management processes that are within the disciplines of safety, quality, security, and environment are harmonized and carefully coordinated by the respective managers. An example of this process is illustrated in Figure 6.6.[6]

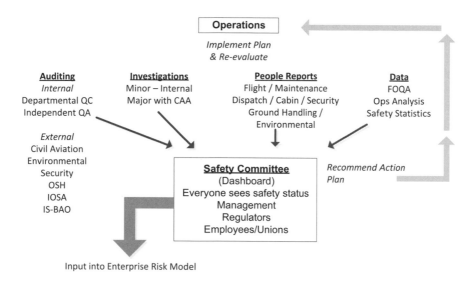

Figure 6.6   Operational risk management

---

6   Captain Henry P. Krakowski, Vice President – Corporate Safety, Security, Quality, and Environment, United Airlines, 2005; Captain William E. Yantiss, Managing Director – Quality Assurance Airline Operations, United Airlines, 2005.

The components of the ORM model are familiar to most safety practitioners and will be reviewed in more detail in the sections addressing safety risk management and safety assurance. Data collection and analysis is conducted primarily by the safety, quality, security and environmental services departments. Data acquisition occurs in four major streams:

- Auditing: independent teams conduct on-site audits to assess the health of each functional area. Regulatory, safety, or operational issues are surfaced and addressed by the functional manager.
- Investigations: investigations occur at many levels in the organization. Minor investigations of significant incidents may be accomplished by the functional manager, division safety departments, or the safety services department.
- People reports: a number of employee reporting systems enable daily collection of operational hazards, minor safety occurrences, safety issues and regulatory violations. Normally there are protective provisions associated with each reporting system to encourage non-punitive reporting.
- Data: this category includes the streams of operational data that are generated by modern jet aircraft. Not only is flight-related data available for periodic download into flight data analysis (FDA) programs, there are hundreds of maintenance parameters that can be downloaded in a similar fashion. Maintenance data enables daily electronic checks of engine health, for example.

In general terms, the six corporate-level services departments collect, assess and report safety issues to the leadership team, employee groups, and regulatory authorities. These processes are clearly defined in SMS, QMS, SeMS and EMS manuals.

For simplicity, we will focus primarily on safety and quality management systems as we delve into the SMS components of safety risk management and safety assurance. The concepts are easily adaptable to security and environmental management systems.

Component #1: Safety Policy and Objectives

Introduction

ICAO has broken down the SMS concept into 4 major components and 12 elements. Each of the elements may have multiple subelements that describe activities that support effective implementation of the individual element. Since many safety and quality practitioners will memorize the SMS outline, we will repeat it for convenience when introducing a new SMS component.

- Component 1: safety policy and objectives
  - Element 1.1   Management commitment and responsibility
  - Element 1.2   Safety accountabilities
  - Element 1.3   Appointment of key safety personnel
  - Element 1.4   Coordination of emergency response planning
  - Element 1.5   SMS documentation
- Component 2: safety risk management
  - Element 2.1   Hazard identification
  - Element 2.2   Risk assessment and mitigation
- Component 3: safety assurance
  - Element 3.1   Safety performance monitoring and measurement
  - Element 3.2   The management of change
  - Element 3.3   Continual improvement of the SMS
- Component 4: safety promotion
  - Element 4.1   Training and education
  - Element 4.2   Safety communication

New Concepts and Definitions

ICAO introduces a number of new concepts to both the regulatory communities and services providers. These terms will be used frequently during any discussion of safety and quality management (ICAO, 2009, 6.4.6).

- Level of safety: the degree of safety within a system expressed through safety indicators.
- Safety indicators: the parameters that characterize and/or typify the level of safety of a system.
- Safety targets: the concrete objectives of the level of safety.
- Acceptable level of safety (ALoS): the minimum degree of safety that must be assured by a system in actual practice.
- Safety indicator value: the quantification of a safety indicator.
- Safety target value: the quantification of a safety target.

The value of understanding these terms is that they may be duplicated in State SMS regulations and therefore will become part of the CAA inspector's vocabulary.

Definition of Safety

As a review, *safety* is the state in which the risk of harm to persons or property damage is reduced to, and maintained at or below, an acceptable level through a continual process of hazard identification and risk management. This ICAO definition implies a system of measurement and evaluation of an organization's safety performance and feedback into the SMS (ICAO, 2009, 2.2). Other valuable

ICAO insights into the changing world of regulatory oversight should be reviewed as a backdrop to our SMS discussion:

- Hazards are integral components of aviation operational contexts.
- Safety encompasses relatives rather than absolutes, whereby safety risks arising from the consequences of hazards in operational contexts must be acceptable in an inherent safety system.
- While the elimination of accidents and/or serious incidents and the achievement of absolute control are certainly desirable, they are unachievable goals in open and dynamic operational contexts.
- Safety is increasingly viewed as the outcome of the management of certain organizational processes.
- Although regulatory compliance is important, its limitations as the mainstay of safety have become increasingly recognized. It is simply impossible to provide guidance on all conceivable operational scenarios in a dynamic system such as aviation.
- Latent conditions contribute to an organizational accident. Most latent failures start with the decision-makers, that is, the management team.

Statistically, millions of operational errors are made on a daily basis in the aviation industry before a major safety breakdown occurs resulting in an accident. One safety expert estimated that at least 15,000 operational procedural deviations occur for each significant safety incident (event) resulting in bent metal or serious personal injury.

All employees will make operational errors – the idea is to reduce the magnitude and frequency of these errors. It is difficult to systematically anticipate typical human frailties such as distraction, inattention, task saturation, fatigue and forgetfulness, and how these factors influence the margin of safety in the operational workplace. The challenge facing the management team is defining the organization's *acceptable level of safety* in terms that are measurable. Examples include tracking the number of operational events, ground damage, engine failures, runway incursions, altitude deviations, or employee injury. Measuring these safety indicators (parameters) and managing them to acceptable levels requires an investment in developing effective systems, policies, procedures and training – all supported with leadership engagement and modeling.

Element 1.1: Management Commitment and Responsibility

One of the more significant characteristics of the ICAO SMS standard is the emphasis on senior management's commitment to operational safety and is clearly reinforced by the term accountable executive. Although this is a familiar concept in the European Union, the introduction of documented safety accountabilities may require some getting used to in some States. For example, the US Federal

Aviation Regulations (FARs) currently say little about a documented management system. However, the new SMS framework creates a significant shift in policy and emphasis and will ultimately be incorporated into new regulations.

Who is the Accountable Executive?

Simply stated, the accountable executive is the senior management official who has overall responsibility for ensuring the safety and security of operations. More specifically, this person has the authority to make policy decisions, provide adequate resources, maintain financial control, lead organizational performance, safety and management reviews, and to accept operational risk. In many large organizations, the accountable executive may be a non-pilot CEO or COO. In contrast, small organizations may identify a director of operations who is assigned additional responsibilities associated with those of an accountable executive.

What the Accountable Executive Needs to Understand

The accountable executive sets the pace for the organization in terms of setting policy, establishing priorities, allocating resources and endorsing initiatives. The commitment to SMS must be endorsed, supported, and communicated by the accountable executive. These leaders do not have to be an expert in safety and quality tools; however, they do need to understand a number of fundamental principles:

- That SMS is a business approach to managing safety.
- The use of risk indices and risk mitigation.
- That the SMS standard requires their personal leadership in the management review process, the identification of operational hazards, the mitigation of recognized hazards and acceptance of predicted residual risk associated with a significant change in operations.
- The requirement to provide adequate resources for the safety and quality services departments that lead SRM and safety assurance (SA) activities.
- That SMS implementation is a major cultural change in terms 'of the way we do business'.
- That the direct responsibility for safety rests with line management and employees, but must be modeled and supported at the senior management level.
- That a healthy corporate culture requires constant nurturing, is composed of multiple components, and that non-punitive methods are necessary to manage human error.
- That there will be individuals throughout the organization that will resist SMS implementation and that the accountable executive must model desired attitudes and behaviors to all employees.

- That there must be continued support for the SMS champion (typically the safety manager) who will lead and communicate the implementation process through the organization.

Safety Policy

An organization's safety policy should reflect the commitment of senior management to safety as a fundamental priority and signed by the accountable executive. Examples of effective safety policies are readily available from ICAO, IATA and regional organizations. The safety policy is the cornerstone document that creates a foundation for a healthy safety culture, and defines specific expectations of the management team to model behaviors expected of all employees.

Management *owns* the resources dedicated to the delivery of products and services and determines the amount of resources required to get the job done – safely. Decisions relating to production versus protection involve trade-offs that ultimately determine the number of hazards in each workplace environment. Management sets an example for adherence to procedures and is a role model for the corporate culture. If management condones or performs activities that deviate from their own published procedures, employees' clearly get the message loud and clear that it's okay to do your own thing and deviate from SOPs when and where you wish. A clear safety policy addresses these issues, establishes safety standards and acceptable performance metrics and defines acceptable employee behavior.

Safety Reporting

A safety policy typically reinforces a commitment to a non-punitive safety reporting process. There's a huge difference between confidential reporting and anonymous reporting. Anonymous safety reports have value in the sense that sensitive safety issues can be surfaced to management without an employee *highlighting* themself. There are several disadvantages to this type of reporting system, such as the inability to provide feedback to the employee and the inability to obtain additional clarifying information from the reporting employee.

Many organizations have adopted a confidential safety reporting system that is based on a principle of safeguarding safety and risk information. The reporter's identity should not generally be available to the Civil Aviation Authority or organizational management. In the United States, the Federal Aviation Administration has pioneered the Aviation Safety Action Program (ASAP) that establishes *incentives* to employees for reporting operational errors, significant safety issues and inadvertent regulatory violations. Although not a get out of jail free card, individual reports are reviewed by an ERC composed of a company management representative, an FAA representative and a safety representative from the employee's professional association. Each report receives appropriate investigation and analysis by the safety services department. Resolution of the reported event involves consensus among the ERC members. The effectiveness of this program varies greatly from operator to operator, as reflected in the number of weekly reports and employee feedback. If an employee feels they will be treated fairly by the ERC and the management team, they will be inclined to report safety issues as a professional obligation. Large operators may receive as many as 10,000 ASAP and hazard reports annually that highlight soft spots in the operation, operational hazards and employee mistakes. Should employee confidence in the safety system be compromised, this stream of reports will diminish accordingly.

Other Policies

A number of international standards and best practices endorse the value of documented security, quality and environmental policies signed by the accountable executive. These policies should be aligned with the safety policy and receive equal commitment by the accountable executive.

Management Review

One of the major responsibilities of the accountable executive is to lead the management review process that is focused on assessment of the various management systems within the enterprise. The objective of this recurring assessment is to verify that each of the component systems is operating to meet expectations in terms of performance, productivity, cost control, customer satisfaction and safety. One of the key objectives of any organization is to thrive in the competitive market and to survive the economic challenges of the aviation industry. Exceptional safety performance is the result of investment in the development and sustainment of solid processes and procedures, reinforced with employee training. Singapore Airlines, one of the most successful airlines in terms of service and profitability, operates with an emphasis on documented procedures reinforced by annual training requirements, particularly for the management team. It's gratifying to see organizations actually increase training when the market gets more competitive, not reduce training to manage expenses. The objective of a management review is to ensure the entire organization and its people are prepared to reengineer operating processes to remain competitive, efficient, effective and safe.

Each enterprise will develop the management review process to meet the desires of the accountable executive. Some organizations review safety performance metrics during a safety review board (may be referred to as a management review) chaired by the accountable executive. Others have combined safety, quality, security and environmental performance into a risk review board concept whereby the managers of these departments provide a summary of the risk management results from their individual perspectives. Obviously, this is most effectively accomplished when they groups operate as a team, focused on holistic operational risk.

The following is a shopping list of topics that may be addressed during periodic management reviews (International Air Transport Association, 2010b, Sec. 2.3):

- results of safety risk assessments
- results of audits, inspections, and investigations
- safety performance results (metrics)
- operational feedback
- changes in regulatory policy or civil aviation legislation
- process performance and organizational conformity
- status of corrective and preventive actions
- follow-up actions from previous management reviews
- feedback and recommendations for management system improvement
- regulatory violations.

Element 1.2: Safety Accountabilities

Any discussion of safety accountabilities involves a clear understanding of three commonly misunderstood terms – responsibility, accountability and authority. The following definitions appear to be the most balanced and useful in understanding the proper use of these terms in safety communications and organizational documentation:

- Responsibility: a specific job function, duty, task, or actions assigned to each line employee by senior management with the power and resources to perform these operational activities. Responsibilities are assigned to people throughout the enterprise to organize and align the work processes, procedures and activities. For example, the responsibility for the safe operation of a flight is assigned to the captain. The responsibility for conducting proper aircraft maintenance is assigned to a specific aircraft maintenance technician (mechanic). Individual employees may be assigned multiple responsibilities that are defined in job descriptions, operational procedures, or organizational manuals.
- Accountability: the term accountability is closely aligned to the term responsibility and is defined as an obligation of an employee to answer for the outcomes of assigned operational work activities. For example, employees may be responsible and held accountable, by a higher authority, for the performance of activities as described in operating manuals. The notion of accountability implies that an employee may be subject to corrective actions by a supervisor or management team members for failure to complete specific work responsibilities, as assigned. There is a difference, however, between responsibility and accountability in that responsibility is the obligation to act, whereas accountability is the obligation to answer

for an action and presumably be subject to consequences from a higher authority for failure to meet assigned responsibilities.

- Authority: authority is a term used to define the power or ability of individual employees to make specific decisions, to give orders, to command, grant permission and/or provide approval. Certain levels of authority may be delegated by senior management to each employee, enabling them to conduct certain work activities without permission. For example, process owners have the authority (power) to approve or change a process or procedure. Another example is a captain that can exercise his emergency authority to divert a flight to an en route alternate airport in the event of a medical emergency. Alternatively, a mechanic may be assigned the responsibility to conduct certain types of maintenance work but may not have been delegated the authority to sign off the work.

ICAO has introduced the concept of *safety services* to suggest that the safety department provides a service to the organization and not directly responsible for safety outcomes. Functionally, the safety services office is a safety data collection and analysis unit using a number of predictive, proactive and reactive methods to provide reliable information to the entire management team. In the same fashion, the quality and security managers utilize similar methods to provide quality and security information to senior management and line personnel who have direct responsibility for risk management and continual improvement. Specifically, the responsibility for managing safety rests with the responsible division (flight operations, ground operations, engineering and maintenance, etc.)

Organizationally, these service departments should remain independent of operational divisions and report directly to the accountable executive. This independence enables these teams to remain unbiased and objective in their analysis of operational issues.

Element 1.3: Appointment of Key Safety Personnel

The safety manager will most likely be the person whom the accountable executive has assigned the responsibility for daily SMS oversight and leading the safety services office. However, some organizations have selected the quality manager to lead the effort, particularly for those organizations that have installed a comprehensive QMS. As an alternative, organizations have the option of implementing SMS as a shared responsibility of the safety and quality managers. The choice will likely depend on the experience, passion, competence and skill set of the existing staff.

The size of the safety staff will vary greatly between organizations. For example, the safety manager in small organizations may be assigned a number of different responsibilities to include flight safety, ground safety, quality assurance and perhaps security. In contrast, large organizations have evolved to a point where the vice president of safety has assumed responsibility for corporate regulatory compliance, which includes the disciplines of quality, security and environment.

Selection Process and Criteria

It is important that accountable executives select their safety manager with great care. The role of the safety manager (to include quality, security and environment) requires an individual with a passion for the work and the ability to find reward in the work itself. As the independent corporate conscience the recommendations to the management team may often be unpopular because they require the expenditure of resources to improve processes or procedures. In an industry that has traditionally lost money, the management team must carefully evaluate every organizational investment in terms of value added, that is, production versus protection. This is a risk-management discussion at its best. The safety manager must have the managerial and business skills to successfully engage with the leadership team and speak with credibility.

The selection of the safety, quality and security managers requires equal attention as they will work together as a team to jointly identify operational safety hazards, assess operational risk, ensure mitigation strategies are appropriate for the identified hazard, ensure compliance with State regulatory requirements and conformance with organizational processes and procedures, and to identify opportunities for continual improvement (IATA, 2010b, Sec. 2.3). If the accountable executive is to receive a comprehensive report on holistic operational risk, it is imperative that the chemistry among these managers fosters effective teamwork through open communication and trust.

There are tools available to assist in the selection in a safety manager. For example, PRISM Solutions, LLC has created a useful candidate screening tool to assist managers in the selection of safety personnel.[7] This tool can easily be adapted to the selection of quality and security managers.

Professional Training

It is common that safety managers are selected on the basis of their experience and leadership skills rather than their experience as a safety practitioner. These individuals typically have earned a strong reputation as an operational expert, but they have received little formal training in the area of systems management, organizational behavior, safety and quality management and human factors. It is absolutely essential that the safety leadership team obtain formal training to have the knowledge to lead the SMS implementation process on behalf of the accountable executive. All professions require solid initial and recurrent training to acquire and maintain the technical skills necessary to be successful. This is

---

7   Professional Resources in System Management, a wholly-owned Subsidiary of ARGUS International, Inc.

particularly true of the domains of safety and quality. Safety experts must understand that quality principles and tools are the enablers of an organization's ability to deliver competitive products and services while preserving assets such as people, equipment, and facilities. This requires continual professional reading and formal training.

Element 1.4: Coordination of Emergency Response Planning

ICAO's *Safety Management Manual* requires that emergency response plans be coordinated between airports, operators, and ATC facilities (ICAO, 2009, 8.7). These plans outline, in writing, specific actions to be taken following an accident and who is responsible for each action. Interestingly, ICAO has provided little guidance in a very important facet of safety management. This should not be interpreted as diminished importance of this important element and function. However, since ICAO is directing the State CAA to provide oversight of the service provider's ERP, all organizations should expect increased interest in the ERP, particularly during the CAA SMS approval process.

ICAO has taken a very realistic approach to describing aviation safety in terms of an ALoS. This new reality has been clearly communicated to the State CAAs who are wrestling with how to manage the communication process. In our complex industry, the obvious goal is to reduce the accident rate to acceptable levels – preferably zero. Although it plays well in the media and with our passengers that 'Safety is our Number One Priority', safety statistics reveal that the industry is approaching the zero accident rate asymptotically through investment in hardware improvements, software development, employee training and infrastructure investments. The accident rate will never be zero. Safety audits reveal that there are typically three threats to every commercial flight that range from weather, equipment anomalies, fatigue, airport congestion, etc. The potential threat list is long! Should there be a major equipment failure or flight crew judgment error during take-off or landing when the risk is highest, there is always the remote possibility of an accident or major incident. Accidents are 'unscheduled' events and always, repeat, always surprise the management team. Management teams who have not recently rehearsed emergency management procedures easily fall victims to confusion, disorganization, indecision and inaction. Therefore, ICAO has emphasized the need for State CAAs to review and approve the operator's SMS, including the emergency response function.

Most organizational leadership teams understand the value of being prepared for the unthinkable accident and have therefore endorsed and supported a solid emergency response program. Here are some keys to success:

- Senior leadership visibility and support of emergency drills is crucial to responding appropriately to an accident or major incident. If senior leadership is involved in the drill, the leadership team will understand that it is an important part of managing the business and being prepared to manage the emergency situation.

- Emergency management skills are perishable and must be refreshed frequently by participation in structured emergency drills.
- Functional managers may resist efforts to train key staff members and dedicate resources to exercise emergency procedures because of mission requirements.

A Regulator's Perspective of Emergency Response Planning

Transport Canada has published excellent material on emergency response in an SMS advisory circular (Transport Canada Advisory Circular, January 2001, Sec. 10). Many of the following techniques and considerations have been extracted from this comprehensive document.

General

Emergency planning is intended to prepare an organization in the event that an emergency situation occurs. It is preferable to use the terms *emergency response* rather than a common term in our industry of *crisis response*. The term emergency response implies that a serious situation confronts the organization; however, processes and procedures are being implemented to effectively manage the situation. In contrast, the term crisis response implies that the management team is reacting to a situation rather than conducting a planned response to a given situation. Well-designed emergency response plans enable the management team to be proactive, methodical, efficient, timely and targeted. In other words, they are doing the right things at the right time to manage the emergency situation.

The maintenance of plans involves more than just their initial preparation. Larger organizations will have a manager or director of emergency response who will ensure the plan is revised and stress tested through emergency drills that ensure each element of the plan will work under adverse conditions.

What Do We Plan For?

The majority of emergency response planning should focus on how to minimize the effects of an emergency by taking prompt action to accomplish a number of objectives:

- provide protection and support to passengers, crew members, and relatives
- protect private, public and operator property
- ensure the legal rights and obligations of the operator and employees are met

- provide timely and accurate information to the operator's leadership team, government authorities, and the media
- properly investigate each incident or mishap.

Although the list appears to be a relatively simple set of tasks, nearly every commercial accident suggests a breakdown in the management team's ability to accomplish these objectives without major blunders, particularly communication. It is important for large commercial airlines to include the media relations team in the planning process. The best media relations teams have developed detailed emergency response plans, tailored to manage the array of potential accident scenarios. These plans may contain 40–50 checklists that orchestrate the response to a vast array of emergencies. Simply stated, a fully prepared organization will ensure a predictable response through a combination of thoughtful planning and frequent practice.

Exercising the Plan

It is unlikely that all primary leadership positions for emergency response will be immediately available at the emergency response center. Therefore, each organization must identify and train back-up staff that has the authority to act on behalf of the primary representative. Inaction or indecision could easily paralyze the emergency response and delay critical actions. A competent staff of trained reserves will ensure that the operator will not find itself surprised when key emergency response positions are unfilled at the most critical time – immediately after the incident.

A number of organizations have discovered that full emergency response drills on an annual schedule are inadequate to maintain the proficiency of key responders. Familiarity with checklists, communication equipment, specific actions and coordination requirements are perishable skills that must be practiced, much like hand-flying instrument approaches or performing cross-wind landings in a commercial aircraft. It is important that each organizational leader receive both initial and recurrent training to maintain familiarity with procedures. It is not uncommon that senior leaders who have not participated in a rehearsal for several months will require special tutoring during the initial stages of the emergency. Obviously this hand holding will diminish the effectiveness of the entire team at critical times.

Quarterly response drills are recommended by a number of organizations. At first blush, this may seem like overkill and unnecessary. However, if a large organization has 2–4 back-up division leaders for each position in the emergency response center, quarterly exercises enable each manager to participate only once or twice each year – minimal training, at best. The frequency of these drills is based on a risk-assessment decision by the accountable executive. A small organization may predetermine that an accident will merely put them out of business and the company will close its doors. Therefore, emergency response training is perceived to be of little value. In contrast, larger organizations will be 'encouraged' by their insurance companies to implement a safety management system with a robust emergency response plan. Don't be surprised if your insurance broker asks to observe one of your emergency response drills and review your plans.

Element 1.5: Documentation of the SMS

Documentation of the SMS is but one of the documentation requirements of any organization. Although ICAO refers to an SMS manual, most organizations cascade safety management information through the documentation management system to ensure safety information is easily available to each employee. Each employee group requires a different level of safety information and this tailored information is provided in manuals that employees utilize in the routine accomplishment of their work activities. For example, safety reporting procedures are described in a flight operations manual (FOM) or general operations manual (GOM) available to all employees.

Large organizations will establish an organized documentation management system that manages manuals, procedures and policy statements into three levels as depicted in Figure 6.7.

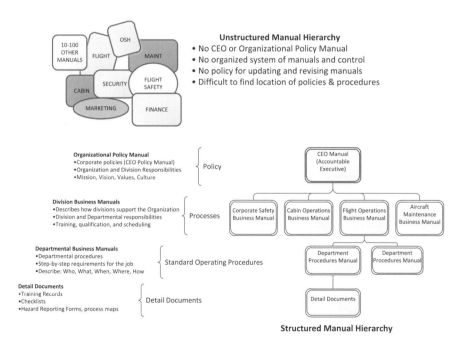

Figure 6.7   Levels of documentation

ICAO has established the requirement to document the structure of the SMS as well as the SMS implementation plan, but again, without specific guidance. To fill this gap, expect the State Civil Aviation Authority to publish specific documentation requirements in the State Safety Program (SSP).

Component #2: Safety Risk Management

Introduction

Operational risk management (ORM) must be differentiated from safety risk management (SRM). Taken holistically, ORM includes the hazards and threats that are surfaced through data collection methods in safety, quality, security, and environmental management systems. For example, a SeMS has embedded risk-management processes that identify and mitigate risks within the discipline of security. Similarly, there are risks associated in the quality system. To illustrate, a set of inadequate tools, improper hangar lights, or cold weather represent hazards to the maintenance technician who is performing engine or airframe maintenance. ORM requires that each of these disciplines be integrated when senior leadership desires an assessment of the holistic, operational risk facing the organization. In contrast, an SMS focuses primarily on the unique risks associated in the discipline of safety.

ICAO describes safety risk management as encompassing two distinct activities or elements: hazard identification and safety risk assessment and mitigation.

- Component 1: safety policy and objectives
    - Element 1.1   Management commitment and responsibility
    - Element 1.2   Safety accountabilities
    - Element 1.3   Appointment of key safety personnel
    - Element 1.4   Coordination of ERP
    - Element 1.5   SMS documentation
- Component 2: safety risk management
    - Element 2.1   Hazard identification
    - Element 2.2   Risk assessment and mitigation
- Component 3: safety assurance
    - Element 3.1   Safety performance monitoring and measurement
    - Element 3.2   The management of change
    - Element 3.3   Continual improvement of the SMS
- Component 4: safety promotion
    - Element 4.1   Training and education
    - Element 4.2   Safety communication

Concepts and Definitions

Safety risk-management processes 'ensure safety risks of the consequences of hazards (threats) in critical activities related to the provision of services are controlled to a level as low as reasonably practicable (ALARP)' (ICAO, 2009, 9.2).

The Australian Civil Aviation Safety Authority (CASA) has described ALARP as meaning 'a risk is low enough that attempting to make it lower, or the cost of assessing the improvement gained in an attempted risk reduction, would actually be more costly than any cost likely to come from the risk itself'. The risk can be said to be ALARP when it can be demonstrated that all justifiable risk reduction measures have been considered and the remaining mitigation strategies cannot be justified. To those of us who have been in the safety world for years this type of thinking is refreshing, but will require the industry to refine existing processes to fully implement this concept, that is, reengineer tools that will quantify justifiable residual risk. Perhaps Figure 6.8 will clarify the ALARP principle (ICAO, 2009):

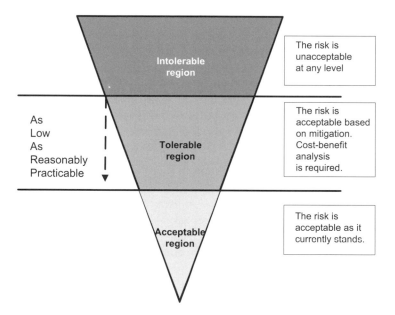

Figure 6.8   ALARP

Transport Canada provides some interesting insight into the SRM process.

Safety oversight or monitoring provides the information required to make an informed judgment on the management of risk in your organization. Additionally, it provides a mechanism for an organization to critically review its existing operations, proposed operational changes... for their safety significance. (Transport Canada Advisory Circular, January 2001, 6.0)

Figure 6.9 depicts that hazard identification occurs primarily through two basic sources: reactive and proactive processes. The reactive process responds to events that have already occurred as reported through flight data analysis, employee reporting, or incident investigation processes. In contrast, proactive processes include on-site audits, formal hazard identification teams and safety surveys. Proactive processes look into the future at what might happen instead of assessing historical events that will likely occur again unless mitigation strategies are implemented.

The most logical place to begin work in risk management is data collection. Actually, data collection has been fundamental to safety systems for years. Do you remember in the mid-1980s the phrase commonly uttered by safety practitioners, 'We are data rich and information poor?' It was clear recognition that aviation safety practitioners were obtaining data but were woefully deficient in distilling data into risk-based information. Reasons for this varied from inadequate staffing, improper

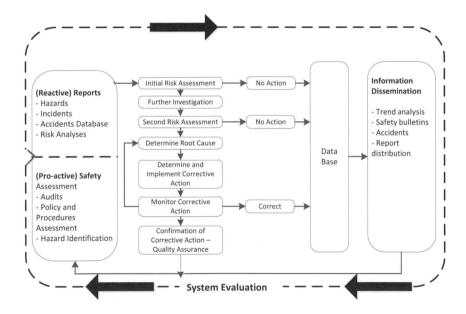

Figure 6.9   Transport Canada SMS process flow

safety training, or lack of analytical tools. Much of the work of the Global Aviation Information Network (GAIN) was focused on developing tools and processes to analyze mountains of data and gleaning nuggets of insight into the most significant safety issues. This process approach is exemplified in Transport Canada's basic process flow chart (see Figure 6.9) that begins with hazard identification. The process then captures the SRM fundamentals of risk assessment, root cause analysis, development of corrective action, implementation of these actions and then continuous monitoring to ensure mitigations were effective. Every safety services department must develop a flow chart that depicts how the organization will track a specific safety issue from discovery to resolution.

What is a Hazard?

A *hazard* is a condition or an object with the potential to cause injuries to personnel, damage to equipment or structures, loss of material, or reduction of ability to perform a prescribed function (ICAO, 2009, 4.2). As a preference, many safety practitioners have inserted the term 'situation' into the definition for clarification, that is, a hazard is a condition, *situation*, or an object with the potential to cause injuries to personnel, damage to equipment or structures.

ICAO groups hazards into three generic families: natural, technical and economic hazards. Operational safety management activities will focus on natural and technical hazards. Economic hazards will be captured and analyzed during the ERM process.

- Natural hazards are a consequence of the environment within which operations take place (for example, in-flight, ground handling or maintenance on an exposed airport tarmac):
  - Severe weather (for example, hurricanes, thunderstorms, wind shear and lightening)
  - Adverse weather (for example, icing, heavy rain and snow)
  - Geophysical events (for example, earthquakes, volcanoes and floods)
  - Environmental (for example, wildlife such as birds, and wildfires)
  - Public health events (for example, epidemics)
- Technical hazards are a result of malfunctions of equipment, software, or sources of energy
  - Aircraft systems
  - Organization's facilities, tools and related equipment
  - Systems and equipment external to the organization, that is, supplier
- Economic hazards are the consequence of the global sociopolitical-economic world in which we exist and operate
  - Global GDP (for example, transition to an integrated global economy)
  - Recession (for example, 2008 global financial crisis)
  - Costs of production (for example, material, equipment, people)

Element 3.1: Hazard Identification

Employee Reports

Aviation organizations are exposed to operational hazards as a normal course of business activity. SRM begins with the identification of these hazards through a formal process of collecting, recording, analyzing, summarizing and providing feedback to operational managers in terms of safety risk. There are four primary sources of safety data available to an organization – external, corporate, functional and individual source:

- *External source.* As a heavily regulated industry, aviation service providers are in continuous contact with regulatory authorities who are conducting routine government oversight to ensure regulatory compliance. In the United States, large organizations will have frequent interface with the Environmental Protection Agency (EPA), Occupational Safety and Health Administration (OSHA), Transportation Security Administration (TSA), and others. For all service providers, feedback from the State's CAA is a valuable source of information regarding the level of regulatory compliance and hazard identification. As State CAAs implement SSPs, there will be a gradual evolution from a compliance-based oversight approach performed primarily through on-site inspections and observations to a performance-based approach that will focus more on proactive hazard identification. Several large operators have used the 'eyes' of the CAA inspectors as an extension of their own staff to collect additional safety data as to the health of their operational functions. This external data is then used to augment data obtained from internal sources, thus creating a more robust database.
- *Corporate or organizational source.* The safety services department typically conducts an independent analysis of all available safety data on behalf of the accountable executive. Hazards or threats to the organization may be identified by the safety, quality, security, or perhaps environmental service departments as they review employee reports, audit results and flight data. It is important that the entire organization use the same risk analysis tools and risk vocabulary to ensure alignment.
- *Functional source.* Process owners are responsible to ensure that departmental activities and operational procedures are planned, trained and executed in a manner to meet organizational goals. Although independent audits are very effective in identifying hazards in the workplace, these audits are typically conducted annually or biennially. New hazards may appear daily and the process owner or line manager supervising the activity is responsible for identifying hazards as they appear. Ensuring an efficient and safe operation is a management responsibility, particularly at the level where the work is taking place.

•   *Individual source.* Workplace hazards exist in every environment and include latent process deficiencies, procedural ambiguities, inadequate training and non-standardized work streams. These hazards are evident on the manufacturing floor, cockpit, maintenance hangar, cabin, air traffic control facility, or airport ramp. This author prefers that each employee think primarily in terms of Threat and Error Management (TEM) while the management team thinks in terms of Hazard Identification and Mitigation. The TEM model is most useful at the functional level while people are performing their duties. Employees find it easy to understand the term *threat* as it relates to their personal safety or ability to safely discharge their duties. As such, threats and personal errors exist in the workplace and must be recognized by each employee as they occur. Examples include freezing, slippery conditions on the ramp, inadequate maintenance tools, inoperative cockpit equipment, inadequate personal protective gear, or schedule pressures. Upon recognition of a threat or personal error, the employee must manage the situation as a threat manager. It is management's responsibility to remove as many of these threats (hazards) as possible; however, it is the employee's responsibility to report these threats (or hazards) for resolution. As threat managers, employees are the last line of defense to keep threats from impacting flight operations (Maurino, 2005).

Additional sources of hazard identification are available to the safety practitioner but may require some personal research:

•   regional safety organizations
•   regulatory advisory bulletins or directives
•   vendor feedback
•   industry safety data sharing organizations
•   industry accident and incident investigation results.

How to Destroy a Safety Reporting Culture?

A discussion of safety reporting would be incomplete without reminding ourselves of the three ways to destroy a safety reporting culture and eliminate or reduce a valuable source of safety data:

•   *Punitive action.* Management responds to safety reports by punitive action or embarrassing the employee. If an employee is to report mistakes and errors, they must feel *safe* to report without jeopardizing their employment. Management must recognize that all employees make mistakes and errors. The only way to make the operational systems safer is to welcome these reports as opportunities for improvement. Punitive action by the management team may be a red flag that reflects a poor safety culture, that is, a culture more interested in taking the easy way out by dismissing an employee rather than taking responsibility for poor system design, procedural structure, or training.

- *No action.* Management responds by taking no action on a significant safety issue. Employees quickly lose faith in a safety reporting system when they observe that no one cares enough to either follow up or resolve their safety concerns. Following report submission, employees are looking for feedback – either an immediate fix to the problem or notification that management is reviewing the issue for resolution.
- *No feedback.* Management ignores the employee report and does not acknowledge receipt. The most important thing a management team can do is to thank the employee for reporting a safety concern and communicating the resolution. This approach builds trust and confidence in the management team and contributes greatly to enhancing the organization's safety culture.

The collection of safety data serves no useful purpose unless there is an effort to proactively identify hazards in terms of risk (frequency and severity) and then prioritize resources for implementation of mitigation actions. Reports of frequent mistakes or errors in a specific work activity suggest improperly designed or trained procedures. An effective SMS contains processes that enable an opportunity to periodically review operational procedures and training programs to ensure they are meeting the needs of the organization.

Accident/Incident Investigation (Reactive Safety Management)

Accident and incident investigation is an important safety program function but is reactive in nature. An accident is a symptom of a system failure at some level(s) in the organization. The purpose of an investigation is simply to understand what happened, how it happened and how to prevent the event from reoccurring. Major accident investigations will obviously be managed by the State accident investigation entity and the affected service provider will participate in the investigative process according to State protocols. However, the same techniques used for a major accident investigation are very effective in internal investigations of significant operational incidents.

The most effective internal investigations occur in an atmosphere of partnership between the corporate safety teams and employee safety representatives. Several airlines have created safety partnerships by drafting a memorandum of understanding (MOU) between the safety services departments and respective employee unions. These MOUs describe investigative processes that define how employees will be interviewed and by whom, how data will be captured and retained, and how the management team will be briefed on investigative results. The following essential elements should be included in an effective organizational accident/incident investigation program (IATA, 2010b, Sec. 3.1):

- Fact finding and data collection:
  - Employee interviews
  - Flight data analysis (digital flight data recorder or cockpit voice recorder)
  - Obtain failed hardware (aircraft parts, etc.) for analysis
  - Conduct a document and procedural review
  - Technical expert interviews (internal and external)
  - Manufacturer participation
  - Ground handling (ramp) video
- Data analysis
  - Assemble SMEs to evaluate the facts
  - Attain consensus on the facts, sequence of events and system deficiencies
  - Ensure that any ambiguities in data are resolved and understood
- Written summary
  - Carefully summarize factual information and analytical methods/ results
  - Explain any gaps in data to enhance the credibility of the report
- Findings, conclusions and recommendations
  - Summarize sequence of events and results of root cause analysis
  - Identify and assign specific corrective actions within the organization to prevent reoccurrence
- Senior management review
  - Provide senior management with a thorough review of the facts and corrective action plan (CAP)
  - Assign and track implementation of corrective actions through a tailored corrective action record or request (CAR)
  - Conduct periodic follow-up assessments to ensure the effectiveness and sustainability of the corrective actions

So what is the value of a reactive incident investigation as it relates to our SMS goal of proactive hazard identification? A nonthreatening employee debriefing conducted in an atmosphere of a classroom seminar will usually yield many additional gems of information relating to procedural and training deficiencies. For example, as a flight or ground crew describes an event chronologically by reviewing their thought and decision processes, it becomes obvious when operational procedures did not work as envisioned by the manufacturer or the operator. It is through these thoughtful conversations with an incident crew that Jim Reason's 'holes in the layers of cheese' become apparent and can be addressed in the investigation's list of recommendations.

As reviewed in previous chapters, ORM also includes the disciplines of safety, quality, security and environment. Crew members are touched by these disciplines on each and every flight. For example, as a flight crew conducts a preflight walk-around aircraft inspection, they may discover a fuel spill or hydraulic leak that leads to activation of an environmental clean-up process. Similarly, as a crew member enters the ramp area, there are a number of security procedures that must be adhered to in terms of ramp access, cargo security and baggage handling. Quality is always an issue in terms of following aircraft maintenance and servicing procedures. ORM may therefore be described as a collage of these activities that are overlapping and integrated.

Let's review an actual real-world example of how the investigation included the disciplines of safety, quality, security and environment. The following event was triggered by a maintenance error during a routine engine overhaul. Once the engine was installed on the airplane, it operated normally for several flights. Ultimately, the maintenance error led to a major fuel leak and an in-flight engine shutdown in the South Pacific at night. Concerned that fuel was still being vented overboard, the crew elected to divert to a remote island airport. Immediately following the safe landing, safety and quality managers began work in partnership to investigate the event and to monitor the repair activity by contract maintenance. While contract maintenance troubleshooting was in progress, the flight crew encountered significant passenger and ramp security issues. The corporate security manager now became a key player in managing the situation, working closely with both the flight crew and ground handling service provider. Additional quality and safety issues surfaced as the contract maintenance staff installed an incorrect part on the engine, thus triggering a massive fuel leak during an engine test run. Fortunately, there was no fire. At this point, the environmental manager became actively involved – first, the crew elected to dump 40,000 pounds of fuel prior to the emergency landing; and second, there was a significant fuel spill on the ramp during the engine test run requiring clean-up.

This situation reminds us that an international divert into a remote airport at night should always be viewed as a high-risk event and as an opportunity to evaluate all operational processes to include divert procedures, emergency checklist structure, crew training, maintenance procedures, etc. The lessons learned and associated opportunities for improvement is the essence of safety risk management.

Flight Data Analysis

Flight data analysis (FDA) is one of the most powerful tools to monitor the health of an operator's daily flight operation. FDA is also known as flight data monitoring (FDM) and Flight Operations Quality Assurance (FOQA).[8] For those

---

8 This author prefers the term flight data analysis (FDA) as it is a more accurate description of the intent and purpose of the analytical process. Unfortunately, the use of the term flight operations quality assurance (FOQA) may be confusing and annoying to some safety practitioners as it is not a quality assurance (QA) program in the traditional sense of the word. In practice, FOQA functions as a quality control (QC) tool to validate the health of flight operations or to identify impending failures of aircraft systems such as hydraulic or electrical systems.

unfamiliar with the FDA concept, today's electronic airplanes produce streams of data from cockpit displays, aircraft systems, engines and flight control inputs that are not always recorded by the crash-proof digital flight data recorder (DFDR). However, operational flight data may be routinely downloaded from an aircraft Quick Access Recorder (QAR) by means of disk exchange, direct download onto a transfer device, or automatic downloads via a wireless network upon arrival at selected locations. Timely downloads of data enables safety teams to analyze the data for undesirable trends and isolate specific events that deserve further investigation. Data is fed through analytical software to identify deviations from accepted performance that may be isolated to a specific fleet, airport, approach procedure, or flight crew. De-identified data from one carrier may be merged with data from other carriers to assess the risks associated with operations from a mountainous airport.

FDA is considered both reactive and proactive. It is reactive in that the data captures a deviation from SOPs that happened in the past, that is, sometime prior to the data download. Similarly, FDA can be proactive in that trend information may show a shift in operational parameters (such as long landings) that may be a leading indicator of heavy brake wear or a runway excursion. FDA programs may also be designed to provide additional training for crew members who have experienced a near miss by operating the airplane at the limit of its performance capabilities in an unsafe manner. Most operators have procedures in place enabling contact with the flight crew to obtain more information, primarily the circumstances contributing to the deviation. Without crew input, the safety practitioner will only know the magnitude of the deviation through data analysis and will not understand how the deviation developed. Through integration of FDA and flight crew reporting programs, safety specialists will understand both the quantitative and qualitative characteristics of a single event.

In many airline and business aviation cultures, the protection and use of flight data is a sensitive issue with flight crews; however, in nearly all cases, stakeholders have recognized the value of this data and collaboratively developed processes to manage data with a focus on safety rather than punitive action. In the United States, de-identification of flight data is particularly important to flight crew representative bodies and an effective FDA program must be founded on a bond of trust between the operator, its crew members and the regulatory authority. In contrast, some States have made FDA mandatory, thus circumventing the sometimes contentious discussions between labor associations and management teams. Regardless of the organization's culture, an FDA program provides accurate data into the risk-management process.

Audits of Operational Functions (Proactive Safety Management)

ICAO describes the audit process as primarily a means to verify the safety performance of the organization and to validate the effectiveness of safety risk controls. Practical experience suggests that on-site audit activity offers much more to the operator. Audits conducted by an independent and fresh set of eyes routinely identifies a number of safety hazards, non-compliance with regulatory requirements, non-conformance to established organizational procedures, latent system deficiencies and opportunities for improvement. These issues often go overlooked by line supervisors. Operational managers may be blind to these safety issues because they have grown accustomed or desensitized to them by overlooking changed work behaviors that have insidiously evolved to counter or circumvent the hazard. In extreme cases, auditors may find it difficult to convince a line supervisor that a hazard actually exists. It will be helpful to collect tangible evidence of the hazard by the use of pictures or employee reports. All too often, this recognition occurs only after a serious incident or accident has occurred.

Many auditing departments have expanded their checklists to include the identification of operational hazards, significant safety issues and opportunities for improvement. The new ICAO SMS standard and FAA SMS framework identifies two levels of audits – internal (system) evaluation and internal (functional) audits. The objective of internal evaluation is to assess operational processes and is conducted by the corporate-level auditing organization – typically the quality assurance group. In contrast, line management conducts internal self audits to ensure that the production system is meeting established standards and expectations. The traditional term used to describe this level of auditing is *quality control*. The good news is that operational hazards will be identified through either type of audit, captured in audit reports and followed up using corrective action requests. The identification of hazards in audit summaries must be shared with the safety staff to ensure a holistic view of operational risks and included in the risk-management process.

The process of auditing will be discussed in more detail in the section on safety assurance.

Existing versus Predicted Hazards

Operational hazards may also be categorized as *existing* or *predicted* hazards. Existing hazards are sand traps that exist in today's operation. Predicted hazards are those that will likely be introduced into the operation as a result of a major operational change (see Figure 6.10).

ICAO and State regulations describe the safety processes used to mitigate existing hazards identified through employee reporting, operational audits, flight data analysis and incident investigation. These activities may be tailored to the size and scope of the organization. One word of caution: the organization must identify someone to summarize and integrate the output of these independent, continuous loop safety processes so as to capture the holistic operational risk facing the organization.

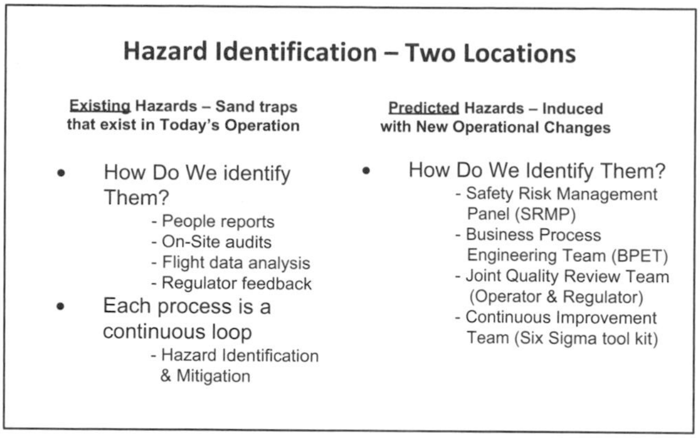

```
┌─────────────────────────────────────────────────────────────┐
│                                                               │
│            Hazard Identification – Two Locations              │
│                                                               │
│   Existing Hazards – Sand traps      Predicted Hazards – Induced │
│   that exist in Today's Operation    with New Operational Changes │
│                                                               │
│   •  How Do We identify          •  How Do We Identify Them?   │
│      Them?                          - Safety Risk Management   │
│          - People reports             Panel (SRMP)            │
│          - On-Site audits           - Business Process        │
│          - Flight data analysis       Engineering Team (BPET) │
│          - Regulator feedback       - Joint Quality Review Team│
│   •  Each process is a                (Operator & Regulator)  │
│      continuous loop                - Continuous Improvement  │
│          - Hazard Identification      Team (Six Sigma tool kit)│
│            & Mitigation                                       │
│                                                               │
└─────────────────────────────────────────────────────────────┘
```

Figure 6.10   Hazard identification

In contrast, predicted hazards are those that a team of subject matter experts and stakeholders believe will be introduced into an operation as a result of a major process or procedural change. Examples include implementation of a new fleet type such as the B787, initiation of passenger service into a new geographical region, transition from old to new computer platforms, or initiating the outsourcing of aircraft maintenance activities. The focus of these discussions is to identify 'what possibly could go wrong' and then develop mitigations to treat the predicted risk associated with each new hazard introduced into the system.

Figure 6.10 introduces several proactive techniques or processes to predict hazards associated with major changes to an existing operation. These will be explained in detail in the chapter on change management and continual improvement.

Element 3.2: Risk Assessment and Mitigation

Safety risk assessment is an analysis of the consequences of the hazards that could threaten the capabilities of an organization. The analysis process breaks down risk into two components – the *probability* of the occurrence of a damaging event or condition, and *severity* of the event or condition, should it occur. A number of matrixes exist in the aviation community such as a 4 × 4 or a 5 × 5 matrix. ICAO recommends that each service organization create its own risk matrix based on the nature and complexity of its operation. However, ICAO also requires the State regulatory authority to agree or approve the service provider's matrix, a process that will be defined in State civil aviation regulations. It is important that your regulatory authority understand the structure of your risk matrix and the definitions of *severity* and *probability*. Figure 6.11 shows the 5 × 5 matrix that may be modified to suit your organization's risk appetite. For example, if the accountable executive is a very conservative individual who is risk averse, the matrix may be modified by changing blocks 2A, 3B, 4C, and 5D to red in color. Similarly, if your organization has a higher risk appetite, blocks 2C, 3D, and 4E may be changed to green in color.

As a reference point, ICAO has provided general definitions of severity (A–E) and probability (1–5). These definitions are summarized in Tables 6.1 and 6.2. Most safety practitioners struggle with how best to use this matrix and have experienced significant differences of opinion as to the assessment of probability and frequency, particularly when briefing an organization's senior leadership team.

Risk Probability	Risk Severity				
	Catastrophic A	Hazardous B	Major C	Minor D	Negligible E
Frequent 5	5A	5B	5C	5D	5E
Occasional 4	4A	4B	4C	4D	4E
Remote 3	3A	3B	3C	3D	3E
Improbable 2	2A	2B	2C	2D	2E
Extremely Impossible 1	1A	1B	1C	1D	1E

Figure 6.11   ICAO Risk Assessment Matrix

Table 6.1   ICAO definitions of probability

Probability	Meaning	Value
Frequent	Likely to occur many times (has occurred frequently)	5
Occasional	Likely to occur sometimes (has occurred infrequently)	4
Remote	Unlikely to occur, but possible (has occurred rarely)	3
Improbable	Very unlikely to occur (not known to have occurred)	2
Extremely improbable	Almost inconceivable that the event will occur	1

Table 6.2   ICAO definitions of severity

Severity	Meaning	Value
Catastrophic	Equipment destroyed	A
	Multiple deaths	
Hazardous	A large reduction in safety margins, physical distress or a workload such that the operators cannot be relied upon to perform their tasks accurately or completely	B
	Serious injury	
	Major equipment damage	
Major	A significant reduction in safety margins, a reduction in the ability of the operators to cope with adverse operating conditions as a result of increased in workload, or as a result of conditions impairing their efficiency	C
	Serious incident	
	Injury to persons	
Severity	**Meaning**	**Value**
Minor	Nuisance	D
	Operating limitations	
	Use of emergency procedures	
	Minor incident	
Negligible	Almost inconceivable that the event will occur	E

A service provider's risk appetite is essentially a business decision that involves allocation of resources to production goals (delivery of services) or protection goals (safety). Conceptually, the management of safety is another organizational process similar to quality or security management. These goals should not be in competition for resources, but should be balanced to ensure the organization is protected during the production of services. For example, most service providers establish annual performance goals that include: financial (shareholder return), employee satisfaction, production, service quality (customer satisfaction), product quality (for example engine shutdown rate), safety (runway incursions, lost time injuries), etc. The budgeting process allocates resources to individual division managers so as to achieve these performance goals. Again, safety and quality are not in competition with other goals – they must be managed as an integrated effort.

Let's look at a typical example of proactive planning to ensure a balance between production and protection. Let's say that your organization was to begin service into a high altitude airport in mountainous terrain. The airport has had a history of unreliable instrument approach systems and struggles to meet international airport standards. Production may be defined as providing customers with reliable and on-time service to the new destination, thus requiring an investment in facilities, personnel, business lounges, maintenance contracts, etc. The issue is how much should be invested? Protection may be defined as the level of support infrastructure and system redundancy to support the operation, such as the amount of additional training provided to flight crews to deal with this special airport, and back-up processes in the event of computer failure, aircraft accident, or significant maintenance issue. Again, the issue is how much should be invested? A company that truly values its reputation and position in the aviation industry may have a low risk appetite and therefore make the necessary investment in facilities and infrastructure to ensure adequate support is available to handle infrequent problems. Additionally, this management team is also willing to make the necessary investment in protection (for example training) to effectively support the new service into a relatively high risk airport.

The risk management discussion revolves around the concept of ALARP. The ICAO Risk Tolerability Matrix in Figure 6.12 may be of value in communicating this concept to organizational leaders and employees.

Suggested Criteria	Assessment Risk Index	Suggested Criteria
Intolerable Region	5A, 5B, 5C, 4A, 4B,3A	Unacceptable under the existing circumstances
Tolerable Region / Acceptable Region	5D, 5E, 4C, 4D, 4E, 3B, 3C, 3D, 2A, 2B, 2C	Acceptable based on risk mitigation. It may require management decision
	3E, 2D, 2E, 1A, 1B, 1C, 1D, 1E	Acceptable

Figure 6.12   ICAO Risk Tolerability Matrix

A Deeper Look into Probability and Severity

Perhaps our discussion of probability and severity as described by ICAO will meet your needs. If so, that's great. However, many safety practitioners strongly desire a single set of definitions of probability and severity that can be used by everyone in the organization including functional managers, division safety representatives and employee associations. You may also discover that the definitions from both ICAO and your State CAA are very general in nature, primarily to ensure flexibility for the service provider. My personal opinion is that each organization must develop more specific definitions to increase the intuitive value of the tool.

To that end, commercial airlines have developed more specific definitions of severity and frequency that are tied to the fundamental concept of achieving a balance between production and protection activities (as defined by ICAO). Further, this type of matrix harmonizes the enterprise risk severity codes for safety, quality, security and environmental disciplines, thus enabling effective integration of risk-management processes.

Development of this type of matrix merely requires a task team to convene and work through a number of operational hazards and develop a suitable set of definitions. The list of definitions may vary greatly from organization to organization depending upon culture and risk appetite. The accountable executive should have final sign-off or approval of the severity matrix. The expanded severity matrix in Table 6.3 may be modified for use by any organization, regardless of size or complexity.

Table 6.3   Expanded severity matrix

Severity of occurrences	Disciplines (all, SMS, SeMS, QMS, or EMS	Value
Catastrophic	**Protection:** SMS, SeMS: equipment destroyed SMS, SeMS: multiple deaths – employee or passenger SMS, SeMS: system-wide shutdown and negative revenue impact All: potential civil penalties (example: >US$1 million/criminal penalties) EMS: large environmental impact (example: >US$2 million/civil penalties) All: loss (or breakdown) of an entire system or subsystem SeMS: security criminal investigations and penalties to groups or individuals All: willful violation of any safety regulation that could result in serious injury or death **Production:** All: severe enterprise risk impact QMS: potential of suspending airline operation (temporary or permanent) All: potential for uncontrollable public relations event(s)	A
Critical	**Protection:** SMS: a large reduction in safety margins, physical distress and/or workload such that operators cannot be relied upon to perform their task accurately or completely SMS, SeMS, EMS: serious injury or death, multiple injuries and personal insurance claims SMS, SeMS: accident or serious incident with injuries and/or major to moderate equipment damage All: civil penalty potential (example: <US$1 million>US$100,000) All: potential criminal penalty EMS: medium environmental impact (example <US$1 million>US$100,000) SMS, QMS: potential moderate damage to an aircraft (out of service ≥5 days)	B

Table 6.3 *Continued*

Severity of occurrences	Disciplines (all, SMS, SeMS, QMS, or EMS	Value
	SMS, EMS: hazardous spill into airplane (cargo compartment or cabin)	
	All: a non-compliance audit finding resulting in major system, process, or operational degradation	
	SeMS: a security audit finding requiring immediate corrective action prior to continued operation	
	SMS: reoccurring violation of any safety regulation resulting in serious injury	
	Production:	
	SMS: employee/customer injury/broken bone. Injury resulting in hospitalization (other than observation)	
	All: moderate enterprise risk where executive management was involved	
	All: very large public relations impact requiring resources to manage information	
	All: system deficiencies leading to poor air carrier performance and chronic schedule disruption	
	All: potential loss (breakdown) of entire subsystem or divisional operation	
	QMS: production errors containing regulatory violations that pose direct consequence to the operation	
	All: shutdown of large operation or hub	
Moderate	**Protection:**	C
	SMS: accident or incident with minor injury and/or minor aircraft damage	
	SMS, SeMS, EMS: non-life threatening employee/customer injury, with recording of LTI	
	All: civil penalty potential (example: <US$100,000)	
	EMS: small environmental impact	
	SeMS: security audit finding requiring a CAP	
	Production:	
	QMS: production element errors that may pose indirect consequences to the operation	
	All: aircraft damage resulting in out of service <5 days	
	QMS: potential to cause sustained irregular operations until issue is resolved	
	All: complete shutdown of small operations or divisions	
	All: system deficiencies leading to poor air carrier performance and/or sporadic disruption to the schedule	
	All: partial breakdown of an air carrier system	
	All: additional public relations efforts and resources required	
Minor	**Protection:**	D
	All: no regulatory action anticipated	
	EMS: no environmental impact anticipated	
	SeMS: no evident security threat affected	
	Production:	
	All: minor errors in completed company policy and procedures	
	QMS: production errors suggest quality system and/or opportunities for improvement	

Table 6.3   *Concluded*

Severity of occurrences	Disciplines (all, SMS, SeMS, QMS, or EMS	Value
	SMS, QMS: no equipment damage to slight damage – outcome deferrable with no operational impact All: US$0 regulatory civil penalty All: no public relations impact All: none to slight aircraft damage – no operational impact	
Negligible	**Protection:** All: no regulatory violation EMS: no environmental impact SeMS: no security element affected **Production:** QMS: continuous improvement (CI) initiative delivered against company policy and procedures All: US$0 regulatory civil penalty All: no public relations impact All: no aircraft damage – no operational impact All: audit findings and observations present limited opportunities for improvement	E

SRM Process Description

Thus far, we have reviewed common sources of safety data and how to classify hazards in terms of probability and severity. Let's now explore the sequence of SRM activities. As an example, the FAA Air Traffic Organization (ATO) has implemented an effective SRM process that can be tailored to any organization and follows the five step process described in FAA SMS framework documents (See Figure 6.13; FAA Air Traffic Organization, 2008, p. 26).

This flowchart requires some study in order to fully understand and appreciate the activities expected for each of the five elements. The process is entered at the *Describe system* element for a new system being designed or prior to implementing a significant change to the operation. This work is predictive in nature because the process focuses on identifying potential hazards and developing mitigations *prior* to implementation of the system or system/procedural change. However, for an existing operation where there are no major changes to processes or procedures, safety data is available from incident investigations, audit reports and employee safety reporting systems. Making changes to a process following a major incident is clearly important work, but reactive in the sense that the event has already happened. Responding to an audit finding that captures a regulatory violation, deviation from established procedures, or identified an operational hazard is proactive work. For hazards identified in a stable operation, enter the process at the *Identify hazards* element.

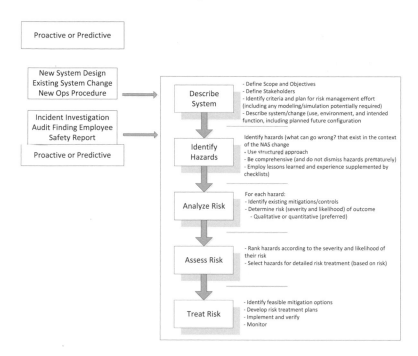

Figure 6.13   SRM analysis process

Risk Analysis vs. Risk Assessment

One of the confusing features of this five step model is the differentiation of *risk analysis* and *risk assessment*. Is there really a functional difference between the two steps or can they be combined effectively because the differences are really semantic and confusing to the management team? ICAO and the FAA do not clearly define the differences between the terms *analysis* and *assessment*. ICAO's approach simply involves a three step analytical process:

- Step 1: complete an overall risk assessment of the safety risk using a safety risk assessment matrix, that is, probability and severity. This risk assessment code may be referred to as the safety risk index or risk assessment code
- Step 2: export the safety risk index obtained in step one to a safety risk tolerability matrix. Is the risk acceptable, tolerable, or intolerable? See Figure 6.12.
- Step 3: develop mitigation strategies to bring the unsafe event (reactive) or condition (proactive) under control.

Transport Canada has perhaps offered the most comprehensive definitions of risk analysis and risk assessment:

- *Risk analysis* is the first element in the risk-management process. It encompasses risk identification and risk estimation. Once a hazard has been identified, the risks associated with the hazard must be identified and the amount of risk estimated.
- *Risk assessment* takes the work completed during the risk analysis and goes one step further by conducting a risk evaluation. Here the probability and severity of the hazard are assessed to determine the level of risk. The risk-assessment matrix is used to determine the level of risk.

In practice, safety practitioners will not waste time and effort to explain to the management team the minimal differences between risk analysis and risk assessment. This author suggests keeping the process simple and straight forward by combining the two steps into a single risk assessment step. Risk assessment should therefore include (1) calculating the safety risk index in terms of probability and severity, and (2) determining if the risk is acceptable or requires mitigation. If the safety issue requires mitigation, the safety practitioner or functional manager moves to the next step of developing mitigation strategies to treat the risk.

Practical Application

The next question is 'How do I perform each of the five steps? Figure 6.14, courtesy of FAA ATO, describes a straightforward process that is based on answering several fundamental questions:

- what can go wrong?
- how big is the risk?
- how can you reduce the risk?
- does the program include mitigation?

The 1E risk index in Figurer 6.14 requires some additional explanation. The FAA ATO considers hazards with catastrophic effects (even though extremely improbable) as high risk (red) if they are caused by:

- single point events or failures
- common cause events or failures
- undetectable latent events in combination with single point or common cause events

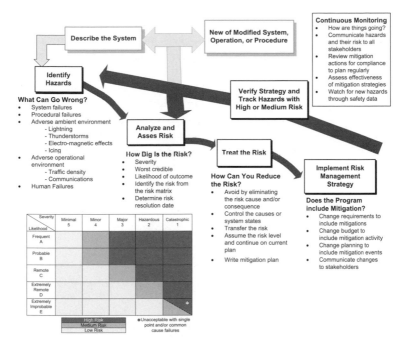

Figure 6.14   How to accomplish a safety analysis

The terms *single point* and *common cause* failures are probably already in your vocabulary, but a review of these definitions may be of value in your assessment of risk (FAA ATO, 2008, p. 43).

- *Single point failure* is defined as a failure of an item that would result in the failure of the system and is not compensated for by redundancy or an alternative operational procedure. In the world of computer-managed air traffic control, an example of a single point failure is a system with redundant hardware in which both pieces of hardware rely on the same battery for power. In this case, if the battery fails, the system will fail.
- *Common cause failure* is defined as a single fault resulting in the corresponding failure of multiple components. An example of a common cause failure is redundant computers running on the same software, which is susceptible to the same software coding errors.

Risk Mitigation

Risk is inherent in nearly all aviation-related activities and the amount of risk with a given activity will vary from day to day as work conditions change. Risk mitigation is nothing more than taking action to manage the potential effects of the operational hazard. Risk mitigation strategies may cross organizational boundaries and therefore must be carefully coordinated among operational managers. This work requires some level of documentation as a communication tool to stakeholders (anyone who is affected or partially accountable for the outcome of a work-related activity) to capture, explain and document the results of the entire risk-management process dealing with a specific hazard. Some organizations call this document the Risk Management Plan and may include mitigation strategies to address multiple risks. The risk management plan is a living document and should be periodically reviewed to avoid the analysis becoming stale by not capturing the changing hazards in each operational environment.

Reactive versus Proactive Risk Mitigation

Risk mitigation activity can occur in one of two situations – reactive risk mitigation (RM) following an accident or proactive RM in advance of a major change to the operation. Reactive risk mitigation is a significant part of the accident or incident investigation process. Remember, a major incident is a symptom of a system breakdown and the investigative process is reactive to an event that is now historical. During the investigative process, the safety investigator and organizational managers will collaboratively develop an action plan to prevent a similar event from reoccurring. This corrective action plan is a form of RM and will contain specific actions that will require management approval, funding and implementation. This risk-assessment process, conducted collaboratively by organizational safety practitioners and functional managers, will capture the predicted residual risk once the mitigation strategy is developed. Stated differently, 'What will be the probability and severity of this event happening again once we implement corrective action?' This residual risk must be accepted by all stakeholders and should be captured in a RM plan.

Proactive RM occurs as part of the planning activity associated with a major system or process change. Examples include implementing a new route, purchasing a new aircraft type, setting up a new process (system) to do internal engine maintenance, converting an entire organization to a new computer platform and the like. Prior to implementation and part of the planning process, all potential hazards are identified by a team of stakeholders with a focus on 'what can go wrong?' The team identifies all of the potential hazards that it can envision based on experience and knowledge. These hazards are evaluated during a risk-assessment process, are prioritized in term of risk, RM strategies are developed, and predicted residual risk is estimated. This information is summarized in a Safety Risk Management Document (SRMD) to capture and communicate the material to senior management. Ideally, the responsible manager for the process must review the material, agree to the risk mitigation strategies developed by the team, fund the strategies and accept the predicted residual risk. This process is simply intended for the management team to think proactively about the potential consequences of changes to systems or processes prior to implementation, that is, instead of a reactive clean-up of mistakes that could have been avoided by an adequate planning process.

Risk Mitigation Categories

Risk mitigation strategies fall into four general categories. Mitigation may include only one of these strategies or be a combination of these options:

- *Accept or assume risk*: the manager or executive team elects to accept the risk – with or without mitigation – as the rewards justify the risk. Accepting risk is merely taking the chance that negative impact will be incurred, the probability and severity being defined in the risk index that was either calculated or estimated during the risk assessment phase. Example: a commercial operator reviews the number of altitude deviations reported by the pilot group and determines that the number of events falls within an ALoS and elects to continue operation with no additional investment in simulator training or revision of procedures. (Note: It is inappropriate to accept or assume high risk as indicated by an ICAO risk index value of 5A, 5B, 5C, 4A, 4B, or 3A.)
- *Avoid risk*: the manager or executive team changes plans in order to prevent the problem from arising. Example: a charter operation may elect to not operate an aircraft into a specific airport because the combination of weather, crew and airplane conditions make it a high-risk operation. Similarly, a commercial air carrier conducts a risk assessment on operating into an airport with the security risks are high to both the aircraft and flight crew and elects to cancel the operation.
- *Mitigate risk*: the manager or executive team approves, funds, schedules, and implements one or more risk mitigation strategies. These strategies are actions or controls that lower or eliminate the risk. Example: a commercial or charter operator revises acceptance procedures to ensure that bogus or unapproved engine parts are not accepted into the parts inventory. Similarly, these operators have elected to install electronic flight bag equipment with moving map displays to increase crew situational awareness during taxi operations, thus reducing the risk of runway incursions.
- *Transfer risk*: the manager or executive team shifts ownership of the risk to another party that can manage the outcome. Example: a commercial operator elects to outsource ground handling activities to a supplier to reduce the organization's LTI rate and costs associated with ground damages. Service contracts are therefore structured so that the supplier pays the associated costs of aircraft repair should there be damage to an aircraft caused by ground equipment operated by the supplier. Furthermore, since the operator no longer employs ground handling employees, the risk associated with lost time injuries is transferred to the supplier.

Value of SRM Documentation

There is immense value for the organization in recording and documenting the results of risk assessment and mitigation activities for each actual or predicted hazard. The documentation process should be designed to meet the needs of the organization. At one extreme, a large organization may actually design an online system of hazard tracking and mitigation templates that are completed for each operational hazard assessed with a risk index of 'high'. For smaller organizations, such as a single-aircraft flight department, the documentation process may be a single paper form that is completed by a small team of internal stakeholders and retained in a paper file or electronically. Regardless of whether the operator uses an electronic online documentation management system or an array of paper forms, the risk-management process remains the same. The value of documentation is summarized as follows:

• Documentation establishes managerial accountability and supports an informed decision-making process.
• Documentation provides a record of risk assessment and proposed mitigation strategies.
• A documentation process gains signatures of process owners and major stakeholders, particularly those who have the authority to commit resources to mitigation strategies.
• Signatures represent acceptance of predicted residual risk and agreement with the analytical process and conclusions.

It's Time to Brief the Accountable Executive

Safety risk management occurs daily throughout the organization – at every level. Data is collected daily by the safety and quality services departments, synthesized into useful information and then periodically briefed to the accountable executive and senior leadership team. This briefing may be referred to as the safety committee, safety and quality review board, or perhaps the risk review board. Nevertheless, this is the point where the safety practitioner's writing and communication skills will be tested.

The information selected to be briefed to the executive team must validated for accuracy, coordinated with appropriate process owners, carefully formatted for clarity and backed-up with irrefutable data. Remember, the safety practitioner is the corporate conscience and the safety issues being surfaced may bring a lot of attention to the deficient process and subsequently to the process owner. In some cases, even the senior team member who owns the functional area may be defensive during the discussion with the accountable executive. He or she may offer a number of excuses as to why an event happened or why there was an unexpected regulatory violation. If the safety practitioner has done a poor job of collecting and understanding the facts, a single error in the presentation will jeopardize the credibility of the entire presentation and analysis. On the other hand, if the safety services department has been open and transparent to the line managers and conducted the safety investigations in a spirit of collaboration, there will be few surprises during the briefing to the accountable executive.

As a technique, the heads of the safety and quality departments must carefully coordinate the content of the quarterly (or periodic) safety briefing with all stakeholders. The point is there should be no surprises. The goal of the safety committee meeting is not only to review the health of the organization from a safety perspective, but to review the safety risks identified during the hazard identification process and approve mitigation strategies that are proposed for implementation.

The chart in Figure 6.15 is an expanded UK CAA process map modified by the author to demonstrate an *integrated* approach to safety and quality management (UK Civil Aviation Authority, October 2008). The quality management process will be thoroughly discussed in the next section on safety assurance.

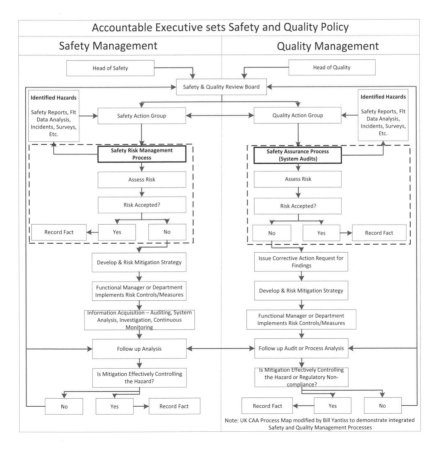

Figure 6.15 Modified UK CAA process map

Component #3: Safety Assurance

Introduction

Safety assurance (SA) is fundamentally a monitoring and feedback system to provide confidence to the entire management team to provide *assurance* as to the performance of operating systems and processes using a number of assessment tools. Safety assurance processes mirrors those of quality assurance, but with a focus on hazard identification, risk assessment and appropriate mitigation. Safety risk management requires an input of safety performance to complete the safety management cycle and make necessary changes to the affected system.

ICAO describes safety assurance as encompassing three distinct activities or elements: safety performance monitoring and measurement, the management of change and continual improvement of the SMS (enterprise management system).

- Component 1: safety policy and objectives
  - Element 1.1   Management commitment and responsibility
  - Element 1.2   Safety accountabilities
  - Element 1.3   Appointment of key safety personnel
  - Element 1.4   Coordination of ERP
  - Element 1.5   SMS documentation
- Component 2: safety risk management
  - Element 2.1   Hazard identification
  - Element 2.2   Risk assessment and mitigation
- Component 3: safety assurance
  - Element 3.1   Safety performance monitoring and measurement
  - Element 3.2   The management of change
  - Element 3.3   Continual improvement of the SMS
- Component 4: safety promotion
  - Element 4.1   Training and education
  - Element 4.2   Safety communication

Concepts and Definitions

A number of States have defined safety assurance in the framework of quality assurance. Regardless of the terminology or organizational structure developed to deliver products and services, the purpose remains the same – ensure procedures are carried out consistently, functional departments are in compliance with State regulations, operational problems (safety hazards) are identified and resolved, continual improvement of processes and procedures and the verification that corrective actions to recognized safety hazards are proven to be effective and sustainable.

Before we delve into the three safety assurance elements, it is important to understand the relationship between the two components of safety risk management and safety assurance. Figure 6.16 should be helpful to visualize the concurrent activities of these two parallel processes (Stolzer, Halford, and Goglia, 2008, p. 79 Fig. 3.9).[9]

Many service providers have elected to perform the safety assurance function within the quality services department, which has expanded quality-oriented checklists with safety and security standards. The value of this integration and partnership is evidenced by the frequent interface between the SRM and the SA activities in Figure 6.16. In practice, the list of operational hazards should be collaboratively managed and monitored by the two services departments.

The components of safety risk management and safety assurance are the areas where both safety and quality practitioners focus much of their effort. For many service providers, these are areas that will also require the most attention to conform to SMS standards. However, there is some good news. Both components are composed of programs detailed in ICAO and State CAA publications. The listing in Table 6.4 provides a few examples of US-oriented reference materials for each of the programs that may in included in an SMS. (Note: It is understood that some service providers will not have a requirement for the following programs.)

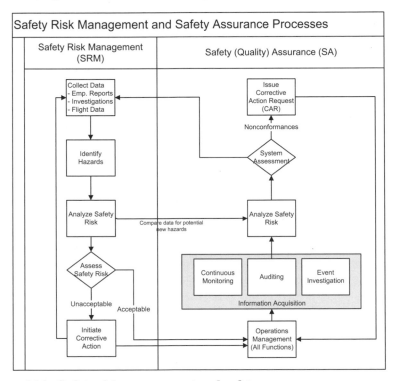

Figure 6.16   Safety risk management and safety assurance processes

---

9   Original flowchart modified by Bill Yantiss to add SRM activities.

Table 6.4   US-oriented reference materials for SMS programs

SMS program	SMS function	Example reference
Aviation Safety Action Program (ASAP) – US	Employee Safety Reporting (Element 2.1)	FAA Advisory Circular 120-66
Flight Operations Quality Assurance (FOQA)	Flight Data Analysis (Element 2.1)	FAA Advisory Circular 120-82
Internal Evaluation Program (IEP)	Management of Change (Element 3.2)	FAA Advisory Circular 120-59A
Continuing analysis and surveillance system	Performance Monitoring and Measurement (Element 3.1)	FAA Advisory Circular 120-79
Accident and incident investigation		ICAO Annex 13

Safety Assurance in More Detail

ICAO describes safety assurance as being composed of three elements supported by a number of subelements. These subelements are very significant in terms of importance and scope. For example, the subelement of internal evaluation is supported by an advisory circular in the United States (FAA Advisory Circular, 2006, p. 114). Figure 6.17 should be useful in visualizing the nature and composition of safety assurance.

Safety Assurance – Parallel and Sequential Methods

Safety (quality) assurance is typically performed using one of two methods: *sequential or parallel*. Should a service provider initiate a new process or revise an operational procedure, the safety risk-management process will proactively be used to identify potential hazards with the change, document the mitigation strategy or controls to manage the associated risk, and gain appropriate management approval of the predicted residual risk. In this case, the safety assurance function takes over at this point (sequential) to ensure that the safety risk controls are being practiced as intended and that they continue to achieve their intended objectives. During the life cycle of the change, safety assurance provides periodic feedback to line managers as to the effectiveness of the controls and overall health of the system.

Similarly, safety assurance is also performed in parallel or simultaneously with the safety risk-management process. Each functional area normally receives annual or biennial audits of the overall health of operational processes and procedures. Checklists are constructed to identify deviations from operational procedures and workplace safety hazards that line personnel have overlooked. Auditors must be trained safety and human factors specialists to gain the most from this independent, operational assessment.

Safety Assurance
An Expanded Definition – 3 Elements

- 3.1–Safety Performance Monitoring & Measurement
  - Reporting systems
  - Investigations
  - Performance metrics

- 3.2–Change Management
  - Process to identify & manage change
  - Identify potential hazards during planning process
  - Assess & mitigate risk prior to change implementation

- 3.3–Continual Improvement
  - Internal Evaluations (IEP & QA)
  - Internal Audits (QC by process owner)
  - External Audits (CAA)

Figure 6.17 Safety assurance elements and subelements

The Value of Alignment

Although ICAO recognizes the benefits in integrating SMS with quality (QMS), security (SeMS), environmental (EMS), and occupational health and safety management systems (OHSMS), the current publications do not provide State regulatory authorities with foundational information on how to complete this alignment (ICAO, 2009, 7.8). This apparent absence of information has generated a debate among safety and quality practitioners, various regulatory bodies and the academic communities as to roles and responsibilities assigned to safety and quality organizations – and how to organize and integrate this work. Since your State CAA is required by ICAO to move forward quickly in the area of safety management, most of the international debate is occurring in this discipline. In contrast, State security and environmental organizations are not feeling the same level of ICAO pressure, as of yet, to establish standardized global security and environmental systems within service provider organizations. This pressure will gradually escalate and hopefully, for the sake of efficiency and alignment, ICAO will require security and environmental management systems to be structured using the same 4 SMS components and 12 elements.

Element 4.1: Safety Performance Monitoring and Measurement

Safety assurance is conducted by monitoring and measuring the outcomes of operational activities that result in the delivery of products and services by the organization. In an SMS, the safety assurance elements are applied to gain an understanding of the human and organizational issues that can impact safety. Safety performance monitoring and measurement may be accomplished by assessing organizational processes through the lens or glasses of each of the following perspectives (ICAO, 2009, 9.6).

- Responsibility: who is accountable for management of the operational activities (planning, organizing, directing, controlling) and its ultimate accomplishment?
- Authority: who can direct, control, or change the procedures and who can make key decisions on issues such as safety risk acceptance?
- Procedures: specified ways to carry out operational activities that translates the 'what' (objectives) into 'how' (practical activities).
- Controls: elements of the system to include hardware, software and procedures designed to keep operational activities on track.
- Interfaces: an examination of such items as lines of authority between departments, lines of communication throughout the organization and consistency or alignment of procedures between work and employee groups.
- Process measures: means of providing feedback to responsible managers to ensure that process outputs and outcomes are being produced as expected.

This tool is useful for the regulatory authority, the quality and safety services departments and functional manager to assess operational processes to ensure they have been designed and operating satisfactorily, that is, meeting expectations in terms of reliability, efficiency, and safety. However, this tool may be challenging to use in the hands of inexperienced practitioners without quality system, ISO 9001:2008, or Six Sigma training. A thorough review of the concepts of accountability, responsibility and authority is absolutely essential to understanding SMS and organizational discipline. Additionally, a thorough understanding of system design, process mapping and procedural development is useful in assessing operational activities.

Data Acquisition Sources

Safety and operational performance data are available from a number of sources. Ideally, this data should be available to the safety and quality services departments enabling a holistic assessment of the entire organization. Safety data may be categorized as reactive, proactive and predictive. Although we will not explore each source in detail, Figure 6.18 highlights the type of available data by category:

Sources of Safety Data

Reactive

- Accident or Incident Investigation

- Employee Safety Reports

- Flight Data Analysis

- Regulatory Violations

Proactive/Predictive

- Onsite Safety/Quality Audits

- Safety Surveys

- Hazard Reports

- Flight Data Analysis (FDA)

- Line Operations Safety Audit (LOSA)

- Safety Risk Management Panels (SRMP)

- Safety Studies

Figure 6.18   Sources of safety data

Measuring Safety Performance

ICAO reinforces the use of three terms that have probably found their way into your safety vocabulary: safety indicator, safety target and safety value. Let's take the time to review definitions and illustrative examples.

- Safety indicator: a parameter that reveals the level of safety in your operation. Examples include:
  - Accidents or incidents
  - Runway incursions
  - Runway excursions
  - Ground collision events
  - Altitude busts (deviations)
  - Number of regulatory violations (during audits, etc.)
  - Rejected take-offs
  - Enhanced ground proximity warning systems (EGPWS) sink rate events
  - Long landings
  - Foreign object damage (engine FOD)
  - Engine shutdowns
  - Unstable approaches
- Safety target: a target is an operator's desired level of safety (or improvement) in this specific indicator (parameter). Examples include:
  - An annual runway incursion rate of X per 10,000 departures

- A reduction in long landings by 30 percent year-over-year
- A reduction in foreign object damage by 20 percent
- Safety value: this term is nothing more than the specific value of the indicator or parameter. Examples may include raw numbers of events or a rate:
  - Lost time injury frequency rate (LTIFR) of 2.37/100 full-time employees. Note that injuries to ground handling personnel are a primary contributor to the rate due to sprains and strains
  - Unstable approach rate of 1.5 percent
  - Ground damage rate of 2.05 (per 10,000 departures)

What is the State's Acceptable Level of Safety?

ICAO has provided additional guidance to the State to complement the traditional management of safety through regulatory compliance, with a performance-based approach. As States respond to recent ICAO guidance, the State's traditional baseline array of regulations will be augmented with the notion of a State ALoS. At the present time, the concept of ALoS applies only to the State within the framework of a SSP. This concept is a work in progress and poses a political dilemma for most governments.

For example, the United States tacitly accepts the fact that there are over 40,000 fatalities on American highways annually. Many safety initiatives have been developed to improve vehicle safety that include side airbags, electronic stability controls and aural lane departure warnings, all of which are aimed at reducing the number of deaths. However, if the US government were to announce a national ALoS goal of 35,000 fatalities as an acceptable annual rate, there would likely be a huge outcry from the media and citizenry. To date, the ALoS for the aviation industry has been described in very general terms without publishing an acceptable level of runway excursions, hull losses, fatalities, etc.

Perhaps the best measure of a State's ALoS is the amount of governmental response to industry safety performance. For example, the amount of investigative effort into the causes of the Colgan Flight 3407 accident on February 12, 2009, and associated media attention reflects an unwritten ALoS that is determined, in part, by the national culture. Simply stated, there is little national tolerance for an aircraft accident that could have been prevented. The national reaction to a single vehicular highway accident involving a bus load of adult passengers is likely to be subdued in comparison.

Service Provider's Safety Performance

Most service providers have already established safety performance indicators and associated targets as a best practice. However, during the SMS approval process, these indicators, targets and values will be shared with State CAA representatives.

Each State will establish SMS approval procedures and will likely follow the general ICAO guidance below:

> Such safety performance must be *agreed* between the State and service providers, as the minimum acceptable the service provider must achieve during the delivery of services. The safety performance of an SMS is thus a reference against which the State can measure the safety performance of the SMS, that is, that the SMS works above and beyond regulatory compliance (ICAO, 2009, 6.6.4).

Element 3.2: Management of Change

A change management process is a documented strategy to proactively identify and manage the safety risks that can accompany a significant change in a service provider, whether operational, technical, or organizational. Changes in the following areas may be excellent candidates for the change management process:

- operational expansion or contraction
- changes to existing systems, equipment, or programs
- new products and services
- introduction of new equipment and procedures
- outsourcing operational activities
- changing business partnerships
- organizational changes to include a merger or reorganization.

ICAO states:

> a formal management of change process should identify changes within the organization which may affect established processes, procedures, products, and services. Prior to implementing changes, a formal management of change process should describe the arrangements to ensure safety performance. (ICAO, 2009, 9.8.4).

The FAA framework document describes the requirement using slightly different language:

> The organization's management will identify and determine acceptable safety risk for changes within the organization that may affect established processes and services by new system design, changes to existing system designs, new operations/procedures or modified operations/procedures. (Federal Aviation Administration, 2010).

Change management, as a formalized way of doing business in the aviation industry, is truly in its infancy. It goes without saying that enterprise change management is absolutely essential to surviving in the aviation industry. Agility and change is simply the nature of our business. All service providers must be agile and unencumbered with bureaucracy in order respond quickly to the changes in the market place, competitor initiatives, fuel prices, etc. We are good at it. However, many organizations have not documented nor standardized change process requirements throughout the enterprise, particularly in operations.

Two Levels of Change – Incremental Change and Deep Change

Within every organization, change is occurring – whether good or bad. An innovative organization is always looking for better ways to do business, to identify new markets and to improve the skills of its people. In contrast, some organizations seem to be satisfied with status quo with little interest for improvement. If one takes a close look at the latter, the best talent is probably looking for more exciting work and is considering a move to another company.

A company is going one of two ways – getting better or getting worse, improving or deteriorating, staying ahead of competitors or slipping behind. No company is standing still. Every company is changing. The question is, 'Is the leadership team guiding this change, encouraging improvement, and investing in employee skills to accelerate change?'

There are two speeds at which change occurs. Most safety and quality practitioners deal with what may be called incremental change. This type of change is perhaps easier to control and relies on a culture of continual improvement with a passion for identifying soft spots or weakness in the operation and getting them 'fixed'. In contrast, Robert Quinn describes *transformational change* in his book *Deep Change,* as being essential to remaining competitive in rapidly changing environments (Quinn, 1996). Perhaps one of the best examples of transformational change occurred at The Boeing Company. Until the early 1950s, Boeing focused on building bombers for the US military – B-17, B-29, and B-47. In 1952, Boeing saw an opportunity to transform the aviation industry by investing 25 percent of the company's net worth into building the B-707. The decision to move forward with the B-707 project represented deep change for both the enterprise in terms of how business is done, but even more so on the personal level of each employee. Let's explore these two levels of change in more detail.

Incremental Change

Incremental change is synonymous with what quality practitioners call *change management*. The aviation industry has evolved over the last 40 years through incremental change and effective change management. For example, aircraft accidents have been reduced through incremental changes in cockpit automation such as EGPWS, Traffic Collision Avoidance Systems (TCAS), weather radar, etc. Flight crew training programs have also been improved through incremental change through the development of full-motion simulators, CRM, electronic flight bag (EFB), evolution of checklists, etc. Each of these innovations was based on a recognized need to change the way we do business in order to improve safety. Each of these initiatives began life as an optional best practice and ultimately evolved into an ICAO minimum requirement.

Change management first begins with identifying an opportunity for improvement. These opportunities may be identified by asking appropriate questions such as: What isn't working and why? What performance measures aren't yielding expected results? Can we reduce the number of hydraulic failures or runway incursions? Why didn't our flight crew handle that emergency as well as we would have liked? Thousands of questions such as these identify opportunities for improvement. Routine safety and quality processes are also a rich source of identifying opportunities for improvement. Examples include:

- employee reports – safety quality, security and environmental
- employee reports of regulatory violations
- employee operational reports – system deficiencies, delays, miscommunication
- audit reports – corporate and functional department audits
- flight data analysis
- regulatory authority feedback
- customer feedback
- performance metrics – missed targets
- functional manager self-assessments – processes, procedures, employee performance.

As discussed in the previous chapter on safety risk management, once an operational hazard has been identified and assigned a risk index of high, the management team designs a mitigation strategy with specific actions to improve the operational process. These actions represent change management.

A word of caution: employees may not always appreciate the value of a steady stream of procedural and process changes. For example, a team of subject matter experts may have redesigned an operational procedure to reduce errors and improve efficiency. Much to the dismay of the well-intended management team, the employees did not welcome the change with open arms but responded with resistance. In this case, the management team made a critical error – it did not explain why the procedure was being revised and how it would benefit all concerned.

Management of Change – Practical Application

The FAA ATO has developed an effective approach to change management to ensure that individual hazards are identified and unacceptable risk is mitigated prior to making a procedural, process, or system change. The process is straightforward and initiated by a change proponent or sponsor who must perform a safety analysis. Three terms are used throughout the process:

- Safety Risk Management Panel (SRMP): An SRMP is a group of carefully selected stakeholders from the various organizations affected by the change. The SRMP identifies potential hazards, conducts risk assessment, develops suggested mitigation strategies and completes the SRMD.
- Safety risk management document: the SRMD is an effective tool to summarize (record) the safety analysis, capture the prioritized hazards and associated risk mitigations and serves to enable the management team to understand the change, associated risks, mitigation strategies and commitment to fund and implement the mitigation strategies.
- Safety risk management decision memo (SRMDM): should the change proponent conclude that no safety risk will be introduced into the system by the change, an SRMDM is prepared to document the justification and rationale (see Figure 6.19).

The value of the SRMP is to bring together subject matter experts (SMEs) to provide senior management with a technical assessment of the proposed changes to existing systems or processes. The size of the SRMP is scalable to the complexity of the organization. In large organizations, the panel may consist of 10–15 stakeholders. In contrast, small corporate operators may conduct the change management process with only 2–3 individuals.

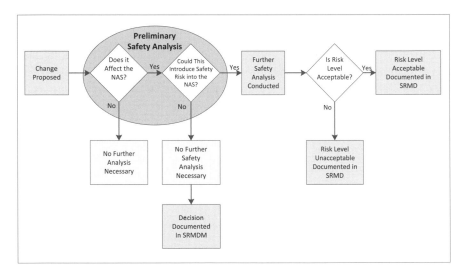

Figure 6.19   SRM decision process

Once the SRMP completes the SRMD, the document is reviewed and signed by appropriate leadership team members. Signatures represent the acceptance of the risk analysis process, the acceptance of predicted residual risk after implementation of mitigation strategies and the commitment of resources to implement the risk mitigation strategy.

An in-depth discussion of this process is described in the IATA SMS Implementation Guide (IATA, April 2010a, 6.2).

Deep Change

Deep change, as described by Robert Quinn, represents a commitment by the organization to escape the slow death to extinction or insignificance. This downward spiral may be evidenced by gradually decreasing market share, a culture of satisfaction with status quo, a staff that is waiting out the last few years to retirement without rocking the boat, or a leader who has effectively lost touch with the people in the organization. Deep change is something that needs to occur periodically, perhaps every 5 or 10 years, to enable the organization to thoroughly reevaluate its vision, mission, strategy, organization and values.

A deep change experience requires strong senior leadership and a commitment to make the necessary personnel, system and process changes required for survival and competitiveness. Strong leadership is required because there is usually grass roots resistance to change because it's threatening to employees – my job may change, I will have to acquire new skills, I may have to move to another department, or my position may be eliminated. Similarly, there may be significant resistance by the recalcitrant management team which openly agrees with the change strategy in group forums but secretly undermines the initiative in private meetings.

The CEO or accountable executive will usually be the driving force behind deep change. The methods used to drive deep change will vary from enterprise to enterprise. At one end of the spectrum, a CEO is hired by a board of directors to initiate deep change in an organization that is lethargic, self-satisfied and non-competitive. The existing senior leadership team has been in place for years and has grown comfortable with inefficient but comfortable ways of doing business. In this case, the CEO may enlist an external consulting firm to help facilitate the deep change process along with the selective retention of leaders who are willing to support the new direction of the company.

At the other end of the spectrum, an organization with an entrepreneurial culture may be eager and emotionally ready to move, as a group, to the next level. The team members are willing to take the risk of change – not knowing exactly what it will require of them on a personal basis. There is a certain anxiety associated with the unknown, but they have extraordinary confidence in their own skills to not only survive the deep change process, but to end up in a much better place at its conclusion. This group is willing to accept both corporate and personal risk to enable the organization to leapfrog its competitors.

Figure 6.20 may be useful in comparing the two levels of change management. The bread-and-butter work of the safety and practitioner will be involved with incremental change. As a highly regulated industry, manufacturers, service providers and industry watchdogs are quick to point out operational weaknesses and opportunities for improvement. Incremental change means to close the holes in the cheese or strengthen the soft spots in the operation through effective change management.

Change Management – Two Levels

Incremental Change

- Focus is on identifying and "fixing" weaknesses in the operation
  - Employee reports
  - Audit reports
  - CAA feedback
- Each process is incrementally improved for reliability & efficiency
- Cultural change is incremental with little employee resistance
- Employees see action to resolve safety concerns (reports)

Deep Change

- Focus is on major change
- New ways of thinking & behaving – transformational
  - Infrequent – not daily
  - Organizational & personal change
  - Involves more risk
- May be required for survival
  - Eliminate organizational 'bloat' or over-staffing
  - Realigns organizational systems
- May need external team to facilitate change process
  - Realigns organizational systems

Figure 6.20   Change management levels

Element 3.3: Continual Improvement of the SMS

Continual Improvement (CI) Tools and Concepts

Every commercial enterprise must embrace the concept of continual improvement in order to survive, to effectively compete in the market place, and to ensure its future. Management systems are never static, but must continually evolve based on lessons learned, remediation of weaknesses in the operation, and identifying opportunities for improvement. Continual improvement is a characteristic of a learning culture that enables proactive risk management through process assessment and improvement.

It has been said that 'The emphasis with assuring quality (and safety) must focus first on process because a stable, repeatable process is one in which quality (and safety) can be an emergent property.' (Transport Canada Advisory Circular,

2001, 3.9). The implication is clear – quality is the key to acceptable safety performance.

Transport Canada, EASA and the FAA have embraced the Shewart Cycle (Figure 6.21) as a relatively *simple*, yet highly *effective* tool to manage change in your organization. Therefore, it is likely your State CAA may expect to see a similar process in your organization during the SMS approval process.

The Shewart cycle was popularized by W. Edward Deming as a way of organizing and executing operational changes. The cycle has also been called the Plan-Do-Study-Act cycle. The repetitive steps may be applied in the following situations:

- as a model for continuous improvement and change management
- when developing a new or improving the design of a process, product, or service
- when defining a repetitive work process
- when planning data collection in order to verify and prioritize problems or root causes
- when implementing any change
- when starting a new improvement project.

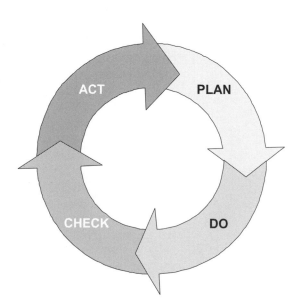

Figure 6.21   The Shewart (PDCA) cycle

EASA integrated Management System Uses Plan-Do-Check-Act

EASA has adopted iMS (see Figure 6.22) as the structure of the new organization. The regulatory agency, much like any service provider, is composed of multiple systems that must be organized, aligned and coordinated for efficiency. A thorough review of EASA organizational documentation reveals the following key points:

- the IMSis focused on organizational risk management
- the IMSincludes all internal business processes
- the organizational structure is certified against the ISO 9001:2008 standard
- a fundamental feature of the EASA PDCA cycle is continual improvement.

As adopted by EASA, the PDCA descriptors can be relabeled to enhance an understanding of each activity:

Plan	Prepare
Do	Execute
Check	Measure
Act	Evaluate

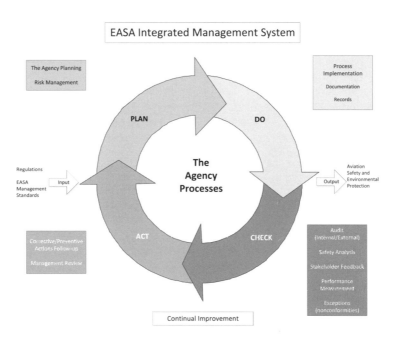

Figure 6.22   EASA integrated Management System

FAA NextGen EMS – PDCA Cycle

During the 2009 National Business Aircraft Association Expo in Orlando, an FAA NextGen EMS briefing included a PDCA diagram (Figure 6.23) that may be linked directly to the SMS four components.

As you look at this diagram, one obvious question comes to mind: 'Is there any link between the four PDCA components and the four SMS components?' The answer to that question is yes, but indirectly. The PDCA cycle is a model used to manage the delivery of products and services for the entire organization, to manage significant operational changes and to continually improve business processes. This is quality management.

In contrast, safety management is the identification and mitigation of hazards during the delivery of these products and services by the quality system. Safety management is a separate and distinct business function that must be considered at the same level and with the same importance as other core business functions (ICAO, 2009, 3.2.5). The bulk of this safety management activity occurs within the components of SRM and safety assurance.

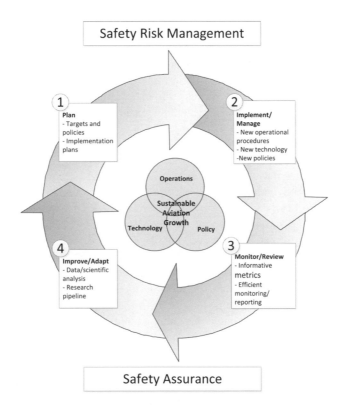

Figure 6.23   FAA NextGen EMS PDCA diagram

In the opinion of this author, the SMS SRM component is most closely aligned with the PDCA plan-do components and the SMS SA component is aligned with PDCA check-act component.

Continual Improvement through Auditing

The operational *assurance* function applies to safety, quality, security and environmental systems that permeate the entire organization. The term *assurance* requires some explanation, particularly in relationship to a separate function called *control*. These terms have been improperly used as synonyms within the aviation industry, thus leading to ongoing confusion as to how these work streams are assigned within an organization. The FAA has given considerable attention to defining the difference between *assuring quality* versus *controlling quality* within an organization. The following definitions are published in Advisory Circular 120-59A:

- Quality assurance: the independent activity of providing the evidence needed to establish confidence, among all concerned, that the quality function is being performed effectively. This activity assures quality through independent evaluation of established processes, procedures and documentation.
- Quality control: the determination of the quality of a product (or service) by inspection and testing to determine compliance with standards. This activity controls quality through the establishment of effective controls, documentation and procedures within specific functional areas.

Quality Assurance

The assurance of safety, security, etc. is an independent confirmation to the accountable executive that the various systems and processes throughout the organization are designed and working as envisioned. In many organizations, the independent assurance function is accomplished by a trained group of auditors reporting directly to the accountable executive. This activity may also be referred to as the organization's IEP (FAA Advisory Circular, 2006, p. 134). This arrangement ensures the independence of the work without inappropriate influence of the functional manager to alter audit results.

A quality assurance group may perform safety, quality, security and environmental assurance through this independent auditing process. However, auditors require the necessary training and checklists to accomplish this work properly. In practice, these auditors are generalists and not necessarily subject matter experts in all areas of the operation; there are simply too many technical areas in which to gain on the job experience and expertise. Should an auditor discover an environmental or security issue, this generalist should have technical experts available to provide technical advice on whether an observation is a serious issue. Large organizations typically have internal experts available within the headquarters staff. Small organizations may have to rely on external technical advice. These efficiencies are being captured as many service providers are consolidating the safety, security, quality and environmental activities into a single, integrated services organization.

Quality Control

In contrast to the assurance function described above, the control function is a responsibility of the functional manager who oversees the day-to-day operational activities. The functional manager desires control over these processes and therefore controls safety, security, quality and environmental outcomes as the process owner. Quality control is a daily activity as the process is operating. For example, the ground handling activities that occur immediately following an aircraft parking at the gate are composed of many independent but integrated processes. The precision and procedural compliance at which this work is done is a quality control activity. Moreover, the latitude that the responsible manager permits the employees to deviate from operational procedures is a quality control decision.

Quality control typically occurs among aviation service providers as inspections of specific work activities. For example, flight crews receive annual training accompanied by performance evaluations to ensure that they possess the required technical knowledge (oral evaluations) and the ability to apply this knowledge in an operational environment (simulator). This validation by the process owner on a frequent basis ensures that the output from training programs meet operational standards, which is a responsibility of the owning organization.

It is important to understand that quality control (for example inspection) is usually conducted by technical experts who can easily identify procedural deviations and errors. In contrast, quality assurance is more of an assessment of the entire system operation, the relationship and interface between systems, and the identification of opportunities for improvement.

ICAO Audit Categories within Safety Assurance

The term *safety assurance* embraces auditing processes to ensure control of safety performance through a layered system of audits. ICAO has organized audit activity into three separate categories (ICAO, 2009, 9.9):

1. Internal evaluations
2. Internal audits
3. External audits

Internal Evaluations (Quality Assurance)

Internal evaluations involve the assessment of operational activities within an organization to include SMS-specific functions. Civil Aviation Authorities may refer to this audit activity in regulations as internal evaluation or quality assurance. These higher-level systems audits provide the service provider with an opportunity to evaluate the selection of performance metrics, the identification of workplace hazards that have been overlooked by employees and the effectiveness of interfaces between work groups.

Internal evaluations are typically conducted by corporate-level teams and focus on the effectiveness of systems, processes and procedures throughout the organization. The implementation of IEP in the United States has required expansion of quality assurance activities from only aircraft maintenance to include all non-maintenance activities. For example, a large airline may have two quality assurance groups – one for operations and one dedicated to aircraft maintenance and engineering.

The effectiveness of the internal evaluation function is dependent on the education, experience and competence of the auditor pool. Auditing is a learned discipline and continues to develop over time. Qualified auditors rarely exist in abundance in any organization. They must be carefully selected and trained over a period of years. Upon selection as an auditor, the organization will have to invest in formal classroom training, a comprehensive on the job certification process, continued professional study and auditor certification exams. It is particularly important for the auditor to understand systems design, ISO 9000 concepts, and Six Sigma tools. Although these tools will not always be used directly by the auditor, it is absolutely essential that the auditor understands how systems are designed and measured in order to be able to perform the internal evaluation function.

A number of techniques are available to the organization to conduct in-depth and comprehensive internal evaluations.

- Business Process Engineering Team (BPET) (IATA, 2010a, Sec. 7.3). Continual improvement of the organization's management system is a never-ending process that presents opportunities for senior management to identify, prioritize and address. Latent deficiencies exist in any process, procedure, or system. A BPET is a team of carefully selected subject matter experts who re-engineer or redesign an existing system that is not meeting expectations or performance targets. Specifically, the team should perform in-depth process assessments to understand the current process, to develop process improvement opportunities, identify potential new hazards introduced into the system associated with change, identify mitigations and prepare an implementation plan. The team may be sponsored by a division head, responsible manager, or head of safety and quality. A facilitator is usually identified from the quality department to organize and direct the activities. This process is similar to the SRMP discussed earlier in this section.

ICAO also requires that independent internal evaluations be performed on the safety management functions, policy making, safety risk management, safety assurance and safety promotion. In most organizations, this work will logically be assigned to those groups or individuals performing the internal evaluation function.

What about security management? It logically follows that independent internal evaluation of security management functions, security risk management, security assurance and security promotion be performed at the functional level during the assessment of safety management functions. For example, when an internal evaluation is conducted of one of many airline operational locations, the most efficient approach is to combine safety, quality, security and environmental checklists to be integrated and thus completed during a single visit to the domestic or international station. These checklists are obviously created by security and environmental experts, but on-site auditing may be completed by a trained generalist.

Internal Audits (Quality Control)

Internal audits are described by ICAO as those performed by the operating department that owns the technical activities supporting the delivery of services. Internal audits are typically performed by trained individuals within the operating division or functional area, not the corporate level. These employees are the technical experts who possess the knowledge and authority to redesign internal processes to meet performance targets.

Internal audits and quality control audits are terms for the same function. These audits may focus on basic regulatory compliance, adequacy of procedures, performance of instructors, potential deficiencies in employee training and adherence to established procedures. Internal audits are conducted between internal evaluation intervals. For example, an organization elects to conduct internal evaluations on a two-year cycle unless performance indicators of a functional area suggest a more frequent interval. Obviously, the performance of a functional area may deteriorate over a matter of weeks and it would be inappropriate to delay an assessment until the next scheduled internal evaluation. Therefore, the internal audit provides a more frequent snapshot of the performance of a functional area and fills the gaps between infrequent internal evaluations.

External Audits

Audits conducted by regulatory agencies, customer organizations, or code-share partners provide the service provider with relatively unbiased information regarding the health of the organization. The role of the regulatory authority is similar to that of the quality assurance department, that is, to ensure compliance with State regulations, internal procedures and continuous improvement of the SMS. Open and candid communication with the regulatory authority is a prerequisite to the sharing of safety information. Obviously, in a world of litigation and inappropriate use of safety information, appropriate safeguards of safety information must be established and agreed upon by the service provider and representatives of the regulatory authority.

- *Joint Quality Review (JQR)*.[10] A number of airlines have improved their safety cultures to such a level that they regularly share safety and audit information with one another on a quarterly basis. The service provider views the findings of the regulator's on-site inspections and observations as rich sources of safety information, that is, system breakdowns, employees deviating from procedures, etc. Regulatory audit results may be used to validate results of internal audits and perhaps identify opportunities for improvement overlooked by internal staff members. The sharing of audit information can lead the organization to take the next step, that is, to conduct system assessments in partnership with the operator. This new concept was successfully beta tested by United Airlines and following several years of process improvement, the joint audit concept was incorporated into the FAA ATOS guidance material. This concept works effectively only when all stakeholders understand systems design, process evaluation, appropriate handling of safety-sensitive information, particularly regulatory violations discovered during the audit process and complete transparency of the corrective action process.[11]

Management Review

Management review is fundamental to enterprise and operational risk management. An acceptable method to satisfy this requirement is a periodic formal meeting of senior corporate executives to review the health of the organization. The output of SRM and SA activities are core metrics indicating opportunities for improvement.

The management review process may include a number of agenda items to assess the health of the organization in terms of regulatory compliance, operational safety, efficiency, productivity and operational risk. Some of the areas for discussion include the following:

- safety and quality performance metrics

---

10   Concept developed by Captain Bill Yantiss, Director Quality Assurance – Airline Operations, 2003.

11   Additional descriptive material on the joint quality review process available in the *IATA SMS Implementation Guide*, April 2010, Section 7.

- operational results that reflect process or system deficiencies
- audit results, to include internal evaluation, internal audits and regulatory feedback
- regulatory violations
- incident and accident investigation results
- status of corrective and preventative actions
- review of prioritized list of operational risks:
  - safety
  - quality
  - security
  - environmental
- organizational effectiveness to include division of work streams
- changes in regulatory policy or legislation.

Overlapping Activities

Any commercial service provider must organize and align safety and quality activities for efficiency, commonality and simplicity. The more that safety, quality, security and environmental management systems are aligned, the easier it will be to explain these processes to managers and employees alike. ICAO has done a terrific job in developing a single SMS model of 4 components and 12 elements that can be effectively applied to all four disciplines. But for the moment, let's only review the critical interface between the safety and quality services departments. Both groups identify hazards, initiate corrective action, and monitor implementation, but use different tool kits. Figure 6.24 highlights this commonality.

Safety & Quality Overlapping Activities

Safety Services

- Identifies hazards
  - Safety Data Analysis
  - Incident Investigation
- Initiates corrective action process
- Monitors implementation of mitigation strategy

Quality Services

- Identifies hazards
  - Internal Evaluation
  - Internal/External Audits
  - Involves more risk
- Initiates correction action request (CAR)
- Monitors Implementation of mitigation strategy

Both departments have some objective – but use different tool kits

Figure 6.24   Overlapping and parallel processes

The principle challenge of the management team is assigning the specific activities within each of the 12 elements to individual departments within the organization. The principle caretakers of the SMS will be the safety and quality managers for the organization. Similarly, the caretakers of the SeMS will be the organizational security manager. Figure 6.25 provides a typical example of how a commercial air carrier may distribute work streams described within an SMS.

The relationship between the safety and quality practitioners continues to evolve. For example, this author visited a large European carrier in 2000 where most of the safety management activity was accomplished by the quality manager, a position within the organization approved by the regulatory authority. The philosophy that *quality is the means to safety* was evident by the organizational structure. The quality services department was robust with resources and much of the work was proactive, that is, with a focus on designing operational systems and processes that result in predictable, reliable and safe performance. The emphasis on quality was equally divided between aircraft maintenance activities and non-maintenance operations. The small safety department functioned primarily in an accident/incident investigation role, collected and processed safety reports and served as a communications conduit to employee groups. Human factors training was being embraced and incorporated into employee training programs. This organizational structure and division of work was highly influenced by the ISO 9000 standard of the day. (Note: ISO has evolved significantly and now recognizes the need for integration of multiple management systems within a complex organization.)

Division of Work – A Synergistic Approach

Safety Services Leadership

- Safety policy
  - Emergency Response
- Safety Risk Management
  - Flight Data Analysis
  - Accident/Incident Investigation
  - Employee Safety Reporting System(s)
- Safety Promotion

Quality Services Leadership

- Quality Policy
  - Organizational Documentation System
- Safety Assurance
  - Regulatory Compliance
  - Change Management
    - Business Process Engineering Team
    - Safety Risk Management Panel
  - Continual Improvement
    - Internal Evaluation
    - Internal/External Audit
- Quality Promotion

Figure 6.25   Sample distribution of work streams

In contrast, the concept of quality assurance or internal evaluation has evolved relatively slowly among US operators. For years, quality assurance was a discipline limited to aircraft maintenance and engineering services, primarily because regulatory requirements didn't require an organization-wide system. The FAA published an Internal Evaluation Program advisory circular in 1992; implementation of internal evaluation was voluntary and very few carriers invested resources to expand the maintenance quality program to include the entire organization. Enterprise QA received a boost when the US Department of Defense added an organization-wide audit requirement into the Quality and Safety Requirements (QSRs) for operators applying for government contracts. Commercial operators quickly complied with new requirement for obvious economic reasons; however, there were significant differences between operators on the conduct of QA because of the lack of detail in the advisory circular and QSRs.

Safety assurance, as described in the ICAO Safety Management Manual, was derived almost directly from ISO 9001:2000. Safety assurance is really nothing more than assuring safety through the use of quality assurance tools and methods, all with a focus on safety. The central message to all of us is to create a simple alignment of the work streams in safety, quality, security and environmental disciplines so that the organization can understand and embrace the processes and tools. Otherwise, the leadership will continue to deal with inefficiency, confusion, and redundancy.

Component #4: Safety Promotion

Introduction

Safety promotion involves the leadership of the accountable executive to advocate safety practices that will improve the organizational (safety) culture, to provide effective safety training for all employees and to communicate safety information that will promote adherence to standard operating procedures and consistent behaviors.

ICAO describes safety promotion as encompassing two specific activities: safety training and education, and safety communication.

- Component 1: safety policy and objectives
    - Element 1.1   Management commitment and responsibility
    - Element 1.2   Safety accountabilities
    - Element 1.3   Appointment of key safety personnel
    - Element 1.4   Coordination of emergency response planning
    - Element 1.5   SMS documentation
- Component 2: safety risk management
    - Element 2.1   Hazard identification

- Element 2.2   Risk assessment and mitigation
- Component 3: safety assurance
  - Element 3.1   Safety performance monitoring and measurement
  - Element 3.2   The management of change
  - Element 3.3   Continual improvement of the SMS
- Component 4: safety promotion
  - Element 4.1   Training and education
  - Element 4.2   Safety communication

Concepts and Definitions

Safety promotion begins with the accountable executive to demonstrate the behaviors and attitudes expected of all employees. These desired behaviors are then reinforced by the entire management team, that is, as they 'walk the talk' during the conduct of day-to-day work activities. New State regulatory standards will require that safety information be included in training programs appropriate for each employee group. For example, the UK CAA has categorized the training into the following levels:

- operational staff
- managers and supervisors
- senior leadership
- accountable manager

Each level of leadership requires a different amount of expertise in safety management. For example, operational employees require information as to when and how to submit hazard reports, what happens to their report and how changes result from their safety suggestions. In contrast, the management team requires training on risk identification, assessment and mitigation. Of course, the accountable executive must understand the safety leadership responsibilities of this position and how to support the safety services department.

Safety promotion involves everyone in the organization. For example, an employee who submits a hazard report is *promoting safety* by identifying a hazard that may result in an employee injury or equipment damage. Although the reporting employee may have identified the threat and managed it without physical harm, perhaps the next employee will not be so fortunate. Similarly, middle managers promote safety through actively identifying hazards in the workplace and ensuring their teams are equipped to manage or mitigate these hazards effectively. Mitigation may include training, protective gear, or improved tools.

Element #1: Training and Education

Safety training must be tailored to each management level and employee group. Several staff organizations, such as the safety and quality departments, will be required to be fluent and knowledgeable of each facet of safety and quality management. In comparison, line employees will not need to be knowledgeable of these detailed processes, but merely aware that they exist and where to find the information in corporate manuals. In general, training and education fall into the following general categories or activities (ICAO, 2009, 9.11):

- documented process to identify training requirements
- a validation process that measures the effectiveness of training
- indoctrination training incorporating SMS, to include human factors
- initial (general safety) job-specific training
- recurrent safety training.

Management personnel that assume a leadership, change management, continuous improvement, or investigative role will require training in the following topics:

- roles and responsibilities relative to SMS, QMS, SeMS, and EMS activities
- event investigation and analysis techniques
- audit principles and techniques
- management system design, analysis and implementation
- root cause analysis
- human and organizational factors
- emergency response preparedness
- communication and training techniques.

Leadership teams have a variety of training delivery options that can be adapted to the size and complexity of the organizations. For small organizations, it may be more efficient to develop a single, comprehensive training program and select specific modules for each level of management. Larger organizations may elect to develop tailored training programs for each employee group and level of management.

Individuals who actively participate in safety risk management and safety assurance processes must receive in-depth training on the use of safety and quality tools. Safety practitioners include members of the airline's management team such as the safety, quality and security services departments, operational managers, process owners and corporate communications.

A terrific technique is to develop an SMS Quick Reference Guide (SMS QRG) intended for distribution to all safety practitioners. The SMS QRG may contain abbreviated process maps, risk matrices, analytical tools, and basic definitions.

Element #2: Safety Communication

Safety communication includes information on safety culture, reporting systems, investigation results, safety lessons learned and management actions taken as a result of employee safety reporting. Tailored communications to each employee group permit an opportunity to meet the interests and needs of each job category, such as pilots, mechanics, dispatchers, ATC controllers, etc. Communication raises employee awareness of operational hazards and how to manage them. Effective communication, both vertically and laterally, enables transparency and builds trust throughout the organization.

Safety Summit

Many organizations schedule an annual safety stand-down or safety summit where senior safety leaders and operational division heads meet to align SMS processes and procedures. Topics of discussion may include the following items:

- review of roles and responsibilities for managers at all levels
- review of organizational cultural characteristics and strategy to improve the culture
- assessment of the relationship between the organization and the regulatory authority
- assessment of existing SMS processes
- root cause analysis
- hazard tracking system
- safety data analysis – techniques, tools, classification and data sharing.

Safety Culture

The study of organizational safety culture requires much more attention than this section permits. As a review, culture has been defined as the values, beliefs and norms shared by a group of people that influence the way they behave. Every organization has a culture that may be described holistically as a *corporate culture* and is composed of multiple components (Stolzer, Halford, and Goglia, 2008, p. 24):

- Informed culture: people are knowledgeable about the human, technical, organization and environmental factors that determine the safety of the system as a whole.
- Flexible culture: people can adapt organizational processes when facing stressful and temporary operations.
- Reporting culture: people have enough trust and confidence in the safety system to respond to hazard safety reports in a way that justifies the time and effort to submit the report.

- Learning culture: people at all levels of the organization continue to identify opportunities for improvement and implement reforms to operational procedures and processes.
- Just culture: a just culture, for the most part, refers to the attitude of the management team to deal rationally, fairly and justly with employees who make unintentional mistakes.

Cultural assessment surveys have proven to be an invaluable tool enabling an understanding into the unique subcultures (work groups) of an organization. Surveys always yield surprises for the management team in terms of new insights as to why people do what they do in the workplace. Dr. Terry von Thaden, University of Illinois, has conducted numerous surveys that provide terrific insight as to employee attitudes, behaviors, loyalty, etc. These surveys provide the necessary insight and understanding of your organization in order to develop a strategy to improve the culture (von Thaden, 2007).

Safety practitioners should ensure that the senior leadership team understands the following fundamentals regarding safety culture:

- the management team must be role models to line employees
- SMS implementation is a change in the way we do business
- corporate culture is composed of many components – safety is one of them
- there is always a safety culture – it may be good or bad
- the management team sets the tone and culture of the organization
- the safety culture is undermined by ignoring hazard reports or punishing the reporting employee for making a mistake
- safety culture is fragile and must be reinforced daily by all managers.

Senior leaders can do a number of things to assume safety leadership. These options are available to every line manager, as well.

- Establish a safety policy, allocate resources to deliver safe products and services and provide support to the safety, quality, and security services departments.
- Encourage all employees to communicate safety issues and to question operational procedures when they appear to affect safety.
- Recognize that all employees will make errors and that to err is a human characteristic.
- Encourage employees to develop and apply their skills and knowledge to enhance organization safety.
- Foster open communications regarding potential safety hazards.
- Present safety lessons learned to all employees.
- Recognize individual and organizational safety accomplishments.

Regulatory Environment

Introduction

As previously mentioned, the ICAO safety management Standards and Recommended Practices (SARPs) are contained in Annexes 1; 6, Parts I and II; 8; 11; 13 and 14. The annexes address the activities of service providers that include the following organizations (ICAO, 2009, 6.2):

- approved training organizations
- international aircraft operators
- approved maintenance organizations
- organizations responsible for type design and/or manufacture of aircraft
- air traffic services providers
- certified aerodromes

The second edition of the ICAO *Safety Management Manual* (SMM – Doc 9859) is aimed at two groups: State civil aviation authorities and individual services providers. Of significance is the new ICAO requirement for States to establish a SSP, defined as a 'management system for the management of safety by the State' (ICAO, 2009, 11.2.1). It may be further defined as an integrated set of regulations and activities aimed at improving safety. Although ICAO continues to receive feedback from State CAAs, the basic SSP requirements are reflected in the following four components and eleven elements (ICAO, 2009, 11.2.5-11.2.8):[12]

- State safety policy and objectives:
  - State safety legislative framework
  - State safety responsibilities and accountabilities
  - Accident and incident investigation
  - Enforcement policy
- State safety risk management:
  - Safety requirement for service provider's SMS
  - Agreement on the service provider's safety performance
- State safety assurance:
  - Safety oversight
  - Safety data collection, analysis, and exchange
  - Safety data-driven targeting of oversight of areas of greater concern or need
- State safety promotion:
  - Internal training, communication and dissemination of safety information

---

12   Each element is composed of subprocesses, specific activities and tools in which the State must utilize in the conduct of service provider oversight.

- External training, communication and dissemination of safety information

It should be recognized that most State CAAs are currently performing the functions and activities embraced by the SSP eleven elements. However, the organization of this work into a more structured system, defined as a SSP, is a new concept. The intent of the SSP structure is to encourage a higher level of standardization among CAAs in each geographical region. Moreover, the SSP represents the transition from a predominantly prescriptive (regulations) approach to an integrated system of prescriptive and performance-based oversight. In defining both the SSP and SMS, ICAO has introduced the concept of ALoS for an SSP and safety performance for an SMS. Both concepts have been discussed in previous sections. Regulatory compliance, however, still remains the foundation of safety management (ICAO, 2009, 6.4.20).

The ICAO SMM identifies a number of philosophical shifts in regulatory oversight application (ICAO, 2009, 6.3, 11.3, 11.4). It is important that the service provider attempts to understand the internal evolution taking place within your oversight organization.

- The SSP considers regulations as safety risk controls and requires that the process of rule-making be done using principles of safety risk management.
- The CAA will monitor the effectiveness and efficiency of regulations as safety risk controls through its safety assurance component.
- The addition of a concept of ALoS to be achieved by the SSP.
- The two core SSP activities are State safety risk management and State safety assurance.
- A recognized requirement for additional skills in such areas as safety risk analysis, system evaluation, management system assessment and general awareness of new technologies enabling service providers to achieve production objectives.
- Organizing the State safety responsibilities and accountabilities in a principled and structured manner.
- Measuring the effectiveness with which safety responsibilities are discharged and safety accountabilities are fulfilled by the State.
- A requirement to promulgate SMS requirements for service providers requiring demonstration of their safety management capability.

The relationship between the SSP and SMS must be clearly understood by both the CAA and individual service provider. This relationship is depicted by an ICAO chart (see Figure 6.26) which may be duplicated in your State regulations (ICAO, 2009, 6.8.3).

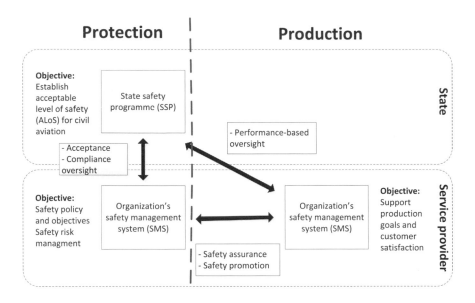

Figure 6.26   SSP and SMS relationship

The introduction of the SSP concept will likely require a significant cultural change for your State regulatory authority, particularly during the transition from a system of predominantly on-site inspections to a system of monitoring the service provider's SMS performance and associated indicators and targets. This will be a challenge in that a full understanding of safety management concepts requires dedicated instruction on fundamental SRM and SA processes reinforced with several years of practical experience to gain confidence that these systems are effective in managing safety. ICAO plans to continue a training program for State CAA representatives to provide additional knowledge of safety management and to fully understand the spirit and intent of ICAO SARPs.

State CAAs will adopt one of two alternative strategies when developing an SSP and defining SMS requirements for service providers. The easiest implementation option is for the State to adopt the ICAO SMS standards as they are and merely duplicate them in State regulations. This may be the preferred option for States with small regulatory bodies with limited resources. IATA has taken this approach during the development of the current IATA IOSA standard, that is, it reflects a basic SMS in compliance with ICAO guidelines. Similarly, EASA has approved the IS-BAO standard as a method for business aviation to meet basic ICAO SMS requirements.

In contrast, State CAAs have the option of adding additional requirements to the ICAO four components and twelve elements. The FAA, for example, is working closely with industry to ensure that the SMS requirements in the US are thoroughly vetted through an extensive and exhaustive validation process. As service providers develop their SMSs during the pilot program, the FAA will gather ideas on how best to design and implement the SSP.

This author believes that one of the thorny issues facing individual State regulatory authorities is defining the ALoS to be achieved by the SSP. What metric(s) does it use to define a State ALoS?

- An acceptable accident rate?
- An acceptable number of fatalities?
- An acceptable number of incidents?
- An acceptable number of employee-reported events, such as an altitude deviation (level bust)?
- An acceptable number of near misses (reported and investigated serious incidents)?
- An acceptable level of safety as measured by a specific parameter in a national database? Example: Aviation Safety Information Analysis and Sharing (ASIAS) system in the United States.

It should be apparent that publishing an acceptable accident rate or specific number of fatalities presents a significant public relations problem for the State. Perhaps the best way to deal with this issue is for the State to emphasize a continual improvement in safety without publishing a specific, acceptable target. Perhaps ICAO will embrace this approach during periodic Universal Safety Oversight Audit Program visits. This is ground yet to be traveled. In the end, each State will find a politically palatable way of dealing with this requirement.

An additional issue facing both the individual service provider and the regulatory authority approving the SMS is gaining consensus as to what an SMS is and what it is not. It is absolutely imperative that we do not take the approach of *I'll know it when I see it*. This will inevitably lead to conflict and disagreement. There are several techniques that will smooth the process of gaining State approval of your SMS. In many States, approval of the SMS may be delegated to a local CAA office that has responsibility for oversight of the service provider's operating certificate. The following techniques may be useful in building a positive relationship with those CAA representatives who will interface directly with the service provider's leadership team and ultimately approve the SMS:

- Partner with the regulatory authority. The value of a collaborative relationship is to ensure that the SMS and SSP are interfaced appropriately during the parallel development process. There is value in exchanging views during the entire developmental process, as the State is required to accept and oversee the development, implementation and operational performance of the service provider's SMS. Partnering provides an opportunity to develop trust, exchange perspectives and to clarify the rationale behind your SMS procedures and processes. Ongoing dialog during the SMS development process provides an opportunity to resolve differences of opinions. Surprising the CAA with your SMS proposal will likely generate delays and the possibility of disapproval.

- Take a leadership role. There is immense value in assuming leadership in building your SMS. Most regulatory officials are comfortable in an oversight or inspector role as this is how they have been trained. Design the SMS to meet your organizational needs and solicit input along the way from the regulatory representative. Avoid asking questions such as 'What does it take to meet the standard?' 'How do you want me to write this?' Again, take the lead and respectfully walk the inspector through your logic and explain how your system meets the regulatory requirements. Leadership means doing your homework and being an expert in understanding the SMS framework (components and elements) and State regulatory guidance and materials. As the service provider's SMS champion, leadership means being an SMS subject matter expert, fluent with SMS terminology and vocabulary, demonstrating an understanding of the integration of SMS processes, and knowledgeable of the State regulations. The regulator needs to have confidence that the service provider knows what they're doing as evidenced by a solid strategic plan.
- Identify an SMS champion. The accountable executive will likely assign day-to-day leadership of SMS design and implementation to the head of safety or quality. The SMS champion may serve as the single point of contact with the CAA on all SMS implementation issues.
- Learn together. Develop an understanding with the regulatory representative that alignment of the service provider's SMS and the regulatory SSP is essential for both organizations. A learning relationship permits a healthy exchange of ideas and alternative techniques that is essential to designing an efficient and effective system. It is to the service provider's advantage to have an input into the design of the local CAA office SMS oversight process. The CAA oversight process needs to be efficient and not burdensome to the service provider. An efficient oversight system will enhance the working relationship over the long term.
- Transparency and openness with the regulatory authority. Regulatory representatives frequently have limited management experience and will therefore be uncomfortable in leading a discussion about the technical aspects of an SMS. Stay in the roles for which each representative team was trained. Again, walk the regulatory representative through applicable process maps, procedures and supporting documentation. Transparency includes acknowledgement of SMS design and implementation challenges. This candid and open communication develops trust, a valuable commodity when both organizations are developing new processes.
- Educational mindset. Regulatory representatives may be unfamiliar with the details of the service provider's management systems and will require appropriate familiarization training so they understand how SMS is

designed into the enterprise management system.

- Attain agreement on performance metrics. Safety performance is to an SMS as ALoS is to an SSP. More specifically, what parameters will be used to measure safety performance and what will be the acceptable targets to the service provider? The regulator will need to agree with the operator on these safety performance indicators and targets as representative of the generic hazards in the operational context of the service provider's delivery of services. This activity is representative of the CAA's transition to a performance-based oversight process and must be fully understood by all stakeholders.

- Communicate. It is imperative that both the service provider and CAA representatives communicate frequently the status of SMS and SSP development, implementation and interface. Challenges to design and development should be shared openly and candidly. A strong sense of teamwork and partnership is invaluable during the implementation phase. A sense of urgency can be established throughout the organization by communicating that the SSP and SMS will support organizational goals of safety, efficiency, profitability and organizational alignment.

- Invitation to service provider's SMS training. ICAO requires the regulatory representative to understand the service provider's management system, that is, a prerequisite to SMS approval by the State. Employee training will include a brief description of the regulator's new role in monitoring the service provider's SMS and safety performance metrics. Active participation, or at least representation during SMS training, will quickly escalate the regulator's comfort level with the service provider's SMS, gain confidence that the SMS is being implemented properly and gain an understanding of the challenges facing the service provider's management team during implementation.

- Anticipate friction and frustration. Service providers should anticipate the possibility that local CAA offices may resist the concept of an SSP in favor of traditional oversight composed of inspections and work observations. Since SMS oversight requires a significant change in CAA thinking and process, there will be pockets of resistance to transition from a familiar culture of traditional regulatory compliance to oversight focused on management system and process effectiveness. Evaluating systems and processes requires a unique skill set that is learned through training and personal experience. Inspectors, particularly those who have not been in a management position, may be well outside their comfort zone and feel that their inability to think in terms of operational systems will result in personal embarrassment. The service provider must deal with these situations very carefully, much like similar challenges within your own organization.

As a final thought, there are those that believe implementing an SMS will relieve some of the State CAA emphasis on regulatory compliance. This is simply not the case. Regulatory compliance will always serve as a minimum standard; however, the SSP methodology of ensuring compliance will evolve into a combination of inspection and performance-based oversight strategy.

References

The Committee of Sponsoring Organizations of the Treadway Commission (2004). *Enterprise Risk Management – Integrated Framework.* Available from http://www.coso.org/documents/COSDO_ERM_ExecutiveSummary.pdf.

Federal Aviation Administration (2006). *Advisory Circular 120-59A, April 17. Air Carrier Internal Evaluation Programs*, July 15 2009, revision 2. http://www.acsf.aero/attachments/files/88/SMS_Framework_Revision2_07-15-09.pdf.

Federal Aviation Administration Air Traffic Organization (2008). *Safety Management System Manual.*

Federal Aviation Administration (2010). *Element 3.2 Management of Change. Safety Management System Framework*, June 1, revision 3, p. 23.

Feynman, R. (1965). *The Character of Physical Law.* London: Cornell University and the BBC.

International Air Transport Association (2007). *Integrated Airline Management System.* Montreal: International Air Transport Association.

International Air Transport Association (2010a). *Safety Management Systems Implementation Guide.*Montreal: International Air Transport Association.

International Air Transport Association (2010b). *Introduction to Safety Management Systems*, 2nd edn. Montreal: International Air Transport Association.

International Civil Aviation Organization (2009). *Safety Management Manual*, 2nd edn. Montreal: International Civil Aviation Organization.

Kaplan, R.S., Mikes, A., Simons, R., Tufano, P. and Hofmann, M. (2009). Managing risk in the real world. *Harvard Business Review*, October, 75.

Maurino, D. (2005). Threat and error management (TEM). Presented to the Canadian Aviation Safety Seminar, April 18–20.

Quinn, R.E. (1996). *Deep Change.* New York: John Wiley and Sons, Inc.

Risk and Insurance Management Society, Inc. (2009). *The 2008 Financial Crisis.* New York: RIMS Executive Report.

Stolzer, H. and Goglia, J.J. (2008). *Safety Management Systems in Aviation.* Ashgate Publishing Company.

Taleb, N.N., Goldstein, D.G. and Spitznagel, M.W. (October 2009). *The six mistakes executives make in risk management.* Harvard Business Review, October, 1–5.

Transport Canada Advisory Circular 107-001 (2001). *Guidance on Safety Management Systems Development.* Ottawa: Transport Canada.

UK Civil Aviation Authority (2008). *Safety Management Systems Guidance to Organizations*. London: UK Civil Aviation Authority.

von Thaden, T.L. (November 2007). Measuring safety culture in commercial aviation operations. Helicopter Association International, Air Tour Summit. Las Vegas, NV.

Von Thaden, T.L. and Wilson, J.K. (2008). The safety culture indicator scale measurement system (SCISMS). Federal Aviation Administration (NTIS) Technical Report HFD-08-03/FAA-08-02. Conference paper.

Yantiss, Capt. B. (2006). An *integrated* airline management system. Presentation ath the FAA Conference- Risk Analysis and Safety Performance in Aviation. Atlantic City, New Jersey, September 19.

Chapter 7

SMS Training: An Overview

Larry McCarroll

The following is written for those that reside or will soon reside in a Safety Management System (SMS). Each resident will have a role to play in managing safety, and the following chapter attempts to assist in the training needs to fulfill those roles.

According to Peter Senge (2006), 'Organizations learn only through individuals who learn. Individual learning does not guarantee organizational learning. But without it, no organizational learning occurs.' If this is so, then an effective SMS is dependent upon being informed, and continuous learning of a safety culture begins with the individual. Under SMS, the individual accepts and uses risk management and quality management tools in the normal performance of duties and tasks. This can only occur through understanding the *why* that comes through training, and demonstrating the utility, successes, need for improvement and commitment through the actions of the accountable executives, managers and supervisors and continuing to the front line employees in the organization.

The most important individual in training your SMS is your SMS champion. They are your chosen advocate to carry the SMS banner and help to persuade, facilitate and train. The choice of a champion is important as their enthusiasm, credibility and knowledge will be a critical component in the building of an effective and embraced SMS.

In reference to knowledge, a caveat from Alexander Pope (1709) seems to suit the SMS champion:

> A little Learning is a dang'rous Thing;
> Drink deep, or taste not the Pierian Spring;
> There shallow Draughts intoxicate the Brain; and
> Drinking largely sobers us again.

This caveat reminds us that to have an effective SMS, it must be embedded within our business as a core component. Accomplishing that goal will require a champion who not only understands the balance of *production* and *protection*, but also one who is consistent and honest in its application. This is the banner the SMS champion carries. Though not alone in this journey, the first steps taken will be theirs, and their knowledge and skills will blaze and ease the trail for the rest that follow.

What You Need to Know

Strategic and Tactile Training

The ICAO *Safety Management Manual 9859* is the primary source document for SMS education and knowledge. To appreciate many of the nuances and supporting philosophy of SMS, texts such as *Managing the Risks of Organizational Accidents* (Reason, 1997), *Safety Management Systems in Aviation* (Stolzer, Halford and Goglia, 2008), and *Just Culture-Balancing Safety and Accountability* (Dekker, 2007) would make an excellent SMS reference library, as will this one. These books are mentioned for the critical concepts they bring to creating an effective SMS. As an example, *Managing the Risks of Organizational Accidents* will assist in avoiding the latent conditions' that organizations unwittingly create or overlook.[1] *Safety Management Systems in Aviation* ties in the concepts of quality management, ISO, and Deming's Plan Do Check Act mantra with the risk management concepts of SMS. These concepts will be critical to the training language and conversations across the various silos where the SMS champion will require a *hook* to sell and educate on the subject of SMS. *Just Culture – Balancing Safety and Accountability* explains that a punitive culture will not beget a *reporting* culture essential to an informed learning SMS. However, a *just culture* is still an accountable one.

Obviously, this is not a comprehensive list of excellent books on SMS; these are just a few that the author believes will assist readers in their SMS journey.

The purpose of this book is SMS application by those who live and work within it. This chapter will emphasize training required to facilitate one's role within an SMS and teach application skills within this management system.

The ICAO *Safety Management Manual* section 9.11.4 (2009) describes the following training recommendations:

> The safety manager provides current information and training related to safety issues relevant to the specific operations and operational units of the organization. The provision of *appropriate training* to all staff, regardless of their level in the organization, is an indication of management's commitment to an effective SMS. Safety training and education on should consist of the following:
>
> a.   A documented process to identify training requirements;
> b.   A validation process that measures the effectiveness of training;
> c.   Initial (general safety) job-specific training;
> d.   Indoctrination/initial training incorporating SMS, including human factors and organizational factors; and
> e.   Recurrent safety training.

---

1   *Managing the Risks of Organizational Accidents* (Reason, 1997) has one of the best descriptive book covers this contributor has seen. After reading a chapter, review the cover. It will reinforce your understanding and it will become a training aid across all silos.

The *Manual* continues:

9.11.7 Safety training should follow a building block approach. Safety training for operational personnel should address safety responsibilities, including following all operating and safety procedures, and recognizing and reporting hazards. The training objectives include the organization's safety policy and SMS fundamentals and overview. The contents include the definition of hazards, consequences and risks, the safety risk management process, including roles and responsibilities and, quite fundamentally, safety reporting and the organization's safety reporting system(s).

9.11.8 Safety training for managers and supervisors should address safety responsibilities, including promoting the SMS and engaging operational personnel in hazard reporting. In addition to the training objectives established for operational personnel, training objectives for managers and supervisors should add a detailed knowledge of the safety process, hazard identification and safety risk assessment and mitigation, and change management. In addition to the contents specified for operational personnel, the training contents for supervisors and managers should include safety data analysis.

9.11.9 Safety training for senior managers should include safety responsibilities including compliance with national and organizational safety requirements, allocation of resources, ensuring effective inter-departmental safety communication and active promotion of the SMS. In addition to the objectives of the two previous employee groups, safety training for senior managers should add safety assurance and safety promotion, safety roles and responsibilities, and establishing acceptable level(s) of safety.

9.11.10 Lastly, safety training should include a special safety training for the Accountable Executive. This training session should be reasonably brief (it should not exceed one-half day), and it should intended to provide the Accountable Executive with a general awareness of the organization's SMS, including SMS roles and responsibilities, safety policy and objectives, safety risk management and safety assurance.

The cited ICAO references above offer a training structure, a syllabus. It gives us a *strategic* structure of 'what subjects need to be trained'.

These topics are excellent strategic training in the concepts and framework of SMS, especially in allocating the minimum subject matter knowledge that individual positions should have. However, these modules do not say what *appropriate training* is for the application of SMS to these positions. They give depth of knowledge, but we are looking for the *skill* to apply in our day-to-day duties. What is the interface? What is the tool that allows us to practice and apply the concepts and dictates of SMS? How does one teach and gain acceptance of identifying hazards and reporting hazards? What do we need to teach? In a speech to the Sixth FAA International Safety Forum, Captain Chelsey B. Sullenberger (Sully) made this comment about SMS and training:

> No matter what prism we use to look at safety – whether it is SMS, CRM, or some other viewpoint or practice – safety always requires a confluence of factors. I'll identify the four I believe are most important:
>
> 1. An absolute commitment to safety at every level of the organization, but especially at the very top;
> 2. A robust system that identifies and mitigates risks;
> 3. A culture that supports safety recording and encourages cooperative behavior; and
> 4. Effective training that provides professionals with well-learned fundamental skills, in-depth knowledge, and the judgment to handle the unanticipated. (Sullenberger, 2009)

Appropriate and effective training is the mandate we have been given in training SMS. How do we reach that appropriate and effective level of knowledge, skill and attitude? The first step is to remember the goal: training personnel to support the SMS by training hazard identification, controlling when able, reporting when needed, questioning process and procedural changes if necessary and continually monitoring. In return, reports will be *justly* treated; managers and supervisors will be *flexible* in receiving input from the *task experts*, and lessons will be *learned* and applied. This is the SMS contract between the organization and the individual. According to James Reason, a safety culture (our SMS) has four critical subcomponents mentioned above: flexible, reporting, learning, and just. These four subcomponents create the informed culture. To support the informed culture, we return to the individual's SMS.

As described in the Prince study (1999)

> The *only group* that can *identify hazards consistently, accurately, and dependably* are those engaged in the operation under study ... it requires the 'E' in SME (subject matter *expert*) be part of the team ... the real world experience of the *frontline employee* to name for us those hazards they live with everyday. (Stolzer, Halford, and Goglia, 2008, p. 115)

> The safety culture of an organization is the product of individual and group values, attitudes, competencies, and *patterns of behavior* that determine the commitment to and the style and proficiency of an organization's health and safety programs. (Reason, 1997, p. 194)

Thus, an effective SMS formula begins to look like this: SMEs + SMS knowledge, skills and attitudes = an informed, reporting and learning culture. However, the culture is not complete; *just* and *flexible* have to be part of the formula and these attributes come from the accountable executive's absolute commitment to the SMS which assure the flexibility and the *just* treatment within a reporting and resultant informed culture. So to reach Dr. Reason's informed safety culture (an effective SMS), the formula might look as follows:

$$\{(Just \times SMEs) + KSAsms\} + (Just + Flexible) \times | CEOsms | = Effective\ SMS$$

Tactile

Above we described an *informed* culture having four subcomponents. Dr. Patrick Hudson of Leiden University states that there is an additional component; '*wary, ready for the unexpected*' (Hudson, 2006).

So how do we make SMS citizens, the ones who *touch* the product, 'wary' without making them weary? We are looking for a sense of awareness (wary) to identify hazards. Each job has unique hazards; each job requires the proper tool to assist in hazard identification, reporting and controlling. As students of SMS, we are familiar with the many variations of risk matrices as shown in Figure 6.11. Perhaps asking the following questions and deciding where the answers would fall within a risk matrix can answer the appropriate level of SMS training. The more risk, the more in-depth training is required.

- What potential risk impact does a particular position have to the organization, to the individual, to fellow workers?
- What tools are given to the individual to manage risks?
- Where would potential impact consequences of actions and decisions fall within the matrix of risk management?
- What are the consequences of poor risk management knowledge or skills in a given position?
- Have we given the individual in this position enough training, and the tools necessary to support the SMS?

The CEO or accountable executive will definitely impact the risk of an organization, after all they decide the degree of acceptance of the SMS culture; strong, tepid, bureaucratic, or otherwise. The SMS buy-in from this accountable executive can be facilitated by emphasizing the quality management aspects of SMS and the application of Deming's concept of Plan Do Check Act. Thus, training of SMS objectives and philosophy are key as the accountable executive will be the one allocating the resources and will need to see the practicality in production versus protection allocation. However, the heavy lifting and supporting of the SMS pillars will lie in the hands of the frontline, high-risk employees that 'touch the product' and those managers that decide the working processes, environment and procedures of those employees. Identifying and calibrating their impact and control of risk dictate the depth of training required. Those that confront hazard the most need the tools the most. However, since hazards are everywhere, we all need tools to shield, identify, report, control and assist in mitigation.

So what are the tools and how do we train for their use? The easiest topics to teach are the ones that have a tangible return to the student. Therefore, we need a *hook*, a hook that appeals to the 'what's in it for me?' syndrome. The hook is something they can use in everyday life; it is not an addendum to the duties of the job, it reinforces and protects the individual and the organization. The hook for the CEO will be slightly different from the flight attendant, the pilot, the cargo worker, flight instructor, etc. However, the right tool for the job is the requirement, a tool simple enough to use, but robust enough to support the SMS pillars and the individual. A practical risk management model is such a tool. In this case, I refer to the 3P model of Aeronautical Decision Making (ADM) as shown in Figure 7.1. The commonality to both the organization and individual is the requirement to perceive (reporting), process (flexible, learn) and perform (informed). These attributes solidify the connection between the organizational SMS (macro) and the individual SMS (micro). This is the teaching and practical application tool.

From Figure 7.1, it can be seen the 3P model connects (interfaces) the organizational need to perceive, process and perform with the individual's need to do the same. So how does this work? In the traditional use of this model, the pilot is used as the frontline employee subject to hazards and in need of this risk management tool giving us the P as in pilot in the PAVE mnemonic. However, it can be used in any position as will be illustrated later when we introduce DAVE and RAVE (see Table 7.1 for definitions).

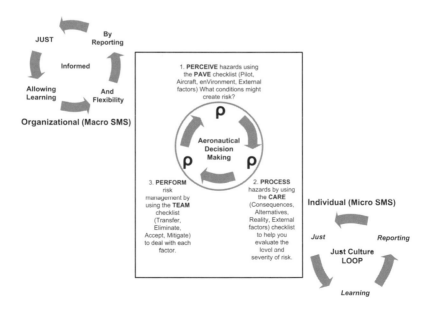

Figure 7.1   The 3P model and the graphics of macro and micro SMS

Table 7.1   The definitions and tools utilized for risk management decision path

Definitions of 3P	Tools of utilization	Acronyms of the tools
Perceive hazards	PAVE	Pilot
		Aircraft
		Environment
		External factors
Process risk level	CARE	Consequences
		Alternatives
		Reality
		External factors
Perform risk management	TEAM	Transfer
		Eliminate
		Accept
		Mitigate

Source: FAA (2008).

In the case of the pilot: perceive hazards with PAVE, process hazards with CARE, perform risk management with TEAM.

Start with the SMS job position in this case, a pilot. The 3P model tells us to perceive with PAVE:

- Do we *perceive* any hazards associated with the *Pilot* (illness, medication, stress, alcohol, fatigue, emotion, experience, currency etc.)?
- Do we perceive any hazards associated with the *Aircraft* (minimum equipment list items, mission capabilities, fuel reserves, unfamiliar avionics, etc.)?
- Do we perceive any hazards associated with the *enVironment* (weather, surface conditions, obstacles, congestion, construction, lighting, facilities, etc.)?
- Do we perceive any hazards associated by *External* factors (economic, appointments, schedules, passengers, ATC, etc.)?

If a hazard is identified, it is then *processed* with CARE to determine risk.

- What are the *Consequences (risks)* if this hazard is encountered?
- What are the *Alternatives*? (Evaluate all available options.)
- What is the *Reality*? (No hoping and wishing.)

- What are the *External* factors? (Be mindful of external factors that may affect safety decisions such as economics, get-home-itis, pride, etc.)

Perform risk management by mitigate or eliminate; how do we *mitigate* (control) or *eliminate* the hazards?

As an example, let's see how the 3P model works for a pilot, a dispatcher and a ramp worker. The model will have PAVE (pilot), DAVE (dispatcher) and RAVE (ramp worker). The potential hazard in this example is the forecast of freezing rain at the departure airport. This example will limit the discussion to ground operations.

The organization is a part 121 air transport company.

How do we perceive, process and perform for each of our SMS-savvy citizens?

Again start with the SMS job position. With each position the 3P questions are similar:

- Do we *perceive* any hazards associated with the Pilot (illness, medication, stress, alcohol, fatigue, emotion, experience, currency etc.)? The dispatcher? The ramp worker? *No, all are fit and ready to work.*
- Do we perceive any hazards associated with the Aircraft (and accessories or equipment utilized by the Dispatcher and Ramp Worker)? *The Ramp Worker has been assigned a belt-loader needing a set of new brakes.*
- Do we perceive any hazards associated with the enVironment (weather, surface conditions, obstacles, congestion, construction, lighting, facilities, etc.)? *Freezing rain is in the forecast.*
- Do we perceive any hazards associated with external factors (economic, appointments, schedules, passengers, ATC, etc.)? The 121 carrier is a scheduled airline and origination delays at the hub can cause accumulating delays throughout the day and missed connections. If there is a long delay, the pilots could miss their deadhead flight home tonight, possibly increasing the pressure to rush to catch up. The dispatcher's work load will increase if delays accumulate and diversions begin creating workload saturation and the possibility of missed or overlooked information. The ramp could become colder, wetter and slippery. Also, an on-time performance bonus could be lost.

In keeping with the 3P model, if a hazard is identified it is then *processed* with CARE to determine risk.

- What are the consequences (risks) if this hazard is encountered? The pilot and ramp worker will encounter slippery ramp and taxiways increasing the likelihood of skidding and an accident with another aircraft, vehicle, or person; especially the ramp worker with the failing brakes.
- What are the alternatives? (Evaluate all available options.) First, exchange

the ramp worker's belt loader with one with better brakes and plan on everything operating slowly today. If braking is inadequate, dispatcher cancels all ground operations.

- What is the reality? (No hoping and wishing.) If delays begin, the pilots will be spending the night away from base again, no hoping and wishing will change that. The dispatcher workload will increase as the day progresses, and they will become more fatigued and task saturated. The ramp worker will be asked to stay overtime and become more fatigued, cold and damp.
- What are the external factors? (Be mindful of external factors that may affect safety decisions such as economics, get-home-itis, pride, etc.). Get-home-itis can affect all three employees and can cause any of them to rush in an attempt to get back to normal; however, the opposite often occurs.

In keeping with the 3P Model of perform with *TEAM*;[2] risk management, how do we *transfer* the risk? Allowing another employee to perform the tasks would transfer the risk from our SMS players, however, that does not resolve the organization's delayed operations and mitigate the hazard. If we can't transfer, ask the following questions.

- How do we *eliminate* the risk? Cancelling all operations would eliminate the risk, but have we balanced production and protection? Can the operation continue safely?
- If we accept the risk what are the controls?
- How do we *mitigate and manage* the risks resulting from freezing rain?

In this example, due to the continuing delays the pilots could mitigate external pressures by calling the hotel for a reservation. This will alleviate the external pressure of get-home-itis. The dispatcher could request assistance to alleviate task saturation and eliminate the risks of fatigue. The dispatcher and pilot will continually monitor conditions and have decided when the risk is not acceptable (nil braking conditions, strong crosswinds, and deicing fluid availability, etc.). For the ramp worker, additional layers of clothing if electing to work overtime or refuse overtime if too fatigued. For all our SMS players, slow down and stay 'wary'.

The above example is a simple demonstration of the FAA's Practical Risk Management Model and there are no doubt additional solutions to the example. However, the purpose of this illustration is to demonstrate the flexibility of a model that can be readily taught to people who may not be accustomed to hazard identification and risk management. It also has the advantage of giving everyone a risk management tool that can be used in all facets of their life. It is a simple tool

---

2   A recent modification of the 3P model has the Perform with TEAM portion reduced to Perform with ME (mitigate or eliminate the risk)

that can enhance the individual's SMS and in the process enhance and support the organizational SMS.

Safety Change Management (The SMS Litmus Test)

Living in an SMS can be challenging. The varying levels of SMS knowledge and application ability, especially in a developing SMS, can create disagreement due mostly to misunderstanding or too vague guidance, especially in the realm of *Safety Change Management*. Thus, 'appropriate and effective training of SMS' becomes even more important in regards to safety change management.

According to Stolzer et al. (2008):

> After this much hard work, the team will be highly motivated to assure that the system (SMS) they so diligently constructed does not decay because of uncoordinated action by individual stakeholders. This is accomplished by building into the process description the steps required to *coordinate changes* with *ALL stakeholders before the change takes effect* and in most cases it is *best to give stakeholders input into the design of the change itself.* (Stolzer, Halford, and Goglia, 2008, p. 192)

In addition, Stolzer et al. state (2008), 'Without the commitment to effective safety change management, no other aspect of SMS matters, regardless of how well-developed it might be.'

Thus, an agreement of what constitutes safety change management should be agreed upon, along with direct input from stakeholders from the various silos and SMEs is obligatory. For expediency or economic savings, it is easy to fall back into the silo mode of thinking 'we know best and we have always done it this way'. However, SMS does a continual gap analysis by continually seeking to learn by being informed. The 'desk-pertise' of SMS decision makers will occasionally be challenged in an informed safety culture if the frontline SMEs' input and data are allowed to flow and be processed. Traditional and tribal knowledge will be challenged or validated by continually renewed data, renewed inputs and renewed information. SMS will be supplying metrics that may not have existed before and now allow measurements of performance not seen before. It could very well rock one's silo. So how does one *decide* to apply safety change management? In the previous section, a familiar risk management model was used to illustrate the commonalities of the organization and the individual via the 3P model. The ADM DECIDE model may also provide a convenient utility. Although designed as a tactical decision tool for aviators, it can be utilized in understanding safety change management. The traditional DECIDE model is as follows:

Detect: the decision maker detects that change has occurred.
Estimate: estimate the need for counter measures or react to the change.
Choose: chooses a desirable outcome.
Identify: identification of actions, which will successfully control the change.
Do: implement the necessary actions.
Evaluate: evaluates the effect of action countering the change.

However, one slight modification is suggested to the DECIDE model to fit into the safety change management process, the addition of an 'I', seeking *Input* from the frontline employee, the subject matter experts (SMEs) creating the DECIIDE model.

The *evaluate* (E) questions to ask in safety change management include: does the decision (control) balance production and protection? Does it introduce any additional hazards? If so, what are the controls? Have the frontline stakeholders' input been sought? If I was the frontline employee directly affected, would I understand why the change was made, how the change is managing risk and how it is enhancing safety, improving the product? Are other departments (silos) affected by this decision? What is their input? These questions are examples of silo-busting questions and the use of the DECIIDE model is recommended to get us to think outside our silo and desk-pertise. The lesson here being, in a systematic approach to safety, changes can have effects across many silos and; once again as described in the Prince study (1999), 'the only group that can identify hazards consistently, accurately and dependably are those engaged in the operation under study'. To get the correct answers, one needs to ask the correct questions to the correct person and or silos.

Common Errors

1. Looking at SMS as a box-checking exercise. Having an SMS means following through, documenting, measuring, monitoring and showing improvement.
2. Not communicating across silos or departments. Not asking the silo-busting questions. (How does this change affect, the dispatcher, the flight attendant, the student, the ramp agent, etc.?)
3. Assuming formal risk management is a normal part of people's life. We all do risk management to an extent (survival instinct); however, many will need to be educated and given a tool or tools to facilitate the understanding.
4. Not agreeing to what safety change management is and when to utilize it.

Summary: Do's and Don'ts

The Do's

1. Find a credible SMS champion who is enthused and willing to spend the time to educate themself on the concepts, goals and application of SMS. Their ability to explain, facilitate (sell) and educate the various silos will ease the training and accelerate SMS acceptance and implementation and integrating silo communication.
2. Involve your local state agency (e.g., FAA, CAA) from the beginning. The training will be symbiotic.
3. Define safety change management process in your organization
4. Involve the labor unions, if any, from the beginning. SMS has to be inclusive; after all, it is 'our SMS'.
5. Have examples of success stories utilizing SMS (for example the oil industry, nuclear power industry, Air Tran-Sat).
6. Stress the quality management side of SMS as well as the safety components.
7. Have specific examples of how it can enhance safety in various silos (for example demonstrating the 3P model to a dispatcher, flight attendant, cargo loader).
8. Emphasize the pillars or portions of the pillars that already exist today as evident by the gap analysis.
9. Follow through. Continuity is the essence of SMS. Emphasize that a hazard identified and not controlled is a risk that may become an incident or accident.
10. Fulfill the SMS contract. Emphasize this is *our SMS*, from the CEO to the entry level position of the company. All must be committed and have the ability and responsibility to assist in risk management in a just and informed culture.
11. Document. When in doubt, document.
12. If attending an SMS school be sure to get more than concepts; make sure there are tools to apply SMS.[3]
13. Use your SMS Tools.[4]
14. Be wary. Be honest.

The Don'ts

1. Don't delay appointing an SMS champion to begin their education process.
2. Don't forget to seek input from the frontline employees (i.e., SMEs).

---

3 IS-BAO has created a comprehensive SMS toolkit and an E learning course for SMS.

4 Various organizations such as IHST (International Helicopter Safety Team) offer SMS tool kits to assist in training SMS.

3. Don't forget to be wary and honest.
4. Don't shoot the messenger.
5. Don't rest on your successes.

Last Word

The following are excellent SMS training resources and associated websites:

- International Civil Aviation Organization http://www.icao.int/anb/safetymanagement/training/training.html
- Federal Aviation Administration (Mitre) http://www.mitrecaasd.org/SMS/documents.html
- Transport Canada http://www.tc.gc.ca/CivilAviation/sms/menu.htm
- Civil Aviation Safety Authority – Australia http://www.casa.gov.au/scripts/nc.dll?WCMS:STANDARD::pc=PC_91436International Business Aviation Operators (IS-BAO) http://www.ibac.org/safety_smslibrary.php

References

Dekker, S. (2007). *Just Culture – Balancing Safety and Accountability*. Farnham: Ashgate Publishing.

Federal Aviation Administration (2008). *Pilot's Handbook of Aeronatuical Knowledge, FAA-H8083-25A*. Washington, DC: Federal Aviation Administration.

Hudson, Dr. P. (2006). Characteristics of a safety culture. Presented at the *ICAO Seminar Baku: Safety Culture; Informed Just and Fair*. April.

International Civil Aviation Organization (2009). *SMS Training Modules*, 2nd edn. Retrieved from http://www.icao.int/anb/safetyManagement/training%5Ctraining.html.

Pope, A. (1709). *Essay on Criticism*.

Prince, A. (1999). The effect of experience on the perception of decision scenarios: A field study. Ph.D. dissertation. University of South Florida. Dissertation Abstract International DAI-3 60/08.

Reason, J. (1997). *Managing the Risks of Organizational Accidents*. Farnham: Ashgate Publishing.

Senge, P.M. (2006). *The Fifth Discipline, the Art and Practice of the Learning Organization*, p. 129. New York: Doubleday.

Stolzer, A.J., Halford, C.D. and Goglia, J.J. (2008). *Safety Management Systems in Aviation*. Farnham: Ashgate Publishing.

Sullenberger, Captain C.B. (2009). Speech to the Sixth Annual FAA International Safety Forum, Washington, DC.

Chapter 8

Quantum Safety Metrics: A Practical Method to Measure Safety

D Smith

The basis and premise for this chapter is best explained by relaying a true account of a revolutionary experience I had many years ago. At the time, I was the Director of Safety for a large international on-demand aviation operator. One of our regional operators experienced three catastrophic accidents within a six-month period. It was after the third accident that the CEO ordered an operational stand-down and full internal audit. The boss directed me to assemble and lead the audit team that consisted of several subject-matter experts from our corporate office and the field. Included were operations specialists, maintenance experts, ramp support personnel and, of course, me. I contacted the regional manager in charge of the operation and informed him of our arrival date and intentions for the audit. I was immediately troubled when he didn't seem to think the audit was necessary, nor was it a big deal. He was more concerned about getting us in and out so he could resume operations. The morning we arrived, I scheduled a desk-side in-brief for him intended to further explain and provide more details about the audit. To put it mildly, the briefing did not go well for me. I barely got a word in edgewise as he explained to me that I had about one or two days to complete the audit and get out of his hair. He further went on to explain that he had already conducted his own internal audit and determined that each of the catastrophic accidents were totally unrelated, coincidental, and in no way connected to one common denominator or causal factor. I politely listened to his brief and secured his agreement to begin my audit immediately so we could wrap it up and get on our way.

I had two very troubling questions, both begging answers and sticking in the pit of my gut like a cheap, greasy two-pound hamburger. First of all, how did this organization pass, and with flying colors I might add, an external third-party audit just six months earlier, and secondly, why the lackluster attitude of the regional manager? I'd chew on those questions day and night for the duration of the audit and beyond. The answers to those questions opened my eyes to some very common fundamental flaws in measuring organizational safety. Those answers and insights are revealed in this chapter. The audit ended up taking the better part of five days as we scrutinized every area of the operation. We looked at their policies, processes, written procedures, pilot training program, maintenance and dispatch operations, ramp and ground operations, quality assurance and, of course, their safety office. At the end of each day, we held a team meeting to discuss team activities and results. It didn't take long to identify a common theme for the organization. Almost every department head, supervisor, and many of the line employees shared the same lackluster attitude as the regional manager. The organization had very sound written policies and standard operating procedures. The problem was they didn't follow or adhere to them. Conducting operations and getting the job done was the one single and most important thing, and it really didn't matter how, as long as the end result was mission completion.

When I looked closer at the previous third-party audit, it was apparent to me that the organization ramped up for the audit. In a manner of speaking, the audit was the mission. It received heavy management emphasis and the order went out to prepare and do well on the audit. Trouble is, once the audit was over, SOPs, processes, policies and procedures were disregarded and forgotten. The one thing that impressed me was the fact that the organization could effectively pass the audit and exert zero effort toward safety, protection, or accident prevention. It seemed to me that we missed the proverbial boat and were measuring the wrong thing. Rather than measuring how well an organization can prepare and negotiate an audit, we should be measuring the current level of effort given to safety, protection, and/or accident prevention.

Many months of pondering and conferring with folks smarter than me, and multiple 'throne of knowledge' sessions yielded the following chapter on measuring organizational safety effort. Quantum safety metrics and measuring accident prevention effort are further explained in this chapter. The bottom line is, if you want to accurately measure safety or accident prevention potential, you must determine the organization's current level of effort directed at safety assurance. You must determine the current level of *accident prevention effort* (APE). Quantum safety metrics provides the means for you to accomplish that.

With little effort, a short amount of planning time and minimal resources, you can easily design your own quantum safety metrics program tailored to meet your organizational safety assurance needs.

What is Safety Assurance

One of the very basic tenants of safety assurance (SA), as related to Safety Management Systems (SMS), is that safety assurance does not measure output; it measures process effectiveness or process goals. The concept of measuring process effectiveness verses output is what separates safety assurance from quality assurance (QA). I've known organizations to assign the SMS safety assurance responsibility to their QA department. I believe there is a fundamental flaw with that logic because QA and SA have two totally separate objectives. Let's further explore the difference between the two. Most of us are familiar with the old adage 'The end justifies the means'. It's a great illustration of how different QA and SA really are. Let's say your organization has a production output objective of producing 100 steel washers per eight hour work shift. The washers begin as molten steel, which is pressed and flat rolled into sheets of metal, stamp cut into the washer, sandblasted smooth, polished, painted and finally packaged for shipment. This is a potentially dangerous process requiring many safety risk reduction control measures designed to prevent accidents. Safety control measures might include things like employee personal protective equipment, machine guards that prevent an employee from assuming an unsafe position, waiting a designated period of time for metal to cool before handling it, or using the right tool for the job. It makes sense that your employees could disregard established safety control measures and continue to make production output goals. And, in fact, that very well might happen unless it's monitored for safety assurance. It's also easy to understand how management might have the false impression that your organization is operating safely simply because production goals are met. Of course, the reality is that every time a safety control measure is eliminated from the production process, the organization takes one step closer to an accident. Although sustained output, meeting production goals and profitability is a primary function and objective of safety controls, safety assurance is not determined by meeting production goals, it is determined by evaluating the process to ensure that established safety controls are in place, that they are adequate, and that they are functioning as designed. The end cannot justify the means if it's setting your organization up for a fall. You, the safety professional, must educate management to resist the temptation of measuring safety performance by output. It's not a job for your QA department; it's a job for a trained, qualified and experienced safety professional.

I also believe it's not so important to get hung up on technical distinctions. It doesn't matter if you call your safety assurance initiative an audit, a survey, an evaluation, or inspection. I've participated in round tables and think-tanks where their meaning and application were argued to death. Of course, there is technical difference between them, but that's not what's really important. What is important is that you establish a meaningful safety assurance initiative, call it what you will.

A common problem for many safety programs and initiatives is associated with gaining management support. The valid question of 'what's in it for me?' often determines the degree of safety commitment from the organization. One method safety managers can use to obtain support is found in the concept of quantum safety metrics.

Measuring safety, however, is not quite as simple as one might think. We can easily measure most types of production in an organization, but in order to quantify safety, we generally find ourselves trying to measure something that is not present: accidents. Even then, the safety professional knows that the absence of accidents does not necessarily mean a solid safety program is in place. Nor does having an accident indicate a malfunctioning or nonexistent safety program. Accident numbers are only part of the equation.

Another (and possibly more critical) element in the formula is something I refer to as *accident prevention effort* (APE). Each organization has a distinct and quantifiable APE that can be measured, evaluated and adjusted to obtain the maximum benefits from the safety program. Additionally, it can be used as a tool to provide management with the necessary information to make informed decisions on safety program outputs, resource needs and areas needing additional emphasis.

In order to fully understand the concept of APE, the following three assumptions are established. These assumptions are based on empirical data with respect to effective and ineffective safety programs.

1. **Assumption 1** – organizations which appoint an auxiliary or collateral duty safety program manager and are actively engaged in loss control and accident prevention efforts are more likely to affect accidental loss than organizations that are not. Example: an organization safety newsletter produced and distributed on a monthly basis that features related safety articles, specific safety guides and encourages employee contributions and participation.
2. **Assumption 2** – an organization's safety program is more effective and has a greater probability of effectiveness when the program manager receives formal training in safety concepts and principles. Example: a facility safety program manager trained and qualified in general industry as an OSHA 10- and 30-hour instructor.
3. **Assumption 3** – properly resourced safety programs are more capable and effective than under-funded or non-resourced programs. Example: a safety program budget is established, approved and reviewed by management on an annual basis.

These assumptions provide the basis for establishing measurable safety criteria within any given organization. Using this concept, a simple formula is used to determine a numerical rating to measure the safety programs and processes in place, the amount of activity within each program, the level of qualified and trained managers, and management support in terms of resources.

The following formula, known as the *Sierra Scale* provides a relative measure by which an organization can evaluate the effectiveness of their safety program:

(Programs × Quantifiable Effect) provides degree of Accident Prevention Effort (APE)

Or

(P × QE) APE

Factor Terms

What Constitutes a Program (P)?

Programs are defined as formal programs, processes, policies, or initiatives that are established and implemented by the organization to enhance or positively impact accident prevention. Examples might include: newsletters or other forms of accident prevention education and awareness, safety committees, accident prevention awards programs, risk management programs, monthly safety meetings, emergency action plans, fire prevention programs, FOQA, or establishing a formal system to track and analyze incidents, such as the Human Factors Analysis and Classification System (HFACS).

The more programs in place within the organization, the greater the opportunity to have a positive impact on accident prevention. Therefore, an increase in programs (*P*) will result in a higher APE. At this point in the formula, the effectiveness of each program is not evaluated. Only the number of formal programs in place is considered. Of course the program must be evaluated for effectiveness, we'll discuss that later.

What Constitutes a Quantifiable Effect (QE)?

Quantifiable effect can be described as the number of tangible effects resulting from any given program. For example, in the risk management program you would count the actual number of hazards identified and corrected for any particular operation. Each hazard controlled or eliminated would be counted as one QE. Likewise, if the organization used the HFACS to track and analyze incidents, each incident that was tracked and analyzed would be considered a quantifiable effect of that particular program.

The following scenarios provide examples of application.

- Your organization is planning a large construction project that will require the displacement of all company aircraft for a six-month period. An alternate non-standard taxiway has been identified for temporary use. The CEO has tasked you to compile an in-depth risk management plan for the operation. You identify and abate two major hazards for the operation. The QE for this process is two.
- You are the safety representative for the companies' quarterly safety committee meeting. During the meeting the committee addresses three major hazards that have been identified in the maintenance shop. Solutions are proposed, discussed and implemented. The QE for this activity would be three.
- The organization's monthly safety newsletter is required reading as established by the employee handbook. First line supervisors are required to conduct routine spot checks and oral examinations to ensure employees

have read and understood pertinent safety information. For the month of February the newsletter provided education and safety guidance on two primary areas of emphasis. Random sampling of the employees validated that they have been read and understood. The QE for the safety awareness program would be two, one for each respective area.

The same QE principle would apply for every accident prevention program established in the organization.

The equation can now be further developed by inserting the number of accident prevention programs established in your organization and the number of QE of each program.

Illustration 1

An organization has three accident prevention related programs and can quantify eight significant effects. The breakdown is as follows:

Programs (P)	Effects (QE)
Active risk management program	3
Monthly newsletter	2
Safety council	3
Totals: 3 programs	8 quantifiable effects

The equation would look like this:

$(P \times QE) APE$

Or

$(3 \times 8) 24$

For illustration purposes, let's say that in the illustration shown above the organization has programs in place, but the safety manager is a part time or collateral duty safety representative and has no formal training. The APE for this scenario is 24. The APE is assessed at the raw value of 24 because the organization does have programs in place and is yielding some degree of APE. However, it makes sense that if the collateral duty safety manager is trained, qualified and or credentialed, resourced and has demonstrated management support, the programs would have a greater potential for success. This rational will be further explained and demonstrated later.

By using this APE formula, cause and effect can be established. An increase in either accident prevention programs or QE will positively impact the APE.

This provides a safety metric by which an increase in APE can be numerically demonstrated to management. Accident prevention progress can be measured and documented, weaknesses can be identified, recommended improvements made and outcomes validated by measurable criteria.

However, two other factors can significantly impact the APE based on the three assumptions made earlier. The first is the degree of skill, training and experience of the safety manager. Obviously, an adequately trained, thus more capable, safety manager has a greater potential to experience better results. The second is program support in the form of demonstrated management commitment and resources. Both of these factors are incorporated into the equation as follows.

Illustration 2

Consider the same organization we used in the first illustration with one change. Let's say the safety manager is properly trained, qualified, and or credentialed to manage the program. In this case, we will increase the value of the program (P) to the second power. Why? The logical result of a more capable safety manager is a more effective program. Therefore, it is reasonable to allow for an increase in APE. For this scenario, the equation would look as follows:

$$(P^2 \times QE)\ APE$$

Or

$$(3^2 \times 8)\ 72$$

In the illustration shown above, the organization has the same three prevention programs in place and the safety manager is full time and properly trained or credentialed. The APE is 72 because it is reasonable to conclude that the program itself will function more efficiently and have a greater effect if the safety program manager is trained and qualified. We assess a greater degree of organizational APE for training the safety manager or the individual managing a given safety program.

Illustration 3

The final illustration depicts the same organization as Illustration 2. Again, we'll make one change and add another factor to the equation. In this illustration, the organization not only has a trained and qualified program manager, but they can also demonstrate management support and operational resources. Based on these factors, the program (P) in the equation is then increased to the third power. Once again, you might ask 'why?' The easiest way for me to answer is by simply referring you back to our third assumption; remember we agreed that an adequately resourced program has much greater potential for positive impact than a non-resourced program. Therefore, once again it is reasonable to allow for an increase in APE. For this scenario, the equation would look as follows:

$$(P^3 \times QE)\, APE$$

Or

$$(3^3 \times 8)\, 216$$

In this illustration, the organization has three programs in place. The program manager is full time, properly trained or credentialed and has management support with proper resourcing. The APE is 216. The P and QE remain the same as the previous illustration. However, the additional increase in APE is a result of management support and resources.

Notice that in each of the illustrations, the number of programs (P) and the number of QEs remained the same. The increase in APE is derived solely from properly trained program managers and adequately resourced programs. A very wise man once said that a good idea without resources, support or training is nothing more than a hallucination. Quantum safety metrics subscribes to and promotes that philosophy.

Several rules apply and require some clarification. They are listed below as follows.

First, it is important to understand that the only way an organization can be awarded a P to the third power is by having both a trained and qualified program manager and demonstrating management support. Both must exist in order to obtain third power APE credit.

Second, is a more descriptive definition of what a trained program manager means. In order to be awarded an APE increase for being trained and qualified, the program manager must have received the initial qualification training and be able to demonstrate a continuing education and professional development training plan. Continuation training must be received at a minimum of every two years to gain and maintain the increased APE benefit. In-house training can qualify for initial and continuation training. You set the rules, it's your organization, however, training and credentialing by an accredited organization is considered optimal.

Third, resourced means having adequate personnel and funds to administer a program demonstrated and documented management emphasis, involvement, participation and oversight of the program.

Fourth, it must be understood that because every organization is unique in size and scope, organizational APE is specific only to your organization and cannot be used to compare one organization to another.

Fifth, every organization must initially determine and define what they will call a program (P) and a QE. Once established, those definitions must then remain constant in order to demonstrate legitimate fluctuation in APE.

Last, your organization must establish criteria for evaluating programs and quantifiable effects. Your evaluation must determine if a program (P) is active and effective, if it will be awarded no credit, the raw value of a P, or an increased value of P^2 or P^3. I urge you to keep it simple.

Evaluations

The one thing we really haven't talked about yet is how to know if a program is designated as a P, P^2, or P^3. That can only be accomplished by looking at the program and determining which category it qualifies for. In order to qualify, it must meet the criteria. Here is your opportunity to tailor quantum safety metrics to your organization. Remember, you determine what constitutes a quantifiable effect. You determine what training qualifies someone to administer or oversee the program in question. You determine if it qualifies as a program in the first place. This makes perfect sense because you are the one designing, developing and implementing policy for the program in the first place. Since you designed the programs, only you can say if they are meeting your expectations. Therefore, it makes perfect sense that you should be the one to determine if the program is functioning as designed. The important thing is to establish the criteria and stick to it.

I do, however, have some suggestions and guiding principles that will help maintain stability and standardization during your evaluation of a program. Break your evaluation down into two separate steps for each program to be evaluated.

Step 1

Evaluate the administration of the program. By that I mean schedule a time and sit down with the individual responsible for administering the program. This portion of the evaluation can easily be accomplished in the office or at a desk. You'll want to look at and discuss the policy, SOP, training records and other established procedures for the program. Be sure to cover the five W's of the program – who does what, where, why, when – how the program is supposed to work at the execution level, and in theory. Be sure to ask the program administrator to explain to you how senior management demonstrates involvement in the program, how the program is monitored and how it is resourced.

Step 2

Get out on the floor and see if the program is actually working as designed. I call this a *performance-based evaluation*. Interview employees and ask managers, supervisors, and line workers to show you how they do things related to the program. Determine beforehand what percentage of folks you need to interview in order to get an accurate accounting of how things are actually working.

I remember one organization where the person in charge of implementing risk management showed me a very comprehensive flight risk management worksheet, policy and SOP. He explained that the policy required the Flight Captain to complete a risk management worksheet prior to every flight. Yet, he could not produce one file copy of a completed worksheet, nor could any flight captain tell me when they had last filled one out. It was a great program on paper but in reality it was nonexistent. I did not give the program credit for a P in my final report.

Quantum safety metrics (QSM) and the *Sierra Scale* provide your organization with leading safety assurance indicators, a method to measure and set goals, and the means to easily predict and project an increase in your organization's APE.

QSM makes safety assurance tangible because it provides management with concise, quick reference cost–benefit analysis information and measurable options to increase their overall APE. The scale can also be used as an internal evaluation tool by comparing current APE to desired APE. This, in turn, assists in the development of strategies to accomplish the desired APE goal. By establishing milestones for increased APE and conducting follow-up evaluations, the organization can demonstrate measurable growth and progression toward program enhancement.

Additionally, I believe that quantifiable effects are essentially accidents prevented and, as such, safety program managers can use the formula to illustrate the number of accidents avoided. Simply by evaluating the organization's historical data and records, one can determine past QE. The number of past QEs can then be used to estimate the number of potential accidents avoided for a given period in the past. The outcome is that ability to illustrate and quantify an actual number of accidents prevented.

Chapter 9

Auditing for the Small Service Provider or New Safety Professional

James Hobart

Audit. What does it mean? To some, it has a connotation of financial peril, but to the trained safety professional, an audit is a valuable tool to assess the state of a particular entity. For smaller aviation service providers and new members of a safety team at larger operations, the safety assurance (SA) component of SMS can be especially daunting, since auditing is an essential part of the SA process. The objective of this section is to de-mystify auditing by describing the elements of any audit, and the variety of audit types that can support an organization's SMS.

What, exactly, is an audit? The American Society of Quality defines an audit as 'a planned, independent and documented assessment to determine whether agreed upon requirements are met'. In this chapter, we will explore how this process works and the key words in that definition. To get started, we'll concentrate on a generalized audit model.

Audits are generally divided into two categories: internal and external. An internal audit is conducted entirely within an organization or department. Trained personnel conduct the audit under the direction of the quality manager or another designated person. Internal audits can be directed at any department within a company, but rarely encompass multiple departments or areas at the same time. An external audit is conducted by a third party consisting of persons specifically trained for that type of audit. External audits usually gather information from multiple departments or divisions for the final tally of findings and observations.

Whether internal or external, audits generally follow a consistent model. Elements within each audit consist of planning, performing, measuring, reporting and closure. Each of these elements will be discussed in detail.

Why do we audit? An audit gives us a 'snapshot' of the current state of affairs within the area or system (company) being audited. Audits, by themselves, do not increase safety. Rather, they tell the auditor the state of things at the moment. This information is very beneficial to the quality managers and safety personnel since it provides near real time information.

Audits can be narrowly focused or very broad in their scope. Defining the actual scope of an audit allows the auditor, and the auditee, to understand just what will be audited and approximately how long that audit will take.

Let's take a moment to review some important aspects of audits, and some definitions. We've already described the two classifications of audits: internal and external. The main difference is the scope of the audit.

Important Definitions

Audit program: structure used to accomplish audits; must be planned, documented and include the methods by which the audit will be conducted

Auditee: an organization or department that is being audited

Auditor: person or organization conducting the audit; if an organization is performing the audit, then it's an audit organization (AO)

Compliance: positive indication that a service supplier has met the requirements of a contract or regulation

Contractor: a person or entity under contract or agreement to furnish services or goods (vendor, supplier, etc.)

Evaluation: act of examining a product or service to a standard

Evidence: verifiable evidence or proof; facts that are verifiable

Finding: results of an examination, specifically to a particular question (IOSA: requirement to close findings)

Key word: something that gives an explanation or identification, or provides a solution

Observation: statement of fact substantiated by objective evidence (IOSA: no requirement to close observation)

Procedure: specific way to do a task; for example a takeoff is a procedure

Process: steps to produce a product, service or outcome; for example the before takeoff checklist

Quality: the degree to which a set of inherent characteristics fulfill requirements

Root cause: the fundamental deficiency that results in a nonconformance and must be corrected to prevent reoccurrence of the same or similar nonconformance

System: a group of processes all working in concert to achieve a universal goal

---

Audit Control System

The audit control system contains the various parts of the audit – planning, performing and measuring the audit. Each part requires thought on what is specifically desired from the audit.

Planning the audit is perhaps the most difficult part of the audit because, without proper planning, the audit might be compromised. Planning requires much forethought in order to make the audit work as it's supposed to.

For example, when do you want to conduct the audit? Should you conduct the audit after a major safety event? Ideally, audits should be conducted on a recurring schedule. Remember, audits are but a snapshot of a point in time. The best audit results come during the normal course of a business, not after a major event, because people are in a more reactive than proactive mode. Remember, things change dramatically after an event, so the answers you get won't be the answers you would have received prior to the event.

Planning takes time and thought. In the planning phase, one must consider what elements to audit, how many people it will take to conduct the audit and what standard will be applied. Elements of the audit might be the flight, maintenance, or dispatch departments. For larger or more diverse companies, the contract administration department or charter department might be an element to audit. Each has different criteria and standards to audit, but all make up the total business model. Of course, if the audit areas expand, it will require more people.

People are just one of the resources to consider when planning an audit. A one-person audit team could not do as much at one time as an audit team of say three or four people. However, no team should conduct more than one element at a time; never audit more than one area at a time. This might lead to confusion on both the audit team and auditee personnel. Other resources include such things as time, money, materials and possibly travel if there is more than one station or location to be audited.

Performing the audit is not as tedious as planning, but can be frustrating for both the auditee and auditor. The auditor must have patience and be able to ask questions to elicit the correct response, and the auditee must have the knowledge to be able to answer questions without any hesitation or concern. An auditee must know the requisite knowledge of their manuals to be able to locate answers quickly or, in lieu of that knowledge, know how to find the requested answer.

Performing the Audit

There are several things to take into consideration when preparing to audit: scope, schedule, purpose and an audit standard. These issues must be fully developed during the preparation phase in order to ensure a smooth audit.

Scope

Know what you are going to audit. Will this be a department audit, or an audit of something within that department? For example, you might not want to audit the entire flight department, but an audit of the training portion of flight department would be entirely realistic. It would take more resources to audit the entire department.

Schedule

Decide on the schedule for your audit. It is much better to devote a couple of full days to the audit than to perform it piecemeal, doing only a couple hours per day over a week's time. This prevents distractions and allows both the auditor and auditee to concentrate on the task at hand.

Purpose

Decide the purpose of the audit during the audit preparation phase. For example, it would be easy to audit training records but difficult to audit simulator training unless you have planned to travel to the simulator to observe the training being conducted. Of course, there might be other reasons to conduct the audit. Have reviews indicated that record-keeping practices have become substandard? Have several people failed checks on one particular area of a check ride?

Another purpose of an audit might be to check on an external supplier. In the above example, you might consider paying a visit to a simulator training facility to observe training being conducted; audit their training records and ensure you are getting the expected product. External suppliers should have measurable standards in their contract or purchase agreement to which they can be audited. The job of the quality department is to ensure those suppliers uphold their part of the contract. Other examples might be to check fuel quality upon delivery, that parts are of acceptable quality, and repair work is being done in accordance with set standards and/or regulations.

Standards

A standard is a reference point against which other things can be evaluated – a checklist, contract, or purchase agreement. In aviation, there are four generally accepted audit standards: IATA's Operational Safety Audit (IOSA), the International Standard for Business Aircraft Operations (IS-BAO), the Air Charter Safety Foundation (ACSF), and the FAA's Air Transport Oversight System (ATOS). Each has unique characteristics and purposes, and each will be more fully discussed later in this chapter.

Regulatory standards also exist that are not usually contained in an audit format. These regulations, domestic (FAA) or international (ICAO), can be audited as necessary and appropriated. Audits help determine a level of compliance with the respective regulatory authority and what, if anything, needs improving or changing.

Another standard also exists – the company operations manual(s). Each manual has processes and procedures that must be adhered to in order to maintain a safe operation. Auditing to this standard should be the least difficult since all persons using the manual should be very familiar with its contents. Unfortunately, it can be the least audited standard for that very reason. Unless an auditor, or audit team, monitors the adherence to company manuals, these manuals are sometimes misinterpreted or, worse yet, unused.

How does one make up an in-house audit checklist? There are several things to consider, but a good foundation is always necessary. The foundation is the main reason for any audit, internal or external. It consists of questions specifically designed to explore the area(s) of concern. One caution, however: do not conduct an audit immediately after an event, as there is too much confusion and/or concern over the event to get accurate answers.

When developing the in-house audit, be sure to consider recent written reports such as safety or incident reports. This provides a good foundation for the audit and gives a good reference if called upon to defend the audit. Always use written reports and not hearsay evidence when developing the audit. However, if the company has a confidential reporting system such as an in-house program or ASAP, take care to preserve the anonymity of the submitter when using those reports.

As mentioned earlier, other areas for the development of the audit are operations or training manuals. These manuals contain a myriad of specifics that should be monitored periodically using audit techniques. How often a specific area is audited depends on the rate, if any, of failures within that department. Of course, in a training environment, repeat check ride items should also be included on an audit. Not only does this provide information on check ride pass/fail rates, but also provides information for use in improving the training events and, quite possibly, to the FAA, should they wish to review training events and scenarios.

Operations and training manual procedures are sometimes best audited during an en route inspection. Results of line checks can be audited not only for completeness of the check form, but for items the check airman deems noteworthy. Results of any check ride are subject to an audit for repeat items, failures, etc.

Often overlooked are standards within a contract or purchase agreement. Flight departments, contrary to what some might believe, have overall responsibility for their training, even if done by an outside entity. For example, a corporate flight department would almost certainly contract simulator events for initial and recurrent training. Do not rely on a successful completion of a check ride as proof that the training is being accomplished according the standards set by a contract. Take the time to visit the training facility and audit a training scenario against the contract you have with that facility. Such audits need not be overly involved, but must be detailed enough to ensure the training facility is upholding its part of the contract. These audits are not only beneficial to the flight department in overseeing the training, but for the contract administration department to ensure the contracts they sign are upheld.

Audits

Now that we've discussed some audit principles, we need to discuss the audits themselves. In SMS, the *gap analysis* is the place to start. Any SMS program has to start with the gap analysis which is a 'how-goes-it' audit. The gap analysis encompasses all aspects of the company regarding how it stands prior to implementing an SMS program. This analysis is performed to look at the 'gap' between what is essential for your particular SMS program and what you presently have in place, and is generally performed by someone in the quality or safety department.

The test for company administration will be how to transition the available information and their understanding of SMS into a functional safety management system for their flight department in the most efficient and effective manner. The gap analysis is performed against generally acknowledged SMS concepts and components and is used to execute your SMS, as well as referenced and updated on a continuing basis. Because it is continually updated, it is a living document. The information gathered from the gap analysis is compared to the ICAO SMS framework in the 4 pillars of safety – safety policy, risk management, assurance and promotion – subdivided into the 14 related elements.

When completing a gap analysis comparison checklist, one only needs to answer 'yes' or 'no' to the analysis questions. Any 'no' answer requires further improvement on that particular SMS process.

Encompassed within the ICAO framework are such areas as management commitment and responsibility, documentation, hazard identification and training and education. Once the gap analysis is completed and fully documented, the items identified as absent or incomplete will form one basis of the SMS completion plan. Each item will be assessed to resolve how the company will form or amend policies, procedures, or processes to integrate the required SMS components and elements. Each component element or task should be assigned milestones, including a suspension date to ensure that completion does not fall outside suitable time limits.

IS-BAO Audit

Designed primarily for the business aircraft operator, the IS-BAO audit focuses on the SMS program instituted by the company. It centers on the SMS and how well the SMS works for the company. The IS-BAO is governed by the International Business Aircraft Council (IBAC) Governing Board which has oversight of the standards and audit processes. IS-BAO encompasses standards for domestic as well as international operations. Therefore, it is important to be familiar with ICAO Annex standards.

Some of the objectives of the IS-BAO audit are: to help the flight department identify any safety lapses, ensure the flight department is in conformity with IS-BAO standards, and make recommendation to IBAC for registration if so desired. The IS-BAO audit allows the successful aviation department to be placed on the IS-BAO Registry signifying compliance with the IS-BAO audit. The registration period is two years and is renewable only upon the completion of another audit. Registration begins with Level 1 and continues through Level 3, signifying the highest level of SMS integration.

The IS-BAO audit must be performed as a system audit – a set of processes all working collectively to reach a common purpose. All IS-BAO audits are conducted by accredited IS-BAO auditors who must comply with the qualifications set forth by the IS-BAO Standards Board. Among these qualifications are: five years experience in aviation operations or maintenance, working knowledge of aviation operations or maintenance and experience in aviation safety management. Auditor candidates must also attend a two-day course on IS-BAO auditing techniques and protocols.

The IS-BAO Audit

The audit protocols assist in leading the auditor through the process of determining whether or not the flight department is in compliance with the standards of the IS-BAO audit. Questions within the protocols are delineated as Standards and Recommended Practices; the difference being that compliance with standards is mandatory, while compliance with recommended practices is not. While these protocols are used to do a comprehensive assessment of the IS-BAO elements, it is important to remember the protocols are paraphrased standards of the IS-BAO. The audit protocols must be completed prior to, and reviewed at, the closing meeting.

The IS-BAO audit does not specifically discuss findings as audits might. It considers any finding either a major or minor nonconformity. A major nonconformity constitutes a significant risk to the safety of the operations while a minor nonconformity does not constitute a significant risk to safety (IBAC, 2010). Minor non-conformities might be isolated instances such as documentation lapses or some similar occurrence.

Now, let's take a couple of examples to demonstrate the difference between the protocol and the actual standard. IS-BAO standards are divided into elements, each having a different number. For example, *Training and Proficiency* is element 5, while item 5.1 is *Training Programmes*.

The Standard reads:

5.1   Training Programmes

5.1.4   No emergency or abnormal situations shall be simulated during flight when passengers are being carried.

Looking at the key words, we can see the standard is very specific – *No* emergency, *shall* be, and *are carried*. This standard is used to ensure, through the operations and/or training manuals, the company does not conduct any emergency or abnormal training while carrying passengers. Of course this is common sense, but, like most safety programs, this restriction must be spelled out in the appropriate manual.

Now, let's see how the protocol is worded:

5.1.4   Does the operator prohibit simulated emergency or abnormal situations in flight with passengers on board? (IBAC, 2010)

In this case, a key word would be *prohibit*. But, just how does the operator go about stating this prohibition in its manuals, and ensuring the flight crews and training department understand and adhere to the restriction?

Another example of an IS-BAO protocol is 5.5.1:

> 5.5.1   Does the operator have a proficiency certification system to ensure that for all required crewmember training courses the training objectives have been met? (IBAC, 2010)

Again, the auditor must interpret what that certification system must contain in order to comply with the standard, which reads:

> 5.5.1   National civil aviation regulations vary in the requirements and processes for proficiency certification for aircraft crew members. Operators must ensure that personnel meet national proficiency requirements but shall at least have in place a process to ensure that for all required crewmember training courses the training objectives of that course were met.

This is where a good working knowledge of certification requirements for various countries comes into play. The auditor must be familiar with these requirements in order to determine the appropriate level of compliance with the IS-BAO protocol. One can see, though, the difference between the protocol and standard. The protocol asks if a proficiency certification system is in place, while the standard only requires there be one.

When conducting the IS-BAO audit, the auditor is required to use the protocols but, should there be any question as to how to interpret the protocol, refer back to the standard for clarification. IS-BAO is designed as a 'one-size fits all' safety program and it is up to the auditor to interpret the protocol and standard and apply it to the particular company operation.

These two examples are indicative of how the IS-BAO audit process functions. Since the IS-BAO is designed to adapt with departments of various sizes and functions, it is, by default, subjective in some areas while specific in others. Auditors must have the requisite knowledge of those interpretive areas and be able to discern what is appropriate for the flight department being audited.

The IOSA Audit

The IOSA audit, like the IS-BAO, was designed by an international council and developed to enhance worldwide safety. Another principal reason for IOSA was cost savings, which is achieved by reducing the number of audits companies must perform of their code-share partners. And, like IS-BAO, the IOSA has its own idiosyncrasies and subjective elements.

IOSA has more rigorous auditor qualification and accreditation requirements than other audit oversight bodies. In addition to aviation and previous audit experience, a prospective IOSA auditor must attend a five-day ground school, including a written test. Furthermore, once the ground school is complete, the auditor candidate must observe an audit for a minimum of one day, participate in an audit under the observation of an accredited auditor for a minimum of three days, and then audit as part of the audit team for five days under evaluation. Also, an IOSA auditor is accredited in areas only where they have experience. For example, an IOSA auditor cannot be accredited in maintenance unless previous experience in maintenance is demonstrated.

As with the IS-BAO, the IOSA audit consists of questions that are Standards and Recommended Practices – these questions are referred to as ISARPS (IATA Standards and Recommended Practices). The IOSA is divided into eight main areas: organization and management (ORG), flight (FLT), operational control (DSP), maintenance (MNT), ground handling (GRH), cabin (CAB), cargo (CGO), and security (SEC). Within these eight areas are some 900 ISARPS, many containing tables with further defined sub-sections. The IOSA is, by far, the most in-depth aviation system audit and is primarily designed for air carrier operations.

Unlike the IS-BAO, however, if an IOSA standard is unable to be answered, it is marked as a finding. Likewise, if a recommended practice is found unanswerable, it is marked as an observation. There are no major or minor nonconformities – only findings or observations. In order to be placed on the IOSA registry, an airline must answer and address all findings to the satisfaction of the audit organization. The difference between a standard and recommended practice is the way the audit question is answered. They are very specific and make *shall* (standard) and *should* (recommend practice) statements.

The IOSA audit also contains what are referred to as conditional questions. Conditional questions usually contain phrases such as *if required*. For example, IOSA Standard FLT 3.4.5 (2010) reads, in part, 'If required by the State' where 'State' would be the regulating authority such as the United States or Great Britain, etc. Alternate means of compliance type questions are also in IOSA. These questions give the operator an alternate means of complying with a standard or recommended practice. Examples might include FOQA programs or emergency locator transmitter (ELT) requirements for certain aircraft types.

The IOSA also has a few subjective questions requiring auditor determination. For example, FLT 1.5.1 states, 'The Operator shall ensure the existence of a physical infrastructure and work environment that satisfies flight operations management system and operational requirements.' Since there are no defined standards stating what, exactly, satisfies flight operations requirements, the auditor is left to his own reasoning to determine whether or not the airline is in compliance with this ISARP.

Another idiosyncrasy of IOSA is the manner an auditor uses in answering the particular standard or recommended practice. While the IS-BAO uses either *Yes* or *No* answers, there are four possible answers to each question in the IOSA:

1. Documented and implemented (*conformity*)

2. Documented not implemented (*finding*)
3. Implemented not documented (*finding*)
4. Not documented or implemented (*finding*)

Obviously, documented and implemented is the only response that fully satisfies the answer to the question. How can an auditor determine the level of compliance with each standard? Documented means published and distributed through a controlled manual system and implemented means the standard is put into practice through the use of that documentation.

Now, let's explore a couple of IOSA audit questions for key words and other elements. FLT 1.5.2: 'The Operator shall ensure operational positions within flight operations are filled by personnel on the basis of knowledge, skills, training and experience appropriate to the position.' Key words in this standard are ensure, operational positions, knowledge, skills, training, experience and appropriate.

When interviewing the auditee, it is important to focus on these key words as they relate to that particular company's manuals and personnel. There are no standard words one must use to write manuals, nor are there penalties for not using a particular phrase. In any audit, it is important to be flexible, yet uphold the integrity of the audit.

In this case, look at what operational positions a company might have – do they have a chief pilot or is that person also the director of operations; is there a manager of the dispatch department ; is there a vice president and a director of operations? These would be among some of the possible *operational positions* described by the question.

The requisite *knowledge, skills, training* and *experience* would be detailed in the job qualification manual or an operational manual (as appropriate). These attributes should outline what skills are necessary to fill the position described in the manual. A college degree or particular experience level, for example, would be satisfactory answers to the question. Keep in mind, however, these attributes are not a job description, but are based on the knowledge or skill a particular individual needs to perform the assigned duties.

Appropriate to the position is a key phrase as well. One would not hire a ground school instructor as a Director of Safety for example, unless that person had the necessary skills, training, or experience to perform the duties appropriate to the task at hand. It is up to the auditor in some cases to determine what is appropriate to a particular position. Knowledge of regulatory requirements is necessary in a few of these instances and can be especially helpful should guidance be needed.

Finally, the auditor must investigate how the company *ensures* these persons hired into operational positions have the knowledge, skills, etc., as called for by the standard. The human resources department is a good resource for this answer. Job prerequisites should be contained in the human resources manuals and expanded upon during the interview process. While it's not necessary to observe an interview, the auditor should research the human resource department manual system to make certain it is up-to-date and matches the qualifications outlined in other manuals.

As seen in this example of an audit question, the auditor must rely on skills, training and experience in order to get to the heart of the question and obtain a sound answer to this standard. When looking deeper at this question, one realizes quickly there is actually more than one question asked. The auditor has to review several factors to obtain a satisfactory response.

Let's take a look at another IOSA question from the operational security section.

SEC 1.1.1   The Operator shall have a security management system that ensures:

i)   supervision and control of functions and activities associated with the Security Programme;

ii)   compliance with standards of the Operator and requirements of the civil aviation security programme of the State of the Operator and other relevant states.

First of all, is this a Standard or Recommended Practice? The word *shall* gives the clue: this is a Standard. As such, if the operator does not fully comply with this question, a finding will be issued. What are the key words of the question? *Security management system, ensures, supervision, control, functions, activities, compliance requirements* and *State* are all key words of this question.

What is the security management system of the operator? How does the operator ensure compliance with this standard? How are supervision, control and compliance requirements met? What are the requirements of the State? It is up to the auditor to answer each of these 'questions within a question' to determine the level of compliance. As with all IOSA standards, this standard will be fully documented and implemented, or a finding shall be issued.

IOSA, as we can see, is a very involved and detailed audit. It is a system audit in the fullest meaning. Auditors wishing to use this audit to gather information about their company's performance or compliance must be well-versed in the IOSA methodologies. They must be willing to devote the time and resources to this audit for it to be meaningful.

The ACSF Audit

The ACSF, in existence for only a few years, has established its own audit designed for shared-owner, or fractional ownership, and FAR Part 135 charter operators. Modeled like the IOSA audit, the ACSF audit is divided into sections and subsections in order to fully explore the compliance with its standards. And, like the IOSA, the ACSF audit has *shall* and *should* questions.

By now, key words and phrases should be rather obvious, but we need to explore a couple of ACSF questions just for safety's sake. Let's start with something from cabin operations:

> Cabin 1.0.2: The Operator shall have documented policies and procedures regarding the duties, responsibilities and authority of managers and cabin personnel. (ACSF, 2010)

Here, key words are *documented, policies, procedures, duties,* and *responsibilities*. Those key words apply to both managers and cabin personnel; so, what we have are ten questions in one. In IOSA, we learned what documented is and we know what policies, procedures, duties and responsibilities are – what the auditor must do is link those together through observation and interviews of appropriate personnel. Conducting interviews and reviewing necessary manuals are the only ways an auditor can be certain how the key words of any questions lead to conformity.

Let's explore another question on the ACSF audit. This particular question involves interfaces and how they relate to the PIC, SIC and any cargo loadmasters.

> Dangerous Goods 1.0.3: The Operator shall have documented policies and procedures regarding responsibilities, authority, and interfaces between the pilot-in-command, second-in-command, and assigned cargo loadmasters.

We now get into a larger area – that of interfaces. What is an interface and how is it accomplished? An interface is the place at which independent and often unrelated systems meet and act on, or communicate with each other. Some methods of interfacing are email, written communication and verbal (as long as verbal is a documented method), among others. The auditor has to review procedures and company standards to determine how those interfaces take place.

As with other questions, this one contains key words – in this example, *documented, policies, procedures, responsibilities, authority, pilot-in-command, second-in-command* and *cargo loadmasters* are the key words. This example also presents several questions within a question.

What this question asks is, simply put, does the flight crew and loadmaster have formal communication standards whereby the loadmaster advises the flight crew of any dangerous goods on board? It is important that the crews understand their individual responsibilities and authority during these types of operations. The operator will document these responsibilities in all operating and training manuals and provide instruction to make certain all personnel understand their respective roles. While specific verbiage is not necessary, it is very helpful to set a standard method of communication (interface) between parties. Interfacing goes much deeper, of course, than only the area of dangerous goods – it must be encompassed throughout the entire operational system.

The ATOS Audit

As described on the FAA website, ATOS implements FAA policy by providing safety controls (that is, regulations and their application) of business organizations and individuals that fall under FAA regulations. Three major functions further define the oversight system: design assessment, performance assessment and risk management.

Design assessment is the ATOS function ensuring an air carrier's operating systems conform to regulations and safety standards. Performance assessments authenticate that an air carrier's operating systems generate intended results, including mitigation or control of hazards and associated risks. Risk management procedures deal with hazards and related risks.

ATOS uses what the FAA describes as data collection tools (DCTs), and are used to evaluate whether the design and performance of an operator's systems meet the standards set by the FARs. The ATOS audit uses seven systems each containing Safety Attribute Inspection (SAI) and Element Performance Inspection (EPI) elements. The SAI is used to collect data about system design and regulatory compliance, while the EPI is used to collect data so the auditor can determine if the operator follows its written processes, and whether the process produces the desired result.

The ATOS SAI's are divided into five main elements – procedures, controls, process measurement, interfaces and management responsibility and authority. EPI's are divided into two elements: performance observables, and management responsibility and authority observables.

The ATOS audit methodology can be quite cumbersome and, some have said, ineffective at performing as intended. It is, however, a very good audit tool, due in part to its in-depth questions, and requires the auditor to dig deep into a company's manual system. When performing an ATOS-type audit, the auditor must read the questions very carefully as some read very much like other questions in the same element.

Unlike other audits, ATOS is not a compliance audit per se, rather an oversight approach used by the FAA when conducting certification and surveillance of Part 121 air carriers. ATOS is not a set of standards or processes an operator must comply with. There is no regulatory order for a carrier to incorporate the principal concepts of system safety or to reference the ATOS surveillance tools in any portion of their operations.

Below are two examples of ATOS questions regarding training of check airmen and instructors:

> 1.12.6.4   The Certificate Holder specifies that initial ground training for check airmen (airplane) must include the approved methods for performing the required abnormal procedures in the airplane.

1.12.6.5 The Certificate Holder specifies that initial ground training for check airmen (airplane) must include the approved procedures for performing the required abnormal procedures in the airplane.

We can see these two questions are very closely related. But, unlike other audit methods, ATOS asks in two questions (*methods* and *procedures*) what other audits often include into one question. Unfortunately, this can lead to misreading questions or needing to take extra time to decipher what the question is really asking.

When using an ATOS checklist for an audit, the auditor quickly notices the repetitive style of ATOS. Depending on the system and elements being used, many questions within a particular element start with or contain the same verbiage, except for one or two words the questions are specifically dealing with. This sometimes contributes to misreading the question. For example, in the two questions above, the phrase '*The Certificate Holder specifies that initial ground training for check airmen*' is repeated 16 times within one question of one element in one system.

Notwithstanding the above, the ATOS method of auditing is very comprehensive and detailed. It provides more than enough specificity in the questions that, through use by a skilled auditor, no aspect of a carrier's system will be left out. It just takes time and resources to complete.

Conclusion

As we've seen in this chapter, auditing is a challenging method of determining a level of compliance or conformity within an organization. Used by a skilled auditor, each of the checklists previously described would work well for the type of operation it is designed for. Of course, one could choose to use a particular audit designed for a different type of operation, which might provide additional oversight and tools for use in the quality department.

In summary, auditors have high personal and qualification standards that must be adhered to and maintained during the course of their duties. Audit training is an essential ingredient for anyone wishing to delve into the rewarding job of auditing. Skills developed during that training and study will be further honed as more audits are accomplished and more audit methods are explored.

References

ACSF (2010). *Operator Standards Flight Manual, March*. Retrieved from http://www.acsf.aero/attachments/files/99/Operator_Standards_Manual (OSM)Rev_3.pdf.

IOSA (2009). *IOSA Standards Manual, July 1, 2nd edn, FLT 31*. Retrieved from http://www.iata.org/SiteCollectionDocuments/Documents/ISMEd2Rev2.pdf.

Chapter 10

The Road from 32 to 320 Reports: A Brief History of How a Large Airline and a Determined Union Representative Introduced Non-punitive Reporting – and Meant it

Nick Seemel

A Short Tale (Tail) of a Lesson Lost

One never knows how these things really start, but I sure remember being impressed the first time I watched the helicopter base operations while living with my family in a small company town in Labrador. I was, perhaps, fourteen years old and, for the most part, bored with school and all things expected of me. Later on, I met one of the helicopter pilots who had spent the day flying my father around the site – this guy was, by far, the coolest person I had ever met.

Ten years later, there I was flying a single-engine Cessna T207a with an overload of canned beer flats filling my nose baggage compartment, every seat filled with a local sport fisherman, beer in hand, heading out to do some spring ice fishing. It could not have been a nicer day for a flight; a welcome treat for a Canadian bush pilot dealing with the demands of seven-day-a-week operations during the unstable late winter weather.

Arriving over the remote lake destination, I could see the remnants of last weekend's ice-fishing holes cut through the ice in a narrow passageway between a small island and the shoreline at the north end of the lake. Without open water or fires, there was no way to read the wind in this hilly part of north-eastern Canada. A well-flown, rate-one turn with a constant bank and speed will tell which way the wind is blowing. This is important, since I will be landing on soft ice without skis so I want to be heading directly into whatever wind there is when I touchdown. There is a chance that the ice will be covered in a layer of water making in much like landing on greased ball-bearings over a sheet of Teflon.

My oblong circle told me that I should land heading north towards the narrow hole-filled passageway. With my best unfazed, bush pilot, poker face I made a gentle descending turn to line-up halfway up the lake landing towards last week's fishing holes. With all of my flaps extended, I was still carrying what seemed like a lot of extra groundspeed; with seven passengers, beer and supplies, I was probably above the legal landing weight of the aircraft. The featureless grey and white surface of the wet frozen lake required a technique where I kept a little power on and raised the nose high while very slowly searching for the first gentle contact of the wheels. The soft snow-drifts along the edges and the fact that the shorelines thaw sooner forced me out towards the middle of the lake, without the shoreline to help me judge the last few feet. This prolonged flare was sure to eat up some lake.

As I entered the flare the hair on the back of my neck stood up. I felt like I was landing a small Cessna jet, not a Cessna single. Where was all this speed coming from and where is the surface – more important, how far away is the island now hidden under my nose? Moments later I felt the relief of a splashdown as we touched the surface but as I dropped the nose, the windscreen was full of island. I jammed on the brakes and couldn't go right; that would be towards the fishing holes and they happen to be the same size as a C-T207 tire! So left it is. Meanwhile, there is a torrent of water coming straight up from the hydroplaning wheels soaking the windscreen, wings and everything else Cessna. I have never pushed harder on toe-brakes and all I was rewarded with was that odd feeling of actually going faster! At that moment, I had learned more than I ever wanted to about landing on wet ice with a tailwind.

As I eased this island torpedo to the left, the airplane started to do a slow-motion drift sideways and suddenly a plan began to form – if I could get it to go around just a little more and if I haven't pulled the throttle out of the panel by then; I could add power and it may help me slow down. Of course, somehow I have to avoid rolling up onto the wing while I am sliding sideways. My seven, slightly drunk, fishermen had gone completely silent as my new plan unfolded and the tail was helped around by my gentle rudder pressure followed by a very un-gentle application of the 310 horsepower available to me. It was the last few moments that put the punctuation on this story as we stopped with the brakes locked, throttle slammed back to idle and the airplane finally into the wind. It rocked hard on its tail and nose-wheel twice with a bang and a crunch that I will never forget.

Thank god for the comedian in the back that said 'that calls for a beer'. Everyone fell out of the airplane laughing as the empties fell into the 4 inches of standing water on the ice. I, however, sat for a moment and took in how close I had just come.

What Does that Have to do with Non-punitive Reporting?

With all that could be wrong with the airplane after a thorough drenching and a couple of very solid impacts on the tail, not to mention that the locked tires had been badly worn with the heat generated while they were locked and boiled in the standing water at 80 knots, I must tell someone that this happened. Well maybe not, the only people I could tell were the owner, the head of maintenance, or perhaps the manager of HR at the small northern Canadian aircraft charter company that I worked for. The problem was that all of those people I could have told were the same person, a fellow who embodied every type of pilot pushing each time he opened his mouth. He constantly reminded us that if we did not want to work or if we screwed up, he could find someone else who would do a better job – 'pilots are a dime a dozen'. After looking for a *real* flying job for two years, it was crystal clear to me that I was not going to jeopardize this one by admitting any failings to him.

The Airlines Must Be Different, Just Look at the Uniforms!

After a few years and a bit of a struggle, I found myself wearing one of those spiffy uniforms and walking through the terminals with my hat slightly cocked to one side, just a little but not enough to attract my chief pilot's attention. I knew it was better to stay below the boss's radar. It had become clear, after a few thousand hours of flying, that the less *they* knew the better.

You see, it is a funny business where, no matter what they say, errors are not really tolerated, and complaining about one item gets you a pat on the back for pointing out a safety problem but another item may get you branded as a complainer. It is hard to know which is which, so it's better to just get along.

Why Would I Tell You What Happened?

Eventually, I figured out that the best way to find out what the company really expects is to ask the captains who had been around for a while. The best advice I ever received was something I had already figured out. If I do it the company way and don't look for trouble, it won't come looking for me. I have spent years trying to get this job. I have a mortgage and a pregnant wife; the last thing I want to do is jeopardize my captain upgrade or my job. There are many ways that management can make my life miserable since they control all the aspects of my employment including recurrent training and upgrades to higher-paying assignments. I have seen the pilots who were asked to leave or the ones who have been banished to the co-pilot seat to pay for their errors. Considering the fact that the company would hire pilots who perhaps had never been captains or had never flown turboprop airplanes and punish them for shortcomings in these very areas of expertise, it all seemed quite unfair at times. The very skills that the company had to teach these pilots, knowing they did not have them previously, were the areas where the company often found fault, and the faults were only identified to be the pilot's. Those of us who were not caught in this twisted web were simply lucky or, perhaps, excellent storytellers.

'Never admit you are wrong' seems to be the mantra of the experienced guys I work with. One day, I messed up some manual weight and balance calculations that we do as first-officers while sitting at the gate in a loaded airplane with everyone, ground guys, passengers, the captain and the whole 'on time performance' gang, in the main office holding their breath and waiting to push the airplane back as soon as the calculations are done. Six or eight times a day and eighteen days a month doing the same little routine and I finally made a mistake. The result was that we appeared to have taken-off over weight. Not realizing my arithmetic error I called in the weight to the dispatcher as I did on every take-off and he accepted the illegal weight, so much for crosschecking. When this error became apparent during the climb out I redid my figures and we were actually within limits so I radioed the change to the dispatcher. My captain said 'write a report on this and keep it in your flight bag in case anyone asks for it'. No one ever did. It ended up in a hotel room trash can along with anything that could have been learned from it.

Defending an Accused Pilot, AKA 'It's Never his Fault!'

Once I got my captain seat and my first child was born with the second one on the way, I settled into the routine of airline flying and raising a family. I found myself feeling that there had to be a way to help improve things.

Thus began years of representing the rank and file front-line pilots of my airline as an Airline Pilots Association (ALPA) representative. After a few different portfolios, I learned that ALPA was not only the bargaining agent for its pilot membership, but it also spent a great deal of its resources on promoting flight safety. Bearing in mind that more than half of the original key men who organized ALPA in the beginning had been killed in aviation accidents, this made sense. I decided to put my energy into the flight safety and accident investigation efforts of the association. It became clear that if an airplane crashes, the investigation leaves no stone unturned and we make industry safety improvements in response to what we learn from the loss of human life and the destruction of valuable property. This struck me as a harsh way to make change but for the most part it was all we had.

Oddly though, I discovered that, more often than not, when a pilot makes an error or is cornered by the system he works in and the result is what we now call an 'undesired state' but not a full-fledged accident, we see fit to punish the pilot. Apparently this offers management and the regulator the opportunity to appear to be doing something and certainly that pilot won't make that mistake again, it could lead to an accident you know …

Well we pilots are generally clever folks. We spend all day navigating through a system that is full of traps, all the while making adjustments to procedures and adapting the rules to satisfy the demands of the airline managers. That can mean any number of things from landing in the path of a thunderstorm to 'troubleshooting' a problem with the airplane for the last two legs so as not to ground it until the day is done. It also often means that when we are singled out as the one who failed to comply with one of the hundreds of air regulations or approved airplane flight manual procedures, many of us are clever enough to only offer the smallest of detail and to deflect whatever blame is being tossed about away from us. It is far more important to us that it not be the pilot's fault and the punishment is reduced than there be any opportunity for others to learn from whatever happened. We will wait for, and only learn from, the next accident as we always have.

No-one Really Knows

As my career moved forward through many stages, I learned another thing that most pilots have figured out. While there may be an enforcement branch within the regulator and a number of ways that pilot errors or contraventions are reported, most of the errors made in the cockpit go unnoticed by the very system that is designed to catch those errors. Once again as a clever pilot, it soon becomes apparent that I need not report myself and look a fool in the eyes of any of the observers. And the pilot in the other seat is part of the brotherhood and in on the game.

There have been efforts to learn from the mistakes of pilots and some have had some limited success. The Aviation Safety Reporting System in the US is a good example, but it had to be connected to a *get out of jail free* card to make it work. The question that remains unanswered is why is a professional culture held to the questionable standard of a Hasbro board game? Are pilots such a wild bunch that we must be held to the fire to behave? It is quite a dichotomy as most of the pilots I know are regular hard-working honest professionals who want to do the best job they can.

The Aviation Safety Action Program came along and has been embraced but while it continues to be successful, it remains locked into the notions of get out of jail, and the regulator and employer cling to the ability to accept or refuse reports. If accepted, they cherish the right to put pseudo-discipline letters on pilots' employment files when they judge the actions to be unacceptable to them as they sit in an office with the bright light of hindsight giving them special powers. Each time I find myself in a group of pilot representatives the subject of 'sole-source' comes up as they struggle to maintain their safety reporting programs. It is often like a big game trying to keep the report in the system and keep the pilot confident that he did the right thing by reporting. The problem is that the game appears to the pilot to be the same as the street corner huckster game where the con-man moves the cups around and around while you try to guess which one the ball is hiding under. How did we ever get to a point where safety is all wrapped up in discipline to the individual employee?

There Must be a Better Way

It is an interesting thing that sometimes it just so happens that the right people are in the right place at the right time. Just when I was feeling strongly that there must be a better way, it turned out that the leadership in the offices that govern air transportation in Canada were facing the same conclusions and realizing that the best opportunity to see any improvement in flight safety in Canada would be to change the core way we manage flight safety. As it turns out, my ALPA staff member, who had a long and significant past as an employee of the government, had been spiriting the same effort for years. Combine those forces with my air safety committee at one of the largest airlines in Canada, and we just may have an opportunity to make something happen.

Transport Canada is Trying to Change the Safety Culture Nationally

With new regulations being crafted and approved, Transport Canada (TC), the Canadian Government air transportation regulator, was making a commitment to add a layer of risk and performance-based regulation into the current proscriptive regulations already in place. This change of culture would make managing safety in an organization the responsibility of the corporate executive leadership. It would also include provisions to have employee reporting of hazards and errors to be handled without the old-school automatic enforcement action taken towards the employee. The mantra was that simple old-school regulation compliance does not mean an organization, or individual for that matter, is safe.

The Enforcement Inspector Says – 'Okay'

While the new rules were being crafted at the highest levels of government, we began to work with our Transport Canada Enforcement Inspectors to allow for the company to be involved in the 'defense' of their professional pilot staff. When things that were not accidents by definition but were occurrences and required some regulator action happened, we would approach our pilots from the safety department and ask for a report. Working together we had agreed that when the enforcement inspectors were seeking out the crew of a certain flight to pursue an alleged contravention of the regulations, we would work together and allow the safety team at the airline to investigate the case and see if there was something to be done organizationally to improve the system and help prevent a recurrence of the same contravention by any of the airline's pilots. This evolved without any tri-party written agreements: we decided to carefully trust each other and as a pilot representative I asked the pilots to tell the whole truth in their reports; a far cry from my advice when we were singled out and facing punitive action such as fines and suspensions. The results were encouraging. The enforcement inspectors were accepting the company solutions based on internal investigations and the company was able to improve some procedures. The pilot's errors were being treated like honest mistakes and opportunities to learn instead of opportunities to punish. Based on a 'just' or 'professional' culture approach, we would accept that professional pilots make mistakes but certain behaviors are well-known as not acceptable.

Let's Learn Together

While this was occurring we, as an industry, were facing a challenging but desired change in the culture at the regulator as the new SMS regulations came on stream. While truly uninhibited employee safety reporting was an important component within SMS, there were many other new and often confusing requirements that had to be managed by the airlines. Not all of these requirements were spelled out in detail as traditional regulations had been. Some things, such as documentation of safety endeavors, were clear, but most of the new regulations asked that an organization have policy and procedure in place to manage safety but did not stipulate exactly how each expectation would be achieved. This was important since the whole idea of evolving management, so that safety is treated much like any other management function, must allow for the size and type of organization involved putting these concepts into practice in a way that works for them. TC managed this with a defined phase-in over three years but there was much learning on the fly throughout the industry.

There is Never a Right Time

I have often heard that the difficulties of changing the corporate culture are such that we should wait for stability in the industry before we take something like this on. Let me take this opportunity to say that if you did not see the humor in the previous sentence than consider this: as an airline, we were also struggling with a multi-airline merger, and operating under bankruptcy protection for a prolonged period of time. If you have not experienced this folly take a moment to consider the amount of confusion and organizational chaos that this fosters within management. Add to that the union issues that would obviously be in play and then consider the struggle that we safety folks were facing in asking for change and trust. I can't say how many times I heard Rodney King's words in my head 'People, I just want to say, you know, can we all get along?'

Union Holds a Couple Training Sessions – Management Shows Up!

The local Airline Pilots Association through my committee took on the effort of calibrating the local association representatives and the company managers on the fundamentals of a safety culture working within an SMS with a special look at employee involvement. We held these three-day courses in company classrooms at no cost to the company. So much for 'we already have SMS we just need to tweak our current practices'. In the beginning, this was probably one of the most important steps we took. The 'learning together' atmosphere set us off on the same journey of change – together – for the most part anyway.

Those sessions included a wide variety of managers and safety folks from our company management. We were fortunate that we had a structured recurrent classroom CRM program that did touch on concepts such as threat and error management and we had been working with TC to develop the organizational approach to safety investigations, thus allowing us to begin these training sessions with some common ground. At this time, we shared a few things: one was a discussion that at the time we had 540 flights per day and we were receiving roughly 32 safety reports per month from pilots. That does not sound so bad until you do the math – 32 reports over 16,000 flights! Another interesting item was the corporate belief that they already had a non-punitive approach to safety reports and felt that it was spelled out sufficiently in the company operations manuals. This, of course, opened a discussion about the past few discipline situations that had occurred when the company managers did not approve of an error or a safety decision made by a crewmember. We also shared in a little classroom game where the results demonstrated that the managers thought we had a much more robust safety culture than the front-line employees.

Project Team is Full of It

The airline then established a 'project team' to help with the implementation of SMS and included all stakeholders in the team. Employee representatives and managers from all departments met regularly to sort out the new policies and procedures. The management leaders of the project team reported to the executive steering committee to keep the process moving. This team started with a gap analysis and went on to develop high-level guidance for every aspect including safety policy, employee reporting, investigation, participation and promotion of a safety culture. Transparency in the process was key to establishing procedures that would facilitate buy-in from the front-line employees. It was hard to get the message out that this was not just another corporate program that in the end would not amount to much.

Day-to-day

In the development of a better safety culture, we felt a need to create more open doors to introduce employee involvement in the culture change that we were trying to nurture. The Safety Management Committee (SMC) was born of this and each month, the local pilot union safety representatives would meet with the company flight operations and flight safety folks in a focus meeting. It turned out to be a powerful tool to build both trust and understanding. When a corporate VP takes the time to hear the concerns of the front-line employees and the union safety representatives listen to the VP's difficulties effective compromise can be reached and trust can begin to permeate the activities. These meetings covered a wide range of topics, some old and some very new, as we strived for an acceptable safety reporting system.

It is important to mention that meetings such as these are requested and managed by the company so they pay the employee flight time credit by putting a daily credit on the pilot's schedule. I agreed to have ALPA pay the related travel expenses and initially we used an ALPA conference room at no direct cost to either side. The concept that pilots volunteer to attend important meetings is a culture issue – the message to the pilot volunteer is clear.

So your Current Program is Working – 32/540?

Who really knows what is going on in your operation? It seems obvious that those front-line employees who actually apply your processes to everyday tasks in all of the pony-express conditions that airlines operate in would know best what works and what does not. So I guess we can assume that if there are only 32 safety and irregularity reports coming in every month while we are operating 540 flights a day (16,000/month) then we must have some solid and relevant procedures. Or maybe not!

An ongoing discussion in SMC was the current so-called immunity reporting policy endorsed by past managements and outlined in the operations manuals. It was agreed that we would do a formal safety risk assessment on the current program and seek to qualify that the current program was effective or propose changes that would fill any gaps. We brought in an outside facilitator and as a result of this assessment we made a number of changes. It was clear that the pilots did not trust the current system and any system, to be effective, has to have ease of access and very little complexity. It was also clear that the only way our employees would believe in the process is if the union representatives endorse the process and are involved in the whole process.

The significant findings were trust and access. A joint endeavor to deal with these issues was undertaken by the SMC and, while the road was anything but smooth, we faced the trust and access issues directly.

Access was more of a technical problem. We had to make it possible for employees to easily report from many locations, and to allow for the union safety team to see the original reports without difficulty. This has been handled by a web-based approach.

Trust is a very difficult thing to understand and I will further explore it – it has a lot to do with power and I find that the managers have a difficult time understanding that the lack of trust goes both ways. In my opinion, we have taken this seriously and have taken it head on with promotion, participation and most importantly transparency with the union flight safety team and front-line employees.

The single most important step regarding trust was that we determined that *all* reports would be included automatically. Not after some sort of vetting process where the opinions of some desk jockeys, with the power of hindsight, are considered, but instead they will all be included and the safety department will forever protect the report. If the company safety investigator uncovers something that may fall under the very clear and well-known exclusions he follows the procedure stipulated in the agreement. The report is sent to higher authorities in the union and corporate safety department. If the report is determined to be excluded the contents remain protected by the safety department. The exclusion is communicated but details are not shared with HR or Flight Operations management. If they determine further investigation is appropriate and are considering discipline they must complete their own investigation and the pilots can avail to themselves all of the normal Union and legal support in a potential discipline case.

'I just don't trust you' are harsh words for the managers to accept but consider this: *progressive discipline* has been a typical way to manage pilots who are involved in cases where the airplane ends up, in the opinion of a desk-jockey manager, in an undesired state. If a manger does not like the way you do something he will determine, on his own, what the punishment will be; this behavior was often confused with safety mitigation. In fact, it was the opposite. To add to the problem, progressive discipline is almost never handed out uniformly and all too often without proper investigation of the matter under question.

Many airline pilots are aware of, or have had an experience such as this, and the idea that somehow the company can now be trusted just because they say they are compliant with SMS is laughable. The problem is that the company has all the power and even if they are wrong at the time, a pilot may spend a long time without pay or under scrutiny while he fights to get reinstated. This can be catastrophic to his personal and professional life.

Historically unions spend a remarkable amount of resources defending the rights of those members who are wronged by a manger reacting to a safety event. Someone must always be held accountable for the error and, in the company consciousness, it certainly is not the company's error; so the pilot is singled out and blamed. A proper investigation may very well prove that the pilot was a small player in a series of events that were made possible by poor company decisions, training or procedures. But that would take time so the pilot is blamed and the punishment is handed out so that the problem is managed and will not happen again – until, of course, the same set of events are faced by another company pilot.

Managers argue that the employees will abuse a true non-punitive environment. The claim is that the weak will simply hide behind a non-punitive reporting process and continue to be incompetent. It is remarkable in these discussions how quickly management forgets that they screened and trained and continue to train all of the employees. More to the point is that if you empower an employee to admit when they don't 'get-it' you stand a much better chance of not suffering a costly error by the so-called incompetent. This level of trust takes a lot of work and real commitment to make it clear over time that the corporate reporting culture is trustworthy, and that any individual employee attention as a result of a report will be generative and not punitive.

Another argument from the pilots is that TC will never agree to relinquish the blame culture. Well, as a matter of fact, they understand that to make this kind of a culture change work, they must make changes internally as well. The result is that the internal manuals for TC inspectors clearly lay out the protection of reports filed in an SMS certified company, and the guidance goes on to say that they will support the company efforts to find organizational safety mitigations. The focus is clearly off of the individual and re-focused on the organizational and policy issues. The organization has to manage the individual, but better through improvement than punishment in a professional and respectful culture.

'We Will Impose a Non-punitive Reporting Policy With or Without You!'

As I continued to push this rock up the hill, it became clear that the elected union representatives had little interest in entering into an agreement on non-punitive reporting where the membership would be expected to testify to the company safety department with a union technical safety volunteer as a representative, especially if the agreement was not part of the collective agreement or contract. The company would not enter into an agreement that was part of the collective agreement and at risk during any future negotiations and felt that legal representation in every interview was nothing more than what they already had. Not to mention that we were also arguing over the inclusion/exclusion issues that must be clear in an agreement such as this.

In frustration at one SMC meeting, the VP of operations of the day said 'We will impose the proposed non-punitive policy on our own and issue it in a future amendment to the company operations manual with or without you!' Hardly a way to make friends, but we were happy to see that the company was 'getting it' on some level.

It is a very difficult pill to swallow when someone tells you that you cannot be trusted. It is clear to the union representatives that the company assurances that they will treat all employees fairly are not that simple. What the union considers fair and what the company considers fair has been proven to be, at times, worlds apart. The only real victim in most of these cases is the employee, who suffers both financially and professionally at the hands of the often less than thorough investigation done by the company. The company on the other hand suffers very little, yet behaves as if it has been significantly damaged by one employee error. The result is an environment where there is very little trust and no real desire to forget the past and begin anew in a relationship where we agree to be transparent and trusting with safety matters.

At this point, as the chairman of the ALPA Air Safety committee at my airline and the accidental agent for change, I was in an 'against all odds' situation as no-one else really wanted to commit to this – or as some folks put it – no-one wanted to jump off of the cliff with me. The details of the proposed agreement were significant as they spoke to what was to be included, such as errors and omissions, and what was to be excluded, such as substance abuse and deliberate misrepresentation. The procedural details of how a report would be excluded and what was expected of the participating employee were all outlined clearly. These are the important things that corporate reporting guidelines all too often do not articulate. The employee must know what will happen if his report is included or excluded and there can be no secrets in how the process is to work. The simple promise from the company that safety reports will be confidential and non-punitive is not enough. Remember that the employee has avoided self-incrimination for a long time and they need some solid evidence that the *new* system is strong enough to protect them. It must have enough detail and clarity so that the employee has confidence in the process and can expect that it will be followed and understood. The key word is transparency; the company the union and the employee must see the whole process to begin to trust it.

As you may recall, I mentioned that we were in merger mode and bankruptcy mode through much of this and one never knows where a lifeline may come from. About the same time that I had managed to convince the elected union representatives that a co-signed policy, instead of a contract provision, will be defendable and is the more appropriate approach, the airline made some leadership changes intended to get things back on track. Once the dust settled I pitched the proposed agreement to the new VP and shortly after that, the Pilots Non-Punitive Reporting Policy was signed by no less than six people and it was in effect immediately with a soft introduction of things like reporting timelines.

The value of the SMC once again became apparent as we developed a joint campaign to have the VP of Operations and the Director of Operations along with the company flight safety team, which included the ALPA safety volunteers, and the chairman of the ALPA executive council for our airline, attend a union membership meeting at every base over a six-month period. I gave a presentation that covered the value of an organization having the safety data that only a field employee can offer and how good reporting can lead to change and benefit the employee and the company. I also presented an overview of the reporting agreement and how the reports are handled and investigated and then we opened the floor for questions. I wish that is all it would take to make a program such as this work, but it takes much more. To put it in perspective, we coined a phrase: 'One pilot at a time' as we endeavored to convince every pilot the flight safety team came into contact with that we were in fact what we said we were – by our actions and our words.

Be Careful What You Ask For

The efforts to increase interest in reporting was a success and as we received more and more reports, we faced two problems. Traditionally, the most sophisticated safety investigations came from the flight safety department and the investigations were done primarily by pilots who, although tasked with safety investigations, still flew the line part-time. Pilots may be involved in many aspects of the operation and able to successfully investigate much of what is reported, but they do not have sufficient insight into every part of the operation to be the sole investigators. So we had to increase our investigator staff with folks from other departments such as dispatch, airport management and maintenance. The other problem was volume – as the number of reports increased, we had to increase the number of investigators. We also adopted a flight operations risk-assessment tool that would triage the incoming reports so that we can apply our resources to the reports that exposed us to the greatest risk and track those that did not yet reach that milestone. The complexities of the fight crew operations would be lost if we did not maintain pilot investigators, so what we have done is maintain a group of pilot investigators managed by the local ALPA Air Safety Chairman. The ALPA investigators are available for questions from corporate investigators anytime and also participate a few days a month directly participating with company safety investigators to assist in understanding the subtleties of the flight deck in the real world. The ALPA investigators also receive original reports and are able to offer some insight into the line operations realities that may be involved in an otherwise mundane report. This direct involvement is especially helpful when there are individual performance concerns that involve a pilot or when the company training or procedures are deficient. There are very few positions within the corporate safety team that require a company line pilot background, so it is important that the investigators have access to unbiased active line pilots.

We also help coordinate interviews of employees involved in occurrences across the operation. A union safety investigator is available at each interview. This union investigator is an active line pilot and he is allowed to participate in the interview to whatever extent necessary, bringing his expertise to the questions so that the facts would be revealed. This, however, does not prevent the company from making contact with a pilot to ask for a more extensive report or to clear up a few facts.

As the investigation file is coming to a close, we review the results and through a joint union and company safety team process develop safety recommendations. These recommendations are presented at a monthly meeting with the executives from each branch of the company. This meeting is also attended by union safety team members.

Now You Know… Accept or Reject

There is a documented process for acceptance, acceptance with alterations, or rejection of the recommendations and thereby acceptance of the risk by the various internal enterprise executives.

This whole process is under the oversight of the Director of Corporate Safety and is managed day-to-day by the Manager of SMS and Reporting.

Safety Review

This monthly meeting is attended by executives from every department of the company. The Manager of SMS and Reporting facilitates the meeting and brings forward the higher-risk safety reports and recommendations. In the interest of transparency and to keep some real front-line input, this meeting is also attended by union safety representatives. We have found that by having employee union safety representation, we can plan joint execution of some recommendations, especially when there is a communication or oversight piece.

Traditionally, the front-line employees figure their reports go nearly unnoticed. Often they say that if it costs anything to fix a safety issue, the company is not going to bother. This is one of the many reasons why we have taken a transparent approach to the investigation and recommendation process.

Reject a Report – You Must be Kidding

Survival by Culture

It is inevitable that at some point the process will be challenged with an excluded report. When it occurs, it is an opportunity to see what your internal culture is really made of. In our case, it was with an incident on a landing with the First Officer at the controls. In my opinion the company was suffering from a bit of retroactive pressure. An airplane was damaged so they felt a strong desire to have someone held accountable – who better than the captain? In the midst of all of this, the captain, who did not completely trust the non-punitive system, decided not to offer any elaboration on the events leading up to the incident. This resulted in some discrepancy between the captain's testimony to the Transportation Safety Board (TSB) and the report filed with the company. The investigation team recommended exclusion due to misrepresentation but the process resulted in a stalemate at the first stage of the exclusion process between the manager of SMS reporting and ALPA. There were reasonable arguments to support both positions, but the real test would be the reaction of the flight operations managers at the next step. As it was this went all the way to the accountable executive (CEO and President) and after some consultation he determined that punishing this individual would not serve any real purpose. The pilot had answered all of the TSB questions, which revealed the undesirable actions that occurred. There were those who felt strongly that this pilot could not remain on our payroll, that somehow this was entirely his fault and it would damage the company to keep him employed. Often it was those who had not bothered to look past the damaged airplane and in almost every case the 'off with his head' crowd were not pilots, his peers. In the end, the accountable executive realized the core values of a professional safety culture and he rose above the old-school punitive response, which may have squashed many potential employee safety reports. We agreed on an internal rehabilitation plan, which included some training for the captain, and today he is still a captain and is uninhibited when it comes to safety reporting.

You either have it or you don't, and we are stronger for this experience. The Master Executive Council (MEC) chairman was directly involved and the flight operations managers had to take a balanced approach and had many very transparent consultations with the union representatives while they considered the facts and the appropriate reaction.

320 Reports – Have We Made It?

Get the message out together.

It always seemed like we could do more, but what we did do, and continue to do, is to keep the reporting process in the limelight with the company safety publications and through the company website. We have also maintained a module in our classroom recurrent ground school that in some way discuses SMS and always highlights reporting.

The company unilaterally decided to put the non-punitive agreement (between ALPA and the company) into the approved company operations manual, verbatim. Thereby putting themselves in a position of having to follow it or suffer the wrath of not only ALPA but TC as well. This was a clear indication of the corporate commitment to follow it.

The most important thing we did was to stick to our commitment to never use the reports in a punitive way. We understood that no matter how good a job we did, if we kept asking for participation and promised a safety umbrella but did not deliver, we would fail.

We also had to develop a way to manage the special reports that require a bit of careful handling. Our robust reporting system gave pilots the confidence to report conflicts that arise in the cockpit. At first, we let the company handle these but it quickly became apparent that we would need to put our heads together and find a more appropriate way to follow-up on these reports. A new ALPA Committee was created after some negotiation with the company. The agreement simply says that the local ALPA representative will have an opportunity to investigate the file and discuss it with the pilots involved. All of the medical, psychological and training resources that may be needed to resolve the situation are made available through this committee and the confidential results remain within this program. The goal is to remedy the situation in a generative manner where, if at all possible, the crew is put back to work as quickly as possible. More often than not, these files do not have to go much further than a confidential peer discussion about the findings.

Once again, the transparent and professional culture prevails, as we are better able to manage the most difficult and sensitive files without falling into the destructive traditional roles where both sides defend their actions and we lose the opportunity to make real improvements.

Fragile – Keep this Side Up

It never ends.

I only trust you a little. For some of our pilots that continues to be the mantra and, in fact, for some they still simply do not trust at all. One must consider that this too has an organizational deficiency underlying within it as we are unable to convince the skeptics that we are a safety department and not a discipline department, so we are not getting all of the information we could get in a perfect world. I suppose that we will always face a certain amount of that and much like a safety risk assessment we must balance the response to any findings with the realities of a viable operation. If we capture most of our daily operational errors and hazards, we have come a long way from capturing almost none of them.

Remember the bush pilot who spun one around on the frozen lake 25 years ago and did not tell a soul about it? Well there I was again, 25 years later, wedged between the airlines' 'on time performance' (OTP) goals and a series of delay causing events. As a captain, to a large degree, I set the pace in the cockpit and, given the situation, I could see that it was very possible that we could pushback late. So I did what we often do – I picked up the pace. We were preparing an airplane that had not yet flown that day. Coming from an overnight in a hotel with a slightly late taxi-cab ride we arrived a few minutes later than we would have liked. I was managing a request from our maintenance department to transport a small part to my next station to help get an aircraft repaired sooner. At the same time I was managing a MEL item that required some interaction from the flight crew each day that it was deferred. I was hung up on a phrase in the text that made me believe that I had some actions required to pre-flight the item, but after discussions over the radio with the maintenance department it was clear that the maintenance department was expected to do it and had. I made a mental note to write a report about the ambiguous text in the on-board MEL manual so that my confusion was perhaps prevented in the future. While all of this was going on the standard requests from the cabin staff were being managed through my position – drinking water, heat and other items. To manage the time, the First Officer, who I had flown with often was handling all of the flight management computer, performance and ATC clearance preparation. Once I had completed the pre-flight checks appropriate for this first flight of the day, we went through the long first flight checklist which includes many of the system tests for the day. We finished all of the performance and navigation briefings that sound like lines from a script as we read off the information entered into the various on-board computers, or had we? I dropped the parking brake on time and we taxied out to the active runway.

As it turned out that morning, one of the runways at the airport had a large portion of it blocked off for resurfacing. It was runway 24 and both my company-prepared flight plan and the ATIS mentioned that the threshold end of runway 24 had a significant closure with equipment on it. We had planned to depart from runway 06 using the full length of the runway and set the thrust computer for that departure. The performance software we use is common in the airlines and it does not default to a maximum power take-off. Instead, it calculates the weight the aircraft can lift at certain reduced thrust and flap setting, possibly using the full length of the chosen runway. There are buffers built in of course, but because the reduced performance plan expects the full length of the runway to be available if an aircraft would have to begin to reject a take-off at the V1 decision speed it very well may require a maximum effort to stop within the available length of the runway. So in the dim morning light, we rolled onto runway 06 with minimum flap and a planned reduced take-off thrust to save engine wear and fuel. As we accelerated our 88,000 pound airplane down the runway with computers managing the engines and a perfect view through the modern heads-up display my 30 years of professional flying experience gave me the confidence that all was well on this fine morning. It turns out my confidence was not unlike the very experienced captain of the Titanic on his fateful night: only a short distance offshore to the right of my aircraft as we hurtled towards the construction equipment on the threshold of runway 24.

This morning, the V1 decision (stop or go speed) and the rotation or lift-off speed were the same, indicating that we had a lot of runway ahead of us. Then why as I was about to raise the nose of the airplane did it seem that if I had rejected the take-off I may well have hit one of the metal icebergs in our path?

Once we finished with the busy after take-off duties, we began to do a little investigation and discovered that we had overlooked a very important item in our haste. We have a little task in the cockpit where we extract the performance data from the flight plan and fold it a certain way so that it displays the appropriate take-off flap, speed, thrust and engine failure route all in one spot for quick reference. The problem is that during our rushed preparation, one of us folded the performance data so that the full length of runway 06 data was displayed and entered into the computers apparently not realizing that the beginning of runway 24 is in fact the end of runway 06 and now the stopping area of our take-off runway may be blocked by resurfacing equipment. If one were to unfold the performance data page, you would see that the next column has a reduced length performance for runway 06 indicated by a convoluted heading that indicates the reduced runway length in code. A cold chill ran down my spine and the cockpit went a bit quiet as we read the numbers – we had taken off incorrectly configured for the for the shortened runway. Had we hit a Canadian goose at what we thought was our decision speed and tried to stop I may have joined Captain Smith of the Titanic in infamy.

So why do I tell you of this error? Well, I wrote a report fully disclosing our mistakes that morning. How could a well-trained well-rested crew make the choices we made, and how could our system of pre-flight checks not be sufficient to trap that mistake? The subsequent investigation has allowed the company managers to see that this error is possible and make changes that may prevent it in the future. They got the luxury of investigating this before it became an accident. We could just blame the crew and punish them for not following proper procedure but one must ask what that will achieve.

Keep in mind that until I wrote the report, the only two people in the world who knew that anything had gone wrong that morning were my First Officer and me, and just a few short years ago, a wink and a nod would have been our agreement that it would stay that way.

Since the difficult days of mergers and bankruptcy protection and I am happy to report that the airline has gone on to thrive and expand. Our well supported safety management system has, in no small part, contributed to the success.

Chapter 11

Practical Risk Management

Kent Lewis

We realize that as tragic as past accidents are, the lessons learned from those disasters must not be forgotten. In order to avert future accidents, all causes of a mishap must be vigorously pursued so that preventative measures can be found and implemented. Wallers and Sumwalt (2000, xxiv)

Introduction

Practical risk management is about realizing that tragic mishaps also lay in our future, unless the multiple hazards that combine to create risk are identified and controlled. The risk management process is the engine that drives a generative safety management system (FAA, 2008). As part of this system, hazards are proactively identified by systems experts prior to mishaps in order to create information about risk. 'Information is best understood within the context where the individuals make choices, i.e., utilize the information' (Jeng, 2009). Armed with this unique perspective, better strategic risk decisions can be made, reducing the need for risky and reactive tactical risk decisions. Through valid risk analysis and assessment, preventative measures are crafted and implemented that eliminate or reduce the severity and frequency of mishaps.

Much has been written about risk management and there are many models from which to choose when embarking upon a quest to place risk at an acceptable level within an organization. Several models will be explored here with examples of each model to demonstrate real world applications, including models from the Department of the Navy, the Federal Aviation Administration (FAA), and the International Civil Aviation Organization (ICAO). These models are part of a systematic process that has integrated key elements. These elements include the definition of a system that is to be studied, identification of a problem or hazards, analysis of risk derived from the data, a quantitative and/or qualitative assessment of risk, decisions on risk controls and monitoring of the system to ensure desired results.

The objective of this chapter is to examine some real world examples of risk management, but it will be beneficial to take a moment to review the foundational concept of risk management. The risk management process may be as simple as a two-step process, such as the operational process used in instrument flying that involves *control* of the aircraft and a *crosscheck* of flight parameters. Control and crosscheck is a very simple, descriptive model, but there are many complex and dynamic factors that must be managed during this process in order to ensure that control of the aircraft and that the desired flight path is maintained at all times. This is one elegant example of risk management and risk assurance working closely in concert. Another model suitable for time critical risk management is the FAA Industry Training Standard (FITS) three-step model – *perceive, process, perform*. A common theme found in all risk management models is the use of looping processes, essential to continuous learning and system improvement. One of the most famous looping models is the *OODA Loop*, a four-step conflict resolution process that consists of Observation, orientation, decision and action. This decision-making model was developed by military strategist and United States Air Force Colonel John Boyd, and has been widely applied as an information management resource to organizational operations and continuous improvement processes (Schecthman, 1996, 32). A model that is widely used in deliberate operational risk management is a five-step process (FAA, 2008). The five steps are:

1. System analysis (design)
2. Hazard identification
3. Risk analysis
4. Risk assessment
5. Risk control.

This model contains the key steps that most risk management models contain and is scalable for use from cradle to grave on any large system or small project.

For consideration of our case studies in this chapter, we will use parts of a five-step model that has been in use by the US Department of the Navy for over 15 years. Its genesis resides in lessons learned from US Army Operational Risk Management and the Systems Safety Program requirements of Military Standard 882. 'This military standard addresses a wide range of safety management and safety engineering issues' divided into the groups of program management and control, design and integration, design evaluation, compliance and verification (Naval Safety Center, 2010). This standard was chosen for use in the Naval Aviation Safety Program because human factors experts recognized the success that was achieved when a systematic approach was taken to reducing material factor contributions to mishaps. At the time, the influence of material factors was decreasing because of systematic improvement in technologies, and human factors were becoming increasingly prevalent in system failures. There was a need to enhance the human factors aspects of accident investigations and it was time to adopt a scientific model for widespread use throughout all spectrums of the system. Not only did a new framework need to be developed for mishap investigations, there were also benefits to be realized from applying this framework as a proactive risk management scheme, as 'such a framework would also serve as a foundation for the development and tracking of intervention strategies so that they can be modified or reinforced to improve safety' (Shappell and Wiegmann, 2003, 19). A new focus emerged in the human factors aspects of system performance that complemented a strong engineering focus, and practitioners began work in earnest to apply this model to improve operational safety. The five steps of the Naval Safety Center (2009) risk management model are:

1. Identify hazards
2. Assess the hazards
3. Make risk decisions
4. Implement controls
5. Supervise and watch for change.

Do you notice any patterns emerging yet? These models are basically adaptations of the scientific method: a hypothesis is formed (a problem identified), data is gathered and evaluated, solutions are implemented and then evaluated. The strength of this approach is to capitalize on the use of common sense within a common strategy and it has widespread system applications (Budd, 2005, 35).

Keep in mind that another common theme in all of the models is that the goal is not necessarily to eliminate risk, but to manage risk so that the operation can be completed with the optimal balance between production and protection. Several key considerations must be factored into risk management, such as making decisions on risk acceptance at the appropriate level, when it is beneficial to use strategic or tactical levels of risk management and prioritization of resources to participate in the process. One additional key to choosing a model for an organization is to choose a process that is simple and can be understood by those who will use it, to ensure that it will actually be used. There is no quicker way to increase financial risk to an organization than to implement an unwieldy risk management process as part of a safety management system. Risk management must be made as simple as possible, because if it is hard, people will avoid doing it.

Before considering some practical examples, remember that risk management is an integral part of a safety management system – a system that consists of written procedures and plans coupled with policy, safety assurance and promotion of safety programs. The goal of SMS is to prevent loss of life and property while conducting daily operations, and this is accomplished through the detection and mitigation of hazards. Risk management forms the foundation for an effective SMS, regardless of size, mission or resources of the organization, team or individual. Using the five-step process as a guide, we'll now review some examples from the past and present, and then look to the future.

Step 1: Identify Hazards

Case Study 1

For this example, picture yourself in the pilot's seat. You are the instructor pilot in a T-34C flying in South Texas in a tandem seat turboprop trainer with a primary flight student under your tutelage. Your mission today is to introduce the student to touch and go landings and your destination is Aransas County Airport, a civilian airfield in Rockport, Texas. It will be a challenging day because the airport is also utilized by other military and civilian aircraft, and visual lookout for other traffic is a high priority. The T-34 has a tandem seating arrangement and the instructor sits in the rear seat, so you will have a great view of the back of your student's head with limited visibility to either side of the aircraft. It is your student's sixth flight ever in an aircraft, so his visual lookout skills are not very developed. Compounding the challenge is the fact that your aircraft radio is designed to work only on military frequencies, so you will not hear radio transmissions from the civilian aircraft operating to and from the uncontrolled airport. But it is a nice day, so off you go for a fun day of flying.

The transit to Aransas County goes smoothly, and you are on a base turn for landing on runway 36 when a helicopter unexpectedly appears below your right wing on a collision course. You immediately add power and climb back to altitude to avoid the collision. What happened, and how did the risk controls in place fail to protect operators at a critical time? How did we almost lose three pilots, four passengers and two aircraft on a clear, sunny day? And given this scenario, do we need to wait until we have another near midair collision, or should we take steps to ensure an appropriate level of safety within the system? What is our next step?

Our next step is to identify hazards. We have learned from the definition that hazards are conditions with the potential to cause personal injury, death, or property damage. These conditions can lay dormant until acted upon by external influences and changes to the environment. Hazards may exist for many reasons, as a result of poor design, improper or unprofessional work or operational practices, inadequate training or preparation for a task or mission, inadequate instructions or publications, or because the environment is demanding and unforgiving (Naval Safety Center, 2009).

Our case involves naval aviation flight training and potential for midair collisions. In the early days, flight training bases were situated in isolated areas and there was not much interaction between civilian and military aircraft. The Naval Air Station in Corpus Christi, Texas, is home to several flight training squadrons, and during the course of training, crews operated in the local area and also flew in and out of civilian airfields. Initially the risks associated with this training and midair collisions were assessed as low, because while the consequence of a collision was unacceptable, the frequency of exposure was very low. Through the years, though, both military and civilian air traffic increased, and reports of near midair collision increased. This was obviously a problem, so the first steps that the Training Command took were to station a Runway Duty Officer at the airfield who would relay information on civilian aircraft position to squadron aircraft using a military and civilian radio. There was also a stated policy that no more than four Training Command aircraft would use the airport if civilian aircraft were present. Problem solved, correct? Possibly not, in fact the addition of an extra person to relay time-delayed communications only added to the already busy workload of an instructor, and degraded training efficiency. A long-term risk control was being developed to mitigate collision risks, but this collision warning system would have limited functionality in the dynamic environment of an airport traffic pattern.

The first element of this risk management model is to identify hazards within the chosen system, and to not only identify all hazards but to take immediate actions necessary to maintain the appropriate level of safety within the system. In this case, you took immediate action as an instructor to avoid the collision, and upon return to the air station it was time to report the near midair collision to the safety department in accordance with both military and civilian transportation safety regulations. The military has a hazard reporting system that is similar in functionality to the National Aeronautics and Space Administration Aviation Safety Reporting System (NASA ASRS), where demographic information and a description of the event are provided for further analysis and assessment. Given the information we had so far, our next step was to identify hazards. What factors are present that could cause injury or damage? In this case, the identified hazard was the lack of a radio that operated on civilian frequencies. Without this equipment, pilots were left to use basic *see and avoid* techniques, a procedural intervention at the lowest level that is subject to variations in human performance. So we had a hazard with potential severe consequence, and through hazard reporting it was discovered that several similar events had been reported at other joint use airfields throughout the training command. Utilizing this information and a risk-assessment matrix, the hazard was assessed as a severe hazard that required system interventions. With this risk-management model, when a severe hazard at level one or two was discovered, a report had to be generated to notify the appropriate level risk managers within 24 hours. This report also included recommendations for corrective actions that were generated with input from the squadron's instructor pilots; in this case we recommended continuing education on the midair collision potential as a short-term control and acquisition of a dual band UHF/VHF radio as a long term system solution.

So what was the resolution? Based on inputs from squadron pilots through the hazard reporting system and advocacy efforts NAVAIR and the Navy's Office of Safety and Survivability, money was appropriated to acquire dual band radios for the entire T-34C fleet. The total cost of the retrofit was under two million dollars and, because the trainer was not a combat aircraft, a commercial off-the-shelf system was immediately available. The upgraded radio system now not only offers an enhanced level of safety, it also increases the quality of training by exposing military flight students to civilian air traffic control communications throughout the entire national airspace system, an increase in both training effectiveness and system safety.

The most important step in risk management is identification of hazards. It is impossible to be proactive and implement risk controls if hazards remain unseen until discovered in a mishap investigation. There are many ways to identify hazards, and the most effective risk-management systems are those where hazards are identified by reporters who are most familiar with the system risks.

> The information we are seeking is stored in human heads and in books and data banks. Moreover, the information in books is also indexed in human heads, so that usually the most expeditious way to find the right book is to ask a human who is an expert on the subject of interest. (Simon, 1997, 243)

Hazards can be identified by individuals or by teams, such as safety councils or audit teams as part of the assurance process, and part of the challenge here is to motivate people to identify and report hazards. There must be an informed, learning, reporting culture supported by a foundational safety policy and an easy to use reporting system. In small organizations, this may be a written report, and in large organizations there will most likely be web-based reporting portals. Throughout the air line industry, there are additional conduits of information through programs such as the Aviation Safety Action Program (ASAP), Air Traffic Safety Action Program (ATSAP), Flight Operations Quality Assurance (FOQA), and Advanced Qualification Program (AQP) that provide quantitative and qualitative data to both the NASA ASRS and FAA Accident Analysis and Prevention (AAP) programs. This information is still valuable in a traditional reactive sense, and is also becoming increasingly more valuable as a knowledge development tool in proactive and generative safety systems.

These systems essentially conduct *pre-mishap* interviews, and predict what *could happen* versus waiting to see what will happen. Text-mining programs are being developed to look for patterns in operational data, signals in the noise. Has it happened before? Could it happen again? What are the potential consequences, or severity? And what can we do within the system to prevent it from happening? Environmental scanning is a key element in any business plan, and what better area to invest time and talent than in areas that discover threats to the system before mishaps occur? To improve our reporting, it is also necessary to train people to think like investigators, so that when they come across hazards they can identify them and begin the risk-management process. It is also necessary to have an individual and organizational culture that embraces generative reporting and learning. Trust must be built and shared between groups with different goals, and a staggered approach. The incentives to encourage reporting are that individuals and organizations will have a good idea of what is going on, people will feel empowered to concentrate on doing a quality job and resources can be prioritized to fix what needs to be fixed (Dekker, 2007, 26).

Future information systems will emerge from simplistic functionality and capture the advantages of collaborative web 2.0 and 3.0 environments. In these environments, people can share and shape information in the metaverse, which is needed for advancement of global safety management systems. Improvement of safety at a systems level is the next imperative; organizations and regulators must reach across boundaries and share standards in an open, learning environment. Valid and timely hazard reporting is necessary to support the next element in risk management, which is risk assessment.

Hazard Identification Snapshots

There are many web-based information services that collect, store and disseminate safety information. Some excellent examples of environmental scanning on the web are the Flight Safety Information newsletter at http://www.fsinfo.org, signing up for Google alert keywords at http://www.google.com/alerts, aviation safety wikispaces such as http://www.signalcharlie.net, the NASA ASRS online databases and reporting system at http://asrs.arc.nasa.gov and SKYbrary, a web reference for aviation safety knowledge at http://www.skybrary.aero.

Factoids

In 2009 NASA's ASRS received over 50,000 reports, the most in its 35-year history. That is a safety report every 10 minutes:

*   a major air line receives close to 10 reports a day
*   a flight museum may have a web-based SMS, and may receive one to two safety reports per year. (Vintage Flying Museum, 2010)

Step 2: Assess Risks

Case Study 2

On April 12, 2007, at about 0043 eastern daylight time (EDT), a Bombardier/ Canadair Regional Jet (CRJ) CL600-2B19, N8905F, operated as Pinnacle Airlines Flight 4712, ran off the departure end of runway 28 after landing at Cherry Capital Airport (TVC), Traverse City, Michigan. There were no injuries among the 49 passengers (including 3 lap-held infants) and 3 crewmembers, and the aircraft was substantially damaged. Weather was reported as snowing. The airplane was being operated under the provisions of 14 Code of Federal Regulations (CFR) Part 121 and had departed from Minneapolis-St. Paul International (Wold-Chamberlain) Airport (MSP), Minneapolis, Minnesota, at about 2153 central daylight time (CDT). Instrument meteorological conditions prevailed at the time of the accident flight, which operated on an instrument flight rules (IFR) flight plan.

The National Transportation Safety Board determines the probable cause(s) of this accident as follows: the pilots' decision to land at Cherry Capital Airport (TVC), Traverse City, Michigan, without performing a landing distance assessment, which was required by company policy because of runway contamination initially reported by TVC ground operations personnel, and continuing reports of deteriorating weather and runway conditions during the approach. This poor decision-making likely reflected the effects of fatigue produced by a long, demanding duty day and, for the captain, the duties associated with check airman functions. Contributing to the accident were: (1) the Federal Aviation Administration pilot flight and duty time regulations that permitted the pilots' long, demanding duty day, and (2) the TVC operations supervisor's use of ambiguous and unspecific radio phraseology in providing runway braking information (NTSB, 2008a).

There were many situational hazards present during this operation, and the confluence of these factors resulted in a mishap. Fortunately, no one was injured and the airplane returned to revenue service after the nose landing gear and pressure bulkhead were repaired. When all was said and done, the passengers were not even significantly delayed and some did not even realize there was a mishap until they were asked to deplane and ride a bus to the terminal. This mishap is a textbook example of why hazards should be identified, risks analyzed and assessed and control decisions made *before* an operation is conducted. Just by reading the terse description of the mishap from the NTSB probable cause statement, we begin to see the precursors – the latent conditions that existed before the mishap crew reported to the airport for duty, and in the case of pilot flight time and duty time regulations, existed even before the captain or first officer were born. Multiple factors were identified in this mishap beginning with crew experience. The captain was a line check airman who was giving initial operating experience training to a new first officer, which happens every day in air carrier operations, but as we will discover, perhaps this was not the best time or place to schedule or conduct this training. The crew was qualified to operate the aircraft and there were no mechanical issues with the airplane, so for initial risk-assessment purposes we have elevated potential for incidents of a new first officer balanced by the presence of an experienced line check airman. The crew had a relatively normal first day and reported for duty on the second day with a challenging full schedule ahead. The crew flew four flights and was in position in Minneapolis for the fifth scheduled flight to Traverse City. At this point in its day, the crew had been on duty for over 12 hours and had flown over 6 hours, and by the time they arrived in TVC it would have been on duty for close to 16 hours, with flight time exceeding 8 hours. Add to this that the flight would be conducted during a known window of circadian low, with estimated time of arrival at the airport set for after midnight.

We'll pause again here to see if there are yet any identified hazards that may be of interest to supervisors, flight operations planners, meteorologists, dispatchers, airport operations personnel and pilots. As it turns out, there were, and while the captain communicated frequently with the dispatcher on rapidly changing weather conditions, the airport personnel in Traverse City continued efforts to keep the runways plowed and the airport open for air traffic. Initial weather and runway condition reports caused the flight to be put on hold for several hours, as reduced visibility required the use of a specific runway with an instrument landing system, but that runway had a tailwind and the braking action was also being reported between good and fair. Some system controls were in place, operational specifications prohibited use of that runway with tailwinds and reduced braking action, and the reduced visibility precluded the use of a visual approach to the airport. Many of these factors are considered during time-critical risk management and risk controls are developed, so these were not extraordinarily unique conditions when considered in isolation. For this event though, there were many hazards beginning to coagulate and the system was not robust enough to predict the increasing level of risk.

To recap, we have a low time first officer on day two of line flying, with the crew approaching both the 16-hour duty day and 8-hour flight time limit, planning a flight into an airport that had battled winter weather conditions throughout the evening and into the early hours of the morning. That is the nature of air transport flight operations and these risks are managed successfully hundreds of times every day throughout our global system. But we are not quite through with the risk assessment for TVC, as we need to also consider the fact that the runway is relatively short for air carrier operations, 6,500 feet with a medium intensity approach light system (MALSR) with runway alignment indicator lights and a four-light Precision Approach Path Indicator (PAPI). And just to round out the evening's festivities, the air traffic control tower was closed and braking action reports would have to be relayed to the inbound aircraft by airport operations personnel. All of these hazards were known before the mishap, there were no surprises here, just people who were used to getting things done and moving people from point A to point B in a safe manner. No one showed up to work that day and decided to see how far off the end of a runway a regional jet could actually travel, and the airport personnel certainly did not have emergency response and extrication of a regional jet (RJ) from the mud high on their list of desired activities. As a matter of fact, the airport operations personnel in TVC had been recognized by the Air Line Pilots Association in previous years for their proactive stance on safety enhancements to the airport ground environment.

With all of this information in hand, a time-critical risk assessment could be conducted and most likely it would score in a category that required actions be taken to reduce the level of risk. There was a known risk for this type of mishap; in fact, hardly a winter season goes by where several similar mishaps do not occur. An important element of any business plan is to scan the environment, and by doing so, we would know that the probability of such an event given these conditions was that it would probably occur or may occur in time. Next, we would consider the severity and arrive at the conclusion that a runway excursion caused by winter conditions would have the potential of causing severe injury and/or severe mission degradation. For this assessment, we are using quantitative data to establish probability and qualitative judgment to place severity. Utilizing the Naval Safety Center risk-assessment matrix, our risk level would be placed at a high or moderate level and we would need to take actions to reduce this level of risk.

Using the TEAM approach – transfer, eliminate, accept, mitigate – found in personal risk management checklists, options to control risk would include transferring, eliminating, accepting or mitigating the risk. We know now that something in the system needed to change, so we'll look at the things that we can control. We can't change the weather, but we can wait until it improves. We also predict the weather, so why not predict the risk factors as part of a predictive risk assessment? We could cancel the flight, or can we call in a properly rested reserve crew? We can't add more concrete to the runway but we can brief the threats of a short, contaminated runway and strive to fly the textbook approach. And upon arrival at the airport, we can get the most current runway condition and braking action information available from which landing or diversion decisions can be made. Some of these controls seem reasonable when we consider the performance of a crew that is not fatigued, so we need to remember to weigh our decisions with the knowledge and consideration to the *anticipated* performance level of our crew.

Unfortunately on this April morning, the mitigations chosen were not sufficient, and the challenges associated with operating this flight were not met. As mentioned in the probable cause statement, the decision to land without valid runway condition information resulted in the aircraft departing the end of the runway at a low speed and striking a beam, which damaged the airplane. Now from the comfort of our living room couch and in the clear 20/20 vision of hindsight, many of us would say we would have conducted the flight differently or even cancelled the flight, but that is not the reality of the events from that evening or the reality in real world operations. This is a case of where reality and rationality were in conflict; production and protection were at odds. Can this be written off as just another case of 'pilot error', or were there regulatory, organizational and supervisory factors in play here as well?

> Is it more important to ask whether isolating human action as the cause of an accident is 1) sufficient, given that causation is often multifactorial, non-linear and complex (Woods and Cook, 1999); and 2) meaningful for the designer or organizational decision maker interested in ensuring safety for the future. (Holden, 2009, 36)

This mishap may have been prevented had there been flight time and duty time regulations in place that reflected the 24-hour, jet-age nature of our transportation system and knowledge of fatigued human performance. Had there been better use of proper radio terminology or a fully operational and staffed air traffic control tower at the airport that night, the captain might still be employed at this airline. And, had there been a longer runway with the recommended runway safety area free of obstacles, specifically designed for use by air carrier turbojet aircraft, money could have been saved on repairs to the jet and loss of utilization for revenue service. The challenge we face with safety management systems is to capture these lessons learned and implement system solutions at all levels that will prevent future mishaps of this nature.

Deliberate and time-critical risk assessment is not just part of a reactive process, rather it has been incorporated by many individuals and organizations as part of a systematic, proactive approach to safely conduct operations. There are many examples of risk-assessment matrices and personal minimum checklists available for use; these can be found on the web and are valuable resources that can be used in the early stages of risk management to prevent mishaps. This is a good time to remember that risk control recommendations should address short-term, mid-term and long-term solutions, and investigators should do so without consideration to cost. That is not because cost is not an issue, but because the cost is an issue that should be decided at the appropriate level. Many times, the recommendations offer long-term cost savings benefits because the hidden costs of a mishap can be three to five times the visible costs of a mishap. There may be damage to the environment, loss of trust and subsequently revenue in a customer base, reduction in revenue from loss of assets, civil and criminal legal fees and awards and potential fines from a regulator.

Investigative teams can be exercised by participating in the risk-management process; there is no need to wait for loss of life or damage to property. Many organizations are creating hybrid programs that combine accident analysis, investigation and prevention teams. The FAA is one example and the Air Line Pilots Association is another of organizations which have done just that. In some organizations, pre-formed mishap investigation teams do not come together until after a mishap; in generative programs these teams will work together with flight operations personnel and use risk management principles to investigate hazards and make risk decisions to reduce risks throughout the system.

Risk Assessment Snapshots

Situational awareness is needed in order to properly assess risks. Posting a flyer that says 'Maintain Situational Awareness' is a lower-level risk control that has limited effect on reducing the level of risk in a system. One example is a poster that depicts numerous airports that have complex runway and taxiway layouts. Perhaps a better long-term approach would be to design user-friendly airports and work collaboratively to eliminate the identified hazards at complex airports. The FAA Runway Safety Program has more information on this challenge at http://www.faa.gov/airports/runway_safety.

Factoids

Automatic Dependent Surveillance – Broadcast (ADS-B) services that provide flight and weather information services to enhance pilot's situational awareness over portions of the Gulf of Mexico were activated in December of 2009.

Step 3: Make Risk Decisions

Case Study 3

On November 26, 2008, at approximately 1930 Coordinated Universal Time (UTC), a Boeing 777-200ER, registration N862DA, serial number 29734, operated by Delta Air Lines as Flight 18, experienced an uncommanded rollback of the right hand (number 2) Rolls-Royce Trent 895 engine during cruise flight at FL390 (approximately 39,000 feet). The flight was a regularly scheduled flight from Pudong Airport, Shanghai, China to Atlanta-Hartsfield International Airport, Atlanta, Georgia. Initial data indicates that following the rollback, the crew descended to FL310 and executed applicable flight manual procedures. The engine recovered and responded normally thereafter. The flight continued to Atlanta where it landed without further incident. Flight data recorders and other applicable data and components were retrieved from the airplane for testing and evaluation (NTSB, 2008b).

The National Transportation Safety Board has not yet determined the probable cause of this accident, which had disturbing similarities to the British Airways crash at Heathrow in January of 2008.

> The flight from Beijing to London (Heathrow) was uneventful and the operation of the engines was normal until the final approach. The aircraft was correctly configured for a landing on Runway 27L and both the autopilot and the autothrottle were engaged. The autothrottles commanded an increase in thrust from both engines and the engines initially responded. However, at a height of about 720 ft the thrust of the right engine reduced to approximately 1.03 EPR (ENGINE PRESSURE RATIO); some seven seconds later the thrust on the left engine reduced to approximately 1.02 EPR. The reduction in thrust on both engines was the result of less than commanded fuel flows and all engine parameters after the thrust reduction were consistent with this. Parameters recorded on the Quick Access Recorder (QAR), Flight Data Recorder (FDR) and Non-Volatile Memory (NVM) from the Electronic Engine Controllers (EECs) indicate that the engine control system detected the reduced fuel flows and commanded the Fuel Metering Valves (FMVs) to open fully. The FMVs responded to this command and opened fully but with no appreciable change in the fuel flow to either engine.

> The aircraft had previously operated a flight on 14 January 2008 from Heathrow to Shanghai, with the return flight arriving on 15 January 2008. The aircraft was on the ground at Heathrow for 20 hours before the departure to Beijing on the 16 January 2008. Prior to these flights, G-YMMM had been in maintenance for two days, during which the left engine EEC was replaced and left engine ground runs carried out. (AAIB Interim Report G-YMMM, 2008)

There was opportunity for risk decisions at many levels and by many different system agents for these events. Flight crews made time-critical decisions and responded admirably in both events to prevent loss of life and minimize damage to aircraft and property. The survival of all passengers and crew onboard G-YMMM is a testament to the effort put into crashworthiness design by the industry, and skill and experience of the crew. As a result of the Heathrow crash, interim procedures that dealt with management of cold fuel and potential lack of engine response were updated and emphasized. The crew on Delta Flight 18 executed the applicable interim cold fuel procedures and still encountered the rollback condition on one engine, and they then executed the appropriate engine response non-normal checklist, restoring fuel flow to the engine back to a normal condition. Analysis of the risks associated with flying at high altitudes and cold temperatures had been ongoing for years, and recent advances in aircraft fuel and propulsion systems are pushing the envelope, incrementally increasing exposure to the risks associated with this type of flying. This is not a new hazard; it has been around since high-altitude flights by strategic bombers experienced problems with fuel filter screens being clogged by ice in the fuel system. The fuel system was redesigned to be more tolerant of the ice and variations in fuel chemistry were explored, but there are still hard limits that remain and cold fuel properties that must be considered when designing systems, planning routes and operating the aircraft. The 777 with Rolls-Royce Trent 895 engines has operated on millions of flights with only a few recorded events of cold fuel induced rollbacks, but there was clearly a hazard that had potential catastrophic consequences. Provided with this information, it was clearly time to make a decision on the level and types of controls that needed to be improved or created to reduce the risks associated with these hazards to the appropriate level. As an immediate short-term control, flight crews were educated on the recent events and cold fuel management procedures were republished. During inspections of fuel tanks after long flights, it was discovered that ice could remain in the system for hours and would not be removed through normal fuel dumping procedures. Maintenance and ramp personnel adopted procedures to remove as much water from the system as possible through amended fuel dumping procedures, and by transferring cold wing tank fuel to the warmer center tank, which helped speed the melting of ice once the airplane was on the ground. The center tank always remains warmer because it is located next to the warm cabin, while the wing tank fuel is exposed for long periods to significantly colder skin temperatures. As part of the investigation, test rigs were built to closely examine the interface between cold fuel, water and the aircraft fuel system components. During the course of this investigation, it was hypothesized that restrictions to fuel flow were occurring at the inlet to the fuel-oil heat exchanger (FOHE), but the conditions of temperature, fuel flow and 'sticky fuel' that lead to excess amounts of ice forming and contaminating the system at predictable times remained elusive. Restriction of fuel at the FOHE is problematic, as this creates a situation where there is potential for a single point failure in the system, with no means of bypass or recovery. Initially, one could argue that if one engine shut down, there was a redundant propulsion system to carry the load, but now there was the Heathrow crash to consider. The hypothesis of fuel flow restrictions at the FOHE could not be conclusively supported, but the testing also did not conclusively rule out problems in this area. What the investigation had done was discover a hazard in the system that needed to be assessed, and from the assessment by the multinational investigative bodies, a decision was made by the regulatory agencies to require a new design and replacement of the FOHEs on all of the affected engines. The FAA issued an Airworthiness Directive and estimated the cost of compliance to be approximately eight million dollars, but when one considers that against the potential loss of life and property, the amount is insignificant. The scope of the Airworthiness Directive is wide, as these airplanes operate globally on a 24-hour basis, but there were also many other risk decisions made as part of this process, both at the micro and the macro level. The flight crews were tasked with both reactive and proactive real-time decisions on the flight deck, and these risk decisions must be made with consideration of other situational factors present that are precursors to mishaps. Ultra long-range flying involves fatigue, limited communications, exposure to inhospitable flight regimes over mountainous terrain, polar regions, ocean crossings and periods of time where suitable divert airfields are three or more hours away. Proactive risk decisions must be made by investigative teams on what hazard information must immediately be made to organizations and regulatory agencies, and generative decisions on risk are essential to ensure not only maintenance of safety systems but also continuous improvement.

Throughout this study, we can also see the various levels of risk controls, changes made at the system level to design the hazard out, barriers and controls put into place to reduce the amount and characteristics of water present in the system, warnings to operators of the system and administrative procedures and training. These hazard controls target appropriate levels, cross boundaries and ensure optimum risk decisions are made at all levels, not just within individual departments. The key to effective risk decisions is to seek broad participation from all stakeholders, there are many identified here, so that all creative solutions can be identified and considered by those who make final decisions. Information drives decision making and is best understood when placed in the context where individuals utilize the information and make decisions (Jeng, 2009). Armed with these various threads of information a multidimensional tapestry can be woven that encases analysis, assessment and decision making.

Another key element is to make sure that risk decisions are made at the appropriate level. Problems can arise when risk decisions are made at low levels and workers assume inappropriate or excessive risks for an organization without consulting supervisors and top leaders. In the case of Delta 18, the decision to continue the flight to destination versus a diversion was made in concert with the flight crew, maintenance, dispatch and flight operations managers. All of the personnel involved in this type of decision have the authority to stop the operation, and in the case of aviation the captain retains the ultimate decision authority for crew and passenger safety. Effective decision making incorporates sound judgment, a common strategy and common goals, based on reliable information. A systematic approach to risk management is part of this strategy, problems are identified and assessed, consequences considered and decisions made. Good decision making is promoted by team approaches, adequate time to make decisions, philosophy, policy, procedure, training and experience. Once a hazard is detected, it is essential to assess its potential affects by considering its influence on all parts of the system's software, hardware and liveware. These effects will vary depending on factors of time and environment, some we can control and some we cannot control, and it is essential to consider this confluence of situational factors in our risk decisions. Decision making can be hampered by lack of time, inaccurate or unreliable information, production pressures, and lack of teamwork. Good decision making is an integral part of effective risk management, without this, we are more subject to risks associated with hazards and the threat of poor system performance due to human error (Naval Aviation Schools Command, 2008). Decision on controls is a necessary and important step in the process, and timely decisions lead to effective implementation plans, the next step to be considered in the risk management process.

Risk Decision Snapshots

Many helicopter emergency medical services and offshore gas production operators utilize a risk-assessment matrix as part of the mission planning process. This deliberative process helps ensure that multiple situational factors are considered in their dynamic context, and appropriate risk decisions can be made before the aircraft ever leave the ground.

Factoids

In January 2005, Era Helicopter adopted a SMS which focuses on safety training, evaluation, and communications (Era Helicopters, 2010).

Step 4: Implement Decisions

Case Study 4

In the reporter's own words and NASA ASRS format:

> We pushed back from the gate and i instructed the fo to start both engs, anticipating a short taxi. We performed the after start chklist and the fo called for taxi. As we started the taxi, i called for the taxi chklist, but immediately became confused about the rte and queried the fo to help me clear up the discrepancy. We discussed the rte and continued the taxi. We were clred to cross rwy 4, and i asked the fo to sit the flt attendants. He made the appropriate pa. We were clred for tkof rwy 1, but the flt attendant call chime wasn't working. I had called for the before tkof chklist, but this was interrupted by the coms glitch. After affirming the flt attendants ready, we verbally confirmed before tkof chklist complete. On tkof, rotation and liftoff were sluggish. At 100–150 ft as i continued to rotate, we got the stick shaker. The fo noticed the no flap condition and placed the flaps to 5 degs. The rest of the flt was uneventful. We wrote up the

tkof warning horn but found the circuit breaker popped at the gate. The cause of this potentially dangerous sit was a breakdown in chklist discipline attributable to cockpit distr. The taxi chklist was interrupted by my taxi rte confusion. The before tkof chklist was interrupted by a flt attendant com prob. And for some reason, the tkof warning horn circuit breaker popped, removing the last chk on this type of thing. Both of us feel ourselves to be highly diligent professionals. We got ourselves in a box by allowing ourselves to be distr from the chklist. From now on, if i am interrupted while performing a chklist, i intend to do the whole thing over again. (NASA, 2010)

hazard control is an all-hands effort. In this case, a flight crew discovered several hazards to both man and machine and reported them via the NASA ASRS. They were not the only flight crew to discover the potential hazards associated with a flaps up take off, but they were fortunate to not suffer the fate of the crew and passengers of Spanair Flight 5022, which crashed on takeoff in Madrid, killing 154 people (Spanair, 2010). Some hazards are readily identifiable and easy to correct and can be done so on the spot, others are more difficult to identify and may be more difficult to control and correct. This is why it is not only important to correct hazards as quickly as possible, but also important to report hazards to agencies that can ensure long-term and widespread controls are put in place throughout the system. There is nothing more distressing than to prevent a mishap in one organization and to subsequently suffer loss of life and property in another organization from similar hazards. Implementing risk decisions is another step in the common strategy to manage risk, and one that requires the understanding that is derived from robust risk assessment. The Spanair mishap was not simply a 'pilot error' mishap where a crew forgot to set the flaps and slats for takeoff, rather it was a compound systems failure where failures in multiple systems design, training and operational procedures came together in one horrible moment to create a disaster. How could so many parts of the system fail, and how do we defend against this severe hazard? We must have a robust process to identify the hazards, and key players in industry are taking an in-depth look at crashes and events that have similar markers. Several major airlines noticed threshold exceedances in FOQA data and received ASAP reports that indicated that procedures and aircraft systems were not as robust as desired in preventing attempts at no flap takeoffs. These teams then went out and searched for additional data within the ASRS database and the ASIAS and found a surprising number of cases, hundreds, in fact, of similar events under similar circumstances. And mind you, these are not the kinds of surprises that help flight operations managers sleep at night.

It was time to start over and look at all the factors involved in these near mishaps and disasters, and identify cross-cutting factors. What the team found was that there were numerous distractions and concurrent tasks to be managed in a very dynamic taxi phase, just as mentioned previously in the NASA report. Pilots have to interleave calls from maintenance, ATC, flight attendants, the company, other aircraft, with checklists, runway crossings, engine starts, a saturated cacophony of communications and tasks that humans are ill-suited to manage. On paper, these communication, navigation and procedural tasks are neatly and rationally scripted, but then the reality of flight operations intrudes and disrupts that scripting. Operational research was conducted with the help of scientists from NASA's Flight Cognition Laboratory, and a top-down approach was taken to write a new script and coordinate task timing at the participating carriers. What evolved from the research was that concurrent task management in dynamic environments was a very difficult ideal to achieve, and this was not a good environment to be conducting critical flight duties such as setting and checking flaps and slats for takeoff. A pattern of perturbations to the flight operations 'ideal' appeared during the course of the research:

1. interruptions and distractions;
2. tasks that cannot be executed in their normal, practiced sequence;
3. unanticipated new task demands arise; and
4. multiple tasks that must be performed correctly (Loukopoulos, Dismukes and Barshi, 2009, 106).

Researchers presented these findings and a decision was made to reduce the number of tasks that were being managed during the critical taxi phase, when crews should be focused on external communications and navigation around the airport, and to move as many checklist items and flight deck tasks out of this phase as possible. Items that had previously been checked with a 'Taxi Checklist' were moved so that they were checked either before taxiing or during an operational pause before taking the runway for takeoff. It is a very simple strategy to reduce the workload on crews during a busy time, create a focused period where critical flight items were checked, and open a window where crews could focus on prevention of ground collisions, runway incursions and other severe hazards to aviation. There was resistance along the way, with people concerned that the extra time would lead to longer taxi times and increased fuel burns. These concerns were soon forgotten as the FOQA data started coming in that showed that crews were no longer taxiing onto the runway or attempting to takeoff with the flaps and slats not properly set. Flight crews subsequently reported that they liked the improved procedures and reports of runway incursions also decreased.

Hazard identification should contribute to analysis and assessments that lead to control implementations that will reduce the likelihood of a mishap occurring or reduce the severity. These implementations were proven to be practical in the real world of dynamic flight operations. Implementation decisions should be based on quantitative and qualitative confidence of not only understanding the problem and the assessed risk, but also on confidence that the control will have the desired effect on the system at the appropriate time and place. In this case, taking 30 seconds to conduct a safety critical checklist before taxiing reduced the consequence of and exposure to risks associated with a no flap takeoff. This was a quality decision implemented and verified using quantitative data, one more step in a common strategy to manage risk.

Decision Implementation Snapshot

An aircraft battery charger was left attached and unattended on the hangar deck of the Vintage Flying Museum in Fort Worth, Texas. A volunteer noticed the charger and identified it as a hazard to the Museum Safety Council. Subject matter experts in aircraft maintenance, firefighting and safety program management on the volunteer staff were available and assessed the risks to personnel and property. While the frequency of reported battery charger fires was difficult to ascertain, they were known to have happened throughout the industry. This, coupled with the potential catastrophic loss of irreplaceable vintage aircraft, led to the Council's recommendation to discontinue battery charging in the hangar unless personnel were present to monitor the charger and maintain a fire watch. The recommendations were presented to the Museum's Directors and approved, and a record of the risk management process will be kept as part of the Museum's Safety Management System.

Factoids

The entire risk management process at the Vintage Flying Museum took approximately 10 minutes, cost nothing (we're volunteers, remember) and helped ensure the preservation of one of the world's oldest flying bombers, the B-17 'Chuckie' (Vintage Flying Museum, 2010).

Step 5 Supervise and Watch for Changes

Case Study 5

On February 12, 2009, about 2217 Eastern Standard Time (EST), a Colgan Air Inc., Bombardier Dash 8-Q400, N200WQ, d.b.a. Continental Connection flight 3407, crashed during an instrument approach to runway 23 at the Buffalo-Niagara International Airport (BUF), Buffalo, New York. The flight was a Code of Federal Regulations (CFR) Part 121 scheduled passenger flight from Liberty International Airport (EWR), Newark, New Jersey to BUF. The crash site was approximately 5 nautical miles northeast of the airport in Clarence Center, New York, and mostly confined to one residential house. The 4 flight crew and 45 passengers were fatally injured and the aircraft was destroyed by impact forces and post-crash fire. There was one ground fatality. Night visual meteorological conditions prevailed at the time of the accident (NTSB, 2009). Operations must be supervised at all levels and not only should we look for changes in the system, we must look for areas where the system is not performing at the desired level, areas where change is not for the good.

Much has yet to be determined, and even more, yet to be written about this devastating mishap. Too often in discussion of SMSs the focus is on the analytical and scientific aspects, but we should pause to reflect on the fact that when we are discussing systems what we really are talking about is the people and their livelihoods. The real world is made up of flight crews, passengers and citizens who view safety of the system as an imperative, not a dot on a scatter diagram. With this in mind, we should always focus on improvement, not blame. Initial media reports were quick to condemn the pilots, in a misguided effort to make sense of the tragedy and restore emotional stability to a weakened psyche. Could it be as simple as pilot error? Those more familiar with the complex dynamics of high-risk organizations and the limits of pilot expertise know that there is more to be learned from this mishap and much needed improvements of the system have yet to be fully realized.

What we have learned from the Colgan mishap so far is that a functioning safety culture is important to an organization, as it forms a foundation for organizational learning and sharing of information. We have also learned that we do not yet understand the chronic and acute effects of fatigue or have suitable awareness tools developed to proactively manage schedules and prevent hazardous fatigue scenarios from developing. Systems must be designed to support humans in task accomplishment versus add to the challenges of instrument flying in adverse weather conditions, and the system failed at multiple stages during the course of this mishap. The 'automation age' is maturing in some parts of the industry to where concern over maintenance of manual flying skills is increasing, and in the meantime, an increasing reliance on automation has left all operators at risk of not fully understanding the various nuances of complex systems or insidious emergence of hazardous system states. These are high-reliability organizations where performance must be at optimal levels, from man, machine and the system. The operators must be highly skilled and experienced in order to make intervention decisions at opportune moments, but this intervention strategy should not serve as a long term defense against organizational, regulatory and manufacturing deficiencies (Reason, 1997). An integrated defense should be developed that is human-centered but at the same time capitalize on the strengths of automated systems.

This is a realm where it would be wise to expand the team and employ the expertise of cognitive engineers. Cognitive engineers are trained in methods that can be brought to bear on solving complex human-centered problems, they understand the rules of scientific observation and most importantly, know how to measure whether an 'implemented solution was indeed a solution' (Cooke and Durso, 2008, 5). The importance of this last step of monitoring the system and gathering data to assess outcomes cannot be understated. It is not easy to do, and good examples of complex problem solving in tightly coupled systems are hard to find. Risk-management strategies are closely tied to the safety assurance process, and we must check on the system to identify residual hazards and initiate corrective actions to improve performance. This is part of the looping process where risk controls are monitored as part of system operations and information is fed back into improvements to future and existing systems design. During this process, we might find ourselves asking if the process itself is working as desired, especially if we consistently identify the same cross-cutting factors in multiple mishaps. During this monitoring step, all agents in the system should be looking for concurrent task management and workload issues, situations requiring rapid response, plan continuation bias, equipment failures or design flaws, misleading or absent cues, inadequate knowledge or experience provided by training or guidance and hidden weaknesses in defenses against error. These factors were identified during a study of mishaps from the NTSB database and represent 'patterns of interaction that appear repeatedly' and 'underlie many of the errors identified by the NTSB' (Dismukes, Berman and Loukopoulos, 2007, 296). Mishaps involving loss of control have been at the top of the list since the dawn of aviation, and while this may just be the nature of aviation, it is necessary to always seek technological, training and procedural improvements in this area. Supervisors monitoring this system will seek changes to design of training programs, flight time and duty time regulations, and improvements to aircraft design as well as advancement of the foundational principles of safety management systems. We must debrief both the strengths and weakness, identify opportunities and threat, understand the nature of human error and take appropriate actions to ensure proper performance (Reason, 1990).

Proper management of safety information is a supervisory function, and each SMS shall include a safety promotion program. This program should ensure proper management of safety information within the organization and education on how to identify, report, and correct hazards at all levels, and shall include the following:

1. Collection and dissemination of safety information – the collection function includes procedures to ensure proper receipt, care of safety message traffic and other safety correspondence, safety publications and safety films/ materials. In this case, investigators past, present and future will consider the standards of how information is handled in premier organizations and the challenges for those striving to reach higher levels of safety and performance. Supervisors are the messengers, and the message must be clear in word and deed that operational safety is the priority over all other organizational objectives.

2. Dissemination of information on all facets of safety education and training; procedures for distribution of safety message traffic and other safety correspondence/material; distribution of safety periodicals and publications; participation in safety conferences, symposia, committees and councils; liaison with subordinate, adjacent and senior commands to exchange safety information; attendance at meetings for safety briefings, lectures and viewing of related films; and training in safety related subjects.

3. Control of safety information – the proper control of certain safety information is critical to the success of any safety management system. The proper distribution, handling, use, retention and release of such information, as prescribed by laws and regulations, is a requirement for regulators, organizations and individuals who operate within the system.

Supervisors should conduct periodic safety surveys to measure the organization's safety posture. They may consist of in-house safety surveys in which organization personnel are used to conduct the survey. They may also consist of external services provided by a regulator, audit team, or other subject-matter experts. The recommended frequency is every two years, a valid survey functions as part of the supervisory and assurance process works in concert with other reporting systems. Supervision and monitoring of the risk-management process completes the loop and encourages future participation in the system, especially when reporters are given feedback and see generative improvements to the system.

Supervision Snapshots

Turbulence injuries to passengers and flight attendants create the highest number of reportable accidents to the NTSB every year. Supervisors in flight operations, meteorology, dispatch and customer service monitor weather conditions and make both strategic and tactical adjustments to flight patterns to avoid areas of known severe turbulence. Flight crews also coordinate in flight activities to ensure passengers and crew are protected when transiting hazardous weather areas.

Factoids

Of the turbulence injuries suffered by passengers, 98 percent were not wearing seat belts and flight attendants were moving around the cabin conducting flight duties. The other 2 percent of the injuries are primarily caused by unrestrained passengers striking those who are properly restrained.

Future Opportunities for Risk Management

'SMS for SMS'

How do we ensure that a SMS does what is supposed to do? The NTSB is beginning to look at the components of SMS during the course of mishap investigations, and increasingly more focus will be placed on the process itself as well as failures of various components. This is not really new, but we need to be sure that the process supports the intent of our evolving safety philosophy. We must scan the environment, develop standards and share recommended best practices to optimize continuous improvement of SMS. In order to do this, we must increase sharing and management of privileged safety information.

Dynamic Risk Management

Organizations are getting better at managing risks in static environments, but there are still plenty of opportunities to proactively and predicatively manage dynamic risk that emerges from organic changes to systems. Risk is like energy, it does not disappear but rather it changes form and moves from one business area to another. Financial risk can easily be translated into risk of physical harm and loss of property, and the factors that exist to help define these interactions already exist, they just need to be captured, stored as information and managed at the appropriate time. Management of corporate knowledge needs to incorporate the common strategies of risk management, with initial emphasis on system design and analysis. If we can't define the system or the process, we have very little chance of controlling it.

'If you can't describe what you're doing as a process, you don't know what you're doing.' W. Edwards Deming (Quotes.net, n.d.).

This also leaves us exposed to the hazard of wasting valuable time, money and brainpower on reinventing the wheel or recreating a mishap. Think about instances in your organization where you have seen this happen and work towards developing information systems that will help in this endeavor.

Conclusion

The most effective safety enhancements have historically come from the investigative process and lessons learned. 'Those who cannot remember the past are condemned to repeat it' George Santayana (Quotes.net, n.d.).

Our chosen field is inherently a collaborative one. We succeed by sharing, whether it is online cataloging, database development, reference networking, or the development of standards and recommended practices. Your colleagues are your best resource, both now and in years to come. These partnerships may be developed with the safety manager at a local helicopter manufacturer, volunteer air safety advocates, fellow employees, the Director of Safety at a large helicopter company, airport personnel and government regulators. Remember that risk management, as well as SMS, is scalable and should be adapted as necessary to address the nature of each unique problem. Whether the discussion revolves around a malfunctioning aircraft tow bar, a tower cab display design, a radio, an airplane, a procedural manual, an air traffic control system, governmental regulation or international law, working together with other agents in the system is the key to effective management of risks.

Terms and Definitions

Control. A mechanism that manages a risk. Risk control options for each identified hazard generally fall into the following categories: engineering (for example, design, tactics, weapons, personnel/material selection, etc.), administrative, for example, instructions, SOPs, letters of intent (LOIs), return on equity (ROE), special instructions (SPINS), etc., or PPE (for example, eye protection, ear protection, body armor, etc.). Some administrative control option methods are accept (accept the hazard's risk), reject (do not accept the hazard's risk), avoid (minimize exposure/effects by different pathway), delay (postpone until another time where risk is less), benchmark (utilize a control from another entity: reinventing the wheel is not necessary), transfer (move hazard to another participant/asset), spread (diminish the hazard's risk by distributing it among multiple participants/assets), compensate (counterbalance the hazard with something that negates its effect), and/or reduce (limit the exposure to a particular hazard).

Cumulative probability. Summation of probabilities of all causation factors and their impact on participants (for example, the more participants exposed to a hazard, the greater the cumulative probability of that hazard leading to a consequential error).

Deliberate risk assessment/ORM. An application of all five steps of the ORM process during planning where hazards are identified and assessed for risks, risk control decisions are made, residual risks are determined, and resources are prioritized based on residual risk. Usually, the risk determination process involves the use of a risk assessment matrix or other means to assign a relative expression of risk.

Hazard. A condition with the potential to cause personal injury or death, property damage, or mission degradation. Also known as a 'threat'.

In-depth risk assessment/ORM. An application of all five steps of the ORM process during planning where time is not generally a factor and an in-depth analysis of the evolution, its hazards and control options is possible. As in the *Deliberate ORM* process, hazards are identified and assessed for risks, risk control decisions are made, residual risks are determined, and resources are prioritized based on residual risk. Usually the risk determination process involves the use of a risk assessment matrix or other means to assign a relative expression of risk.

Operational analysis. A process to determine the specific and implied tasks of an evolution as well as the specific actions needed to complete the evolution. Ideally, the evolution should be broken down into distinct manageable segments based on either their time sequence (that is, discrete steps from beginning to end) or functional area (for example, ASW, ASUW, AAW).

Operator. An individual who has the operational experience, technical expertise, and/or capability to accomplish one or more of the specific or implied tasks of an evolution.

ORM. Operational risk management.

Residual risk. An expression of loss in terms of probability and severity after control measures are applied. Simply put, this is the hazard's post-control expression of risk (that is, risk assessment code [RAC] or other expression of risk).

Risk assessment. A process to determine risk for a hazard based on its possible loss in terms of probability and severity. A hazard's severity should be determined from its impact on mission, people, and things (that is, material, facilities, and environment). A hazard's probability should be determined from the cumulative probability of all causation factors (for example, more assets involved may increase overall exposure to a particular hazard). Ideally, experiential data (that is, hazard/mishap statistics) should be utilized during the hazard assessment process to assist in determining hazard probability.

Risk-assessment code. A numerical expression of relative risk (for example, RAC 1 = critical risk/threat, RAC 2 = serious risk/threat, RAC 3 = moderate risk/threat, RAC 4 = minor risk/threat, and RAC 5 = negligible risk/threat). See *Risk-assessment matrix.*

Risk-assessment matrix. A tool used to determine a relative expression of risk for a hazard by means of a matrix based on its severity and probability. Typically, a numerical RAC is assigned to each hazard to represent its relative risk (1 = critical, 2 = serious, 3 = moderate, 4 = minor, and 5 = negligible).

Risk decision. A determination of which risk controls to implement to mitigate or manage a particular risk. During the risk decision process, a risk control options' effects should be considered before passing recommendations to the appropriate level for making risk decisions. Risk control options' effects should be determined from their impact on probability of the hazard, impact on severity of the hazard, impact of the risk control cost (what's being sacrificed), and impact of them working with other controls (impedance vs. reinforcement). When cost-effective, multiple risk control options (that is, layered or overlapping controls) should be considered. Risk control options should be chosen to enhance their impact on either probability and/ or severity (for example, goggle-use impacts on both probability and severity of eyes being injured) and chosen in the most mission supportive combination (that is, when one set of controls is more supportive of the mission than another set with the same effect, choose the controls that support the mission).

Risk. An expression of possible loss in terms of severity and probability.

Time-critical ORM. A risk management process that is limited by time constraints, which precludes using a deliberate or in-depth approach. One exemplar time critical ORM process consists of four steps:

1. Assess (your situation, your potential for error)
2. Balance resources (to prevent and trap errors)
3. Communicate (risks and intentions), and
4. Do and debrief (take action and monitor for change).

Threat. See *Hazard.*

References

Budd, J.M. (2005). *The Changing Academic Library: Operations, Culture and Environment.* Chicago, IL: American Library Association.

Cooke, N. and Durso, F. (2008). *Stories of Modern Technology Failures and Cognitive Engineering Success.* Boca Raton, FL: CRC Press.

Dekker, S. (2008). *Just Culture: Balancing Safety and Accountability.* Burlington, VT. Ashgate Publishing Company.

Deming, W.Edwards (n.d.) Quotes.net. Retrieved from Quotes.net http://www.quotes.net/quote/8537.

Dismukes, K. R., Berman, B. A. and Loukopoulos, L. D. (2007). *The Limits of Expertise. Rethinking Pilot Error and the Causes of Airline Accidents.* Burlington, VT. Ashgate Publishing Company.

Era Helicopters (2010). *Safety.* Retrieved on December 20, 2009 from http://www. erahelicopters.com/content/e3/index_eng.html.

FAA (Federal Aviation Administration) (2007). *Flight Standards SMS Standardization Manual.* Washington, DC: Federal Aviation Administration.

Holden, R. J. (2009). People or systems. *Professional Safety*, December, 34–41.

Transportation Safety Institute (2005). *Human Factors in Accident Investigation Manual (HFIAI).* Oklahoma City, OK: Transportation Safety Institute.

Jeng, L.H. (2009). *Texas Woman's University. School of Library and Information Studies: Welcome form the Director.* Retrieved December 12, 2009 from https://www.twu.edu/library-studies/welcome.asp.

Loukoplous, L.D., Dismukes, R.K. and Barshi, I. (2009). *The Multitasking myth. Handling Complexity in Real-world Operations.* Burlington, VT. Ashgate Publishing Company.

NASA (National Aeronautics and Space Administration) (2010). *Aviation Safety Reporting System Database Online.* Retrieved January 16, 2010 at http:// akama.arc.nasa.gov/ASRSDBOnline/QueryWizard_Filter.aspx, NASA ASRS CAN 658970.

Naval Aviation Schools Command (2009). *Decision Making.* Retrieved December 8, 2008 from https://www.netc.navy.mil/nascweb/crm/standmat/seven_skills/ DM.htm.

Naval Safety Center (2009). *Naval Aviation Safety Program*, Op NavInst 3750.6.. Retrieved on December 3, 2009 from http://www.safetycenter.navy.mil/ instructions/aviation/opnav3750/index.asp.

Naval Safety Center (2010). *DOD Standard Practice for System Safety.* Retrieved on January 17, 2010 from http://safetycenter.navy.mil/instructions/osh/ milstd882d.pdf.

NTSB (National Transportation Safety Board) (2008a). *Aircraft Accident Report NTSB/AAR-08-02.* NTSB Identification: DCA07FA037 Thursday, April 12, 2207 in Traverse City, MI, Bombardier CL-600-2B19 Registration: N8905F. Retrieved December 15, 2009 from http://www.ntsb.gov/ntsb/brief.asp?ev_ id=20070501X00494&key=1.

NTSB (National Transportation Safety Board) (2008b). *Uncommanded Rollback of the Right Hand (number 2) Rolls Royce Trent 895 Engine During Cruise Flight.* NTSB Identification: DAC09IA014. Wednesday, November 26, 2008 in Atlanta, GA. Boeing 777, registration N862DA. Retrieved January 16, 2009 from http://www.ntsb.gov/ntsb/brief.asp?ev_id=20081201X44308&key=1.

NTSB (National Transportation Safety Board) (2009). *Crash of Colgan Air Bombardier Dash 8-Q400 during instrument approach to Buffalo-Niagra International Airport.* NTSB Identification: DCA09MA027. Thursday, February 12, 2009 in Clarence Center, NY. Bombardier INC DHC-8-402, registration N200WQ. Retrieved December 17, 2009 from http://www.ntsb. gov/ntsb/brief.asp?ev_id=20090213X13613&key=1.

Reason, J. (1990). *Human Error*. New York: Cambridge University Press.

Reason, J. (1997). *Managing the Risks of Organizational Accidents*. Burlington, VT. Ashgate Publishing Ltd.

Santayana, George (n.d.) Retrieved from Quotes.net http://www.quotes.net/ quote/34044.

Schechtman, G.M. (1996). *Manipulating the OODA Loop: The Overlooked Role of Information Resource Management in Information Warfare*. Retrieved on January 9, 2010 *from* www.au.af.mil/au/awc/awcgate/afit/schec_gm.pdf.

Shappell, S.A. and Wiegmann, D.A. (2003). *A Human Error Approach to Aviation Accident Analysis*. Burlington, VT. Ashgate.

Simon, H.A. (1997). *Administrative Behavior*. New York: The Free Press.

Spanair (2010). *Spanair Flight 502*. http://en.wikipedia.org/wiki/Spanair_ Flight_5022.

Vintage Flying Museum (2010). About Vintage Flying Museum. Retrieved on January 16, 2010 from http://www.vintageflyingmuseum.org/.

Vintage Flying Museum (2010). *Safety Management System*. Retrieved on January 7, 2010 at http://groups.google.com/group/ftwasp/web/vintage-flying-museum-safety-management-system,

Wallers, J. M. and Sumwalt, R. L. (2000). *Aircraft Accident Analysis: Final Reports*. New York: McGraw-Hill

Chapter 12

Integrating SMS into Emergency Planning and Incident Command

Jack Kreckie

Introduction

Safety means different things to different people. In *Webster's New World Dictionary*, the definition for safety is: 'the quality or condition of being safe; freedom from danger, injury, or damage; security'. In the fire service, when we think of safety, the first thing that comes to mind is that important component of every command structure and incident action plan. The role of a *safety officer* in incident command is taken very seriously. In fact, the safety officer wields the power to stop an emergency operation without consulting the incident commander if they deem the conditions or the operation to be unsafe.

Aircraft rescue fire fighters (ARFF), airport operations specialists and airport law enforcement officers (LEOs) are charged with the safety of our airport community and the flying public that choose to fly through our jurisdiction. To those people, there is an expectation of safety. They simply desire and expect that they can reach their final destinations safely. They expect that our airport roadways, parking structures, terminals and walkways will be safe and efficient. They understand that there is some risk involved in air travel, albeit a minor risk; there is a chance that a person will be injured or killed while traveling. They typically accept that risk, because it is so small. If, in fact, an accident or injury occurs, they have another expectation that qualified, competent responders will quickly respond to assist them. Members of the airport community are those that work there. They provide the services necessary to support the overall operation. They come from a variety of business segments, diverse backgrounds and cultures, yet, on a very basic level, they share those same expectations. They expect that those responsible for safety, security, and emergency response plan together and work together.

Compartmentalized Awareness

The airport ramp is a unique environment. After developing a comfort level on the ramp, one becomes somewhat oblivious to the activities that are not part of one's responsibility. The changes in ramp service over the years have undoubtedly contributed to this effect. At one time, nearly all of the services being performed on an aircraft at the gate were handled by the airline. The employees knew one another, their procedures were harmonized and they all had the same priority. Awareness under the aircraft was, in general, at a higher level than it is today. In order to remain competitive, many of these services are now contracted out by the airline. It is not uncommon to have many different companies involved in the effort to re-service and turn around the aircraft. Although the services are, in many cases, performed simultaneously and within close proximity to each other, they are often like 'two ships passing in the night'. There are fewer relationships built, and often lacking is a mutual understanding of the task that the other person is accomplishing.

One of the results of this lack of understanding is the development of a competitive rather than a cooperative nature. Instead of coordinating arrival times and positions under the aircraft, vehicles jockey for position. It may not be based on the need for the service to be performed first, but rather to avenge the last time when the 'other guy' blocked the best access point to the service.

Using multiple providers to perform service to the aircraft may be the most cost-effective method, but it will only prove to be safe and efficient if the efforts are harmonized and the entire operation properly choreographed. The harmonized plan needs to satisfy the objectives of each of the individual providers, as well as the overall mission. By satisfying individual needs, the buy-in of the stakeholders will be achieved. Increased safety is part of the overall benefit, but it is not likely sufficient to convince people to change their ways. One must be prepared to answer the question, 'What's in it for me?'

Maslow's Theory

The American psychologist Abraham Maslow is noted for his conceptualization of a hierarchy of human needs. Maslow is considered the father of humanistic psychology. His explanations of basic needs are difficult to challenge, however they are often not considered in the planning process. Maslow's theory first identifies the levels of the five basic needs. A person does not feel the second need until the demands of the first have been satisfied, or the third until the second has been satisfied, and so on.

For aviation safety, planning must satisfy needs in the first two levels of the pyramid of all those involved. The success of the plans will be measured by how many of the relative portions of the three upper levels of the pyramid are satisfied.

Physiological needs are essentially the very basic components of survival. The human body requires air, water, nutrition, rest, homeostatic regulation and excretion to live. Our need for air to breathe is one to which we can all relate. Most of us have found ourselves in a situation where breathing became difficult. We either had the wind knocked out of us playing sports, or perhaps had our airway blocked momentarily. Perhaps, we were thrown unexpectedly into the deep end of the swimming pool and, for just a moment, felt the fear of the inability to fill our lungs with something other than water. The moment our next breath is threatened, we react through survival instinct, striving to take a breath. It is the threat to this basic need that leads sane, rational, thoughtful people to panic and trample those in their way, trying to escape the poisons of smoke filling the space that they occupy. Throughout history, we have seen far too many tragedies wherein the death toll was significantly increased as a result of that reaction and behavior to a threat which takes away one of our basic physiological needs.

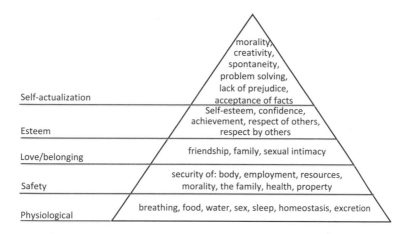

Figure 12.1   Maslow's hierarchy of needs

At the famous Cocoanut Grove nightclub fire in Boston, Massachusetts, US, on November 28, 1942, the doorways were cluttered with the bodies of revelers who climbed over each other to try to make it to the exit. Revolving doors were jammed with bodies, and had to be dismantled from the outside in order to gain access. Other doors were welded shut to prevent people from sneaking into the club. Exit doors swung in, instead of out. Authorities estimate that 300 of the 492 that perished that night would have been saved if the doors only swung out. The panic created by that need for survival piled the fleeing occupants against the door, preventing those doors from opening. Few of these people, if any, walked into the Cocoanut Grove that night thinking about their escape route in case 'something bad happened'. Undoubtedly, they all had an expectation that if something bad did happen, there were systems in place and people trained to keep them safe.

The Cocoanut Grove nightclub fire was over 65 years ago. We've learned a great deal since then about creating and maintaining safe occupancies in areas of public assembly, like airport terminals. Life safety codes have been rewritten and strengthened. Our public is more aware, and our emergency responders better trained and equipped. The things that haven't changed are the basic hierarchy of needs identified by Maslow in 1943. Our very basic needs for survival will, when challenged, overcome our allegiance for rules, laws, social expectations, training and obedience. Unfortunately, we have numerous examples since Cocoanut Grove that prove that our attempts to regulate, equip and train to prevent such catastrophes have missed the mark.

- Cocoanut Grove nightclub fire 28 November 1942; Boston, Massachusetts; 492 dead.
- Club Cinq-Sept fire 1 November 1970; Saint-Laurent-du-Pont, France; 146 dead.
- Beverly Hills Supper Club fire 28 May 1977; Southgate, Kentucky; 165 dead.
- Stardust fire 14 February 1981; Dublin, Ireland; 48 dead.
- Alcalá 20 nightclub fire 17 December 1983; Madrid, Spain; 82 dead.
- HappyLand fire 25 March 1990; New York City, New York; 87 dead.
- Kheyvis nightclub fire 20 December 1993; Olivos, Buenos Aires, Argentina; 17 dead.
- Ozone disco club fire 18 March 1996; Quezon City, Philippines; 162 dead.
- Gothenburg nightclub fire 29 October 1998; Gothenburg, Sweden; 63 dead.
- Luoyang Christmas fire 25 December 2000; Luoyang, People's Republic of China; 309 dead.
- E2 nightclub stampede 17 February 2003; Chicago, Illinois; 21 dead.
- The Station nightclub fire 20 February 2003; West Warwick, Rhode Island; 100 dead.
- Cro-magnon Republic nightclub fire 30 December 2004; Buenos Aires, Argentina; 194 dead.
- Wuwang club fire 21 September 2008; Shenzhen, People's Republic of China; 43 dead.
- Bangkok nightclub fire 1 January 2009; Watthana, Bangkok, Thailand; 61 dead.
- Lame Horse club fire 5 December 2009; Perm, Russian Federation; 149 dead.

The point of all of these examples and the associated death toll is to illustrate our continuous failure to plan, using all of the tools, skills and knowledge available to us. We, in aviation, are indeed more aware than many other industries of the threat posed by terrorism. The war on terror launched by the United States on September 11, 2001 was retaliation to an attack on US soil (and airspace) which turned air carrier aircraft into weapons of mass destruction. These acts need no further description, as the images of that day are indelibly burned into our memories. That day not only took away thousands of lives, but changed our way of life forever.

The aviation industry reacted by developing new security procedures, building better walls, launching an entirely new federal agency and trying to fill all the cracks in a system that allowed these unthinkable acts to occur. We continue to underestimate the desire and 'need to succeed' by our enemies. Our planning and preparation is limited by our imagination and the expertise and experience of our planners. Our plans for dealing with hijackers prior to 911 were not based on the worst-case scenario as contemplated by 'imagineers' on our planning groups, but instead driven by the reactions necessary during historic acts. Typically, these acts were committed by hijackers trying to divert an aircraft to their homeland or to gain attention to their cause. This is not to minimize the violence inflicted, the fears of their victims, or the outcome of these acts of piracy, but rather to identify our reactive planning tendencies which would better suit our needs if they were proactive instead.

Significant changes were enacted immediately after the crimes committed on September 11 2001. Certificated airports in the United States and their territories were not allowed to re-open until all of the new security procedures had been satisfied. But with all the new procedures, who boarded AA Flight 63 at Charles De Gaulle International Airport bound for Miami International Airport on December 22, 2001? Richard Reid's need to succeed in blowing up the Boeing 767 that he was flying on exceeded the basic physiological needs for survival. He was overpowered by passengers and crewmembers while trying to light a fuse attached to one of his shoes. Investigators later found that the fuse Reid was attempting to light led to pentaerythritol tetranitrate (PETN), one of the most powerful high explosives known to man with a triacetone triperoxide (TATP) detonator hidden in the lining of his shoes. Planners in the United States appropriately reacted to this event by requiring the shoes of passengers to be screened at security checkpoints, while the travelers pass through without shoes on their feet.

Further significant changes to passenger screening regulations in the United States occurred after a terrorist plot to detonate liquid explosives carried on board ten different aircraft was discovered by police in the United Kingdom. The targeted flights were to originate from points in the UK and were destined for the United States and Canada. Our planners again reacted appropriately, but followed our template of reactive planning.

On December 25, 2009, another serious incident caused a reactive change to security procedures and regulations. An apparent malfunction in an explosive device designed to detonate the same high explosive used by Richard Reid, PETN, may have been all that prevented tragedy for 278 passengers and the crew aboard Northwest Flight 253. Umar Farouk Abdulmutallab, 23, of Nigeria was on the 'Watch List' as he had known ties to terrorism, but the information was not firm enough to add him to the 'No Fly List'. Once again, regulators reacted to this new threat with additional security screening measures at US airports. More reactive planning! The model we seem to follow requires a specific threat rather than a potential threat in order to factor in procedures to protect against it.

Security planners have a distinct advantage over safety planners. On the federal level, they are able to implement new procedures very quickly across the entire system. Safety planners at airports struggle to implement procedures, and often have to depend on voluntary implementation.

It appears that terrorists have read Maslow's theory. They attack and threaten that which is most precious to us, life. Their planning does not stop there, however, as they use their understanding of our needs and the anticipated reactions of their intended victims who escape their initial attack to satisfy their own goals. Earlier, we discussed the devastating effects of a number of fires in areas of public assembly that collectively caused thousands of deaths. In each of these events, some of the deaths were caused by proximity to the fire, the physical limitations of people who were not able to help themselves, occupancy exceeding the legal limit and poor judgment by event coordinators and facility managers. The vast majority though were caused by predictable human behavior and the human desire to survive. *Terrorist planners* consider these reactions, plant secondary devices and increase their death toll by studying the humanistic psychology of their enemies.

The October 12, 2002, Bali bombings are not the only events which prove these planning strategies exist, but serve as an excellent example. On the Indonesian Island of Bali, in the tourist district of Kuta, members of a violent Islamist Group known as Jemaah Islamiyah launched an attack in and near the popular nightclubs in Kuta. On Saturday night in the popular tourist district at 11:00 p.m., anyone could predict that there would be big crowds. A suicide bomber inside Paddy's Pub detonated a bomb in his backpack. As anticipated, those that were able fled through the door that they entered through into the streets. Fifteen seconds later, a much more powerful bomb hidden inside a white Mitsubishi van parked near their escape path was detonated by another suicide bomber waiting for the crowd to enter the range of the deadly blast he controlled. Research into other attacks shows that terrorists routinely plan primary explosions in areas heavily occupied areas with secondary explosions targeting fleeing survivors, and follow-up devices targeted for the entry points of emergency responders.

For the purpose of this study, we should walk away with some lessons important to our emergency planning process.

- It has been said that terrorist events, that is explosions, fires, sabotage, releases of toxic chemicals, etc., are no different from the types of emergency events for which planners prepare, but in these cases, are intentionally created. This analogy falls short of the level of thinking that is required to properly plan and prepare for new age threats, both intentional and accidental.
- Successful proactive emergency planning and preparation must be mission-driven. It cannot be created around a template imposed by budget or available resources. The draft plan should provide for every possible contingency and need based on the anticipated worst-case scenario considered in the imagination of the planning team.
- The planning team members should represent a true cross-section of the stakeholders based on the event for which preparations are underway. Each offers different perspective, expertise and resources. As each layer of the plan is developed, the review of the various stakeholders provides a reality check and considers the effects of the action and uncovers potential collateral issues. The draft plan(s) should then be evaluated through a

tabletop exercise involving each of the stakeholders. If conducted correctly, this exercise will identify additional issues with which to deal.

ICS/SMS Comparisons

The Incident Command System (ICS) is arguably the greatest management plan in use today. Although typically only used in emergency management, it has realistic applications in nearly any business segment. Let's examine the components of ICS and evaluate the potential applications in an SMS environment.

Definition

The ICS is a standardized, on-scene, all-hazard incident management concept. It allows its users to adopt an integrated organizational structure to match the complexities and demands of single or multiple incidents without being hindered by jurisdictional boundaries.

Key points

By using management best practices, ICS helps to ensure:

- The safety of responders and others.
- The achievement of tactical objectives.
- The efficient use of resources.
- ICS has considerable internal flexibility. It can grow or shrink to meet different needs. This flexibility makes it a very cost-effective and efficient management approach for both small and large situations.

Origin

The ICS was developed in the 1970s. A review of existing incident management policies was called for after a number of catastrophic fires in California's urban interfaces with wild lands (forest fires). In a period of less than two weeks in 1970, 16 lives were lost, 700 structures were destroyed, and over 500,000 acres burned. The calculated cost and loss associated with these fires reached US$18 million per day (1970 dollars). Although all of the responding agencies cooperated to the best of their ability, numerous problems with communication and coordination hampered their effectiveness.

Reviews of these disasters and case histories found that response problems were rarely attributed to a lack of resources or poor tactics. Instead, it was found that response problems were more significantly attributed to inadequate incident management than from any other single reason.

Identified weaknesses:

- lack of accountability
- poor communication
- lack of a planning process
- overloaded incident commanders
- no method to integrate interagency requirements.

A review of these weaknesses in management resulted in the development of ICS. One could say that this too was a reactive rather than a proactive response to an identified weakness or problem. The difference was that the solution did not simply address the specific problem identified, but rather built a system which provided the tools for incident commanders to more effectively deal with all incidents.

Safety Management Systems

The definition of safety as per ICAO is 'the state in which the risk of harm to persons or of property damage is reduced to, and maintained at or below, an acceptable level through a continuing process of hazard identification and risk management'. The ICAO definition seems to specifically relate to safety management. This definition recognizes that the activity (unsafe acts) is not reduced to zero, but instead to an acceptable level. It clearly suggests that safety is measured against the established acceptable level.

The FAA defines the SMS as the formal, top-down business-like approach to managing safety risk. It includes systematic procedures, practices and policies for the management of safety (including safety risk management, safety policy, safety assurance and safety promotion).

Neither of these definitions significantly differs from how we in the fire service define safety and implement safety systems. We have our own language, of course, but the underlying mission is the same. We understand that providing emergency services comes with a level of risk. We use that acceptable level of risk in our risk analysis, which is performed routinely in the development of Incident Action Plans (IAPs). Those of us in emergency services have accepted the fact that there is risk in our profession. We are called upon to mitigate emergencies and even with good training, equipment and procedures, there is a certain amount of risk that comes with the territory. The unofficial rule of thumb is: 'We will risk a lot to save a lot. We will risk little to save little. We will risk nothing to save nothing.'

Safety systems, like safety rules and procedures, are only as good as the level to which they are followed. That typically correlates directly to the level of importance placed on safety by the highest levels of organizational management. Everyone shares the desire to be safe, but routinely, society accepts cutting corners on safety. Examples of this might be rolling slowly through a stop sign, or not buttoning the top button on our fire fighting PPE. It doesn't seem like a big deal and is not likely that anybody will notice or care; nothing terrible is likely to occur as a result. The greatest failure identified by this mini-risk assessment is the conclusion that our safety systems are tolerant of cutting corners on safety.

Failures of Safety Systems

On the airport ramp, there should be no acceptable level of risk in accomplishing routine tasks involved with ground service. We see accidents occur every day which cause damage to equipment, aircraft and other assets. Some of these accidents cause injuries and loss of life. Many are never reported or investigated. Of those ramp accidents and incidents that have been investigated, there is a common factor involved in the cause. There will always be a violation of a procedure, protocol, or instruction. Aircraft pavement markings, cones, delineators, lights, jet bridge restrictions, speed limits, and PPE requirements are all in place to give us visual references, restrictions, rules and precautions governing our actions on the ramp.

Tell-tale signs of a failure of safety practices are generally evident during investigations. A collision involving a piece of ramp equipment and an aircraft is an excellent example. Pictures taken during follow-up investigations will invariably show that the aircraft was not set on the T, or the vehicle was operating inside the restricted pavement lines. There is simply no other way it can happen. A review of policy and procedure during these incidents includes an assessment of the training policy for each position that was involved in activity from the time that the aircraft came on the gate. The requirement for the aircraft to be specifically on the lead-in line, stopped on the T will be seen in each procedure. If safety was a true priority, each of those people would have identified the error and associated risk with the aircraft being in the wrong position. Interviews will usually reveal that people are accustomed to seeing aircraft on that gate parked askew, but they choose not to report it because they feel that their supervisors see the same condition and do not act. They also report that they would be criticized for causing the extra effort in relocating the aircraft, which will be leaving in a few hours anyway.

The current safety culture, all too often, works in the opposite way that we would like. Having the training procedures which emphasize safety does not contribute to safety unless the culture in place not only embraces safety, but demands it at the highest levels.

When asked about an accident, most people will blame the equipment or someone else. It is much easier than accepting blame. When the baggage cart gets blown across the ramp striking an aircraft, everyone throws their hands in the air and says, 'It was the wind. I was not even there.' Both are true statements. However, how many of those same people noticed the broken brake cables dragging under the cart as they drive across the ramp. The cable dragging on the ground serves as a 'cow bell' to alert everyone in the area that this cart has no brakes. Is it reported? Is it immediately taken out of service? Not likely, because the safety culture in place does not start at the highest level. Employees too often feel that management's expectation is that you do your job, use the equipment that is provided, and don't make waves.

Figure 12.2   Baggage cart with broken brake cable blown into B-1900 by wind gust (Jack Kreckie Photo)

Figure 12.3   Tell-tale broken brake cable on same cart (Jack Kreckie Photo)

Maintenance, or the lack thereof, is part of the cause of many accidents on the ramp. Reductions in budgets and personnel cause airlines and FBOs to cut back where possible to protect the bottom line. Cutbacks that affect safety are never acceptable. In the aviation industry, the direct and indirect costs of ramp accidents are significant. In addition to the financial hit, consumer confidence has a huge impact on filling seats on airplanes.

Jetbridges are a critical component of the business of moving passengers. They require skilled operators, routine maintenance and overall awareness by all persons working in the area while operating. Is there a requirement for audible and visual alarms on jetbridges at your airport? Does anybody check those devices? Is routine maintenance performed, or is it simply fixed when broken? A failed jetbridge can cause tremendous damage to aircraft. People walking through jetbridges when they fail are subject to serious injuries. Routine inspection and maintenance of jetbridges are not regulated at most airports. They generally don't get much thought at all unless the air conditioning stops working or an interior light burns out. Most of us are not thinking about when it was lubricated, if the bolts have been torqued, or if an inspection of any kind has been performed. It is outside of our 'compartment' of thinking and the condition is not currently affecting any primary needs, so it does not earn our attention. The passenger that is simply looking to reach their ultimate destination has the level of expectation that they will arrive there safely. They are not thinking that the jetbridge bolts are about to fail, and that the collapse may cause the bridge to drop down on the L-1 door hanging on the aircraft, resulting in the structure coming through the very floor on which they are walking.

Figure 12.4   Jetbridge failure (Jack Kreckie Photo)

Figure 12.5   Jetbridge hanging on door, structure came up through jetbridge floor (Jack Kreckie Photo)

Safety Culture

We recognize the need for safety and, in general, follow most of the safety rules of which we are aware. Each of us can think about examples within our own organization that we follow selectively. If we intend to truly embrace safety, and to integrate safety, we need to raise the level of safety emphasis within our organization and establish a true safety culture. The best illustration of a success in establishing a culture is the security culture that exists at airports today. Those of us who work at airports know the security rules and typically would not violate them. One would not show up for work without one's security badge and expect to get on the airport operations area (AOA). One would never leave a security door open or let a stranger piggyback through a security identification display area (SIDA) door. We adapt to new security regulations as they are implemented and accept these changes as being necessary for the greater good. When everybody puts the same value on safety rules, we may well have been successful in developing a safety culture. If we can achieve a balance which combines our safety culture with our security culture, we will see great dividends for our efforts.

Emergency Management at Airports in an SMS Environment

Effective and efficient management is best achieved when the stakeholders themselves are involved in developing, testing, and implementing a plan. This is true in every case, whether it is management of an aircraft arrival at the gate, an aircraft accident, an airport security plan, or a SMS.

When a team approach is used for developing safety policy and objectives a number of benefits are quickly realized, including the development of important relationships. Aviation is highly competitive, but when these business adversaries meet at the safety table, there is one primary goal. The stakeholders in safety share methods and ideas, working together to develop inter-operative plans, which better serve the airport community as a whole. The benefits to this approach include:

1. The development of a 'safety culture', which universally raises safety awareness and results in a reduction of accidents and injuries.
2. The tendency of the *safety culture* to interface naturally with the *security culture*, which already exists at most airports.
3. Enhancement of emergency operations through established relationships, communications, asset sharing and a spirit of cooperation.
4. Achievement of effective proactive plans and procedures satisfying the needs of those responsible for safety, security, operations and planning.

The most significant lessons learned through the management of the most challenging events in the last decade relate to interoperable emergency management models. The terrorist events of September 11 2001, and the devastating havoc caused by hurricanes Katrina and Rita clearly outlined the need for a standard template for emergency management and incident command, which resulted in the National Incident Management System (NIMS). This system requires that all emergency responders are trained to the same incident management model, that is, everyone is readily reading from the same script and speaking the same language during the active phase of emergency management.

For a number of reasons, airlines and airport support companies deal with rapid turnover of employees. This increases the challenge in maintaining updated contact lists, and keeping employees trained in emergency procedures. A successful SMS requires constant interaction with representatives from every employer. Safety committees conducting risk analysis, training, safety audits and safety promotion continuously keep a line of communications open throughout the airport community, working together in problem solving. Each group learns and better understands the culture of the other stakeholders. Representatives from each airline, FBO and tenant, the police, fire and airport operations departments are critical members of this team.

Similar to NIMS on the national level, an effective SMS provides an environment for the stakeholders to reap benefits for emergency managers at the airport level, who depend upon instant communications, immediate response for information and assets and 'mutual aid' from other airport stakeholders.

ARFF Role in Developing Safety Culture

The role of ARFF in developing a safety culture at your airport is a natural extension of the normal role of a progressive emergency service provider. ARFF should be taking the lead in a proactive approach to community safety services. The airport is a community and, like every community, there is a transient population as well as a resident population. The transient population in the airport community is the flying public and the visitors who pass through the airport. These numbers vary from airport to airport, but are proportionate to response activity, emergency services personnel staffing levels and assets on hand. The resident population is the working community at the airport. These are the people who make their living at the airport as providers of goods or services for the airport operator, the tenants, or contractors. Those in the latter group are the community's stakeholders. They have a vested interest in making the airport a safer workplace.

Fire prevention programs have been launched by nearly every fire department in the world. It is a natural, based on their primary mission – firefighting. Response to accidents typically consumes a larger statistical portion of emergency response than actual fire calls. Incidents causing injuries and damage on our airport ramps occur every day. Shouldn't accident prevention and safety programs be one of the standard services provided by fire departments?

Ramp accidents and incidents causing injuries and damage are a problem that airports need to take more seriously. Industry experts estimate that the airlines lose more than US$5 billion a year to ramp damage, typically collisions between ground service vehicles and parked aircraft or jet ways. Because accidents also result in cancelled flights, lost ticket revenue, added costs for passenger lodging and overtime for repairs, even minor ramp incidents can cost airlines in excess of US$250,000. Indeed, the Flight Safety Foundation estimates that, for every dollar of aircraft damage, the actual cost to airlines is minimally five times that amount (Vandel, 2004). One airline reported that US$77 million in aircraft damage from ramp operations produced about US$540 million in actual lost revenue.

Even worse, according to experts, is the number of ramp workers injured every year. At 14 in 100, the rate of injuries to ramp workers is far higher than many other industries (ARFF Professional Services LLC). Human error is the primary cause of ramp accidents as reported by the International Air Transport Association. Approximately 92 percent of incidents can be traced to the failure to follow procedures, lack of adequate training and airfield congestion. Reducing ramp damage contributes to a positive bottom line in an industry that is struggling financially.

There are certain occupations that hold a natural attraction and are highly regarded by the general population. If we are truly 'marketing' an SMS to which we are committed, we need to use every effective available marketing tool. In the big picture, everyone is trying to 'sell' something. It may be a specific product like the latest fragrance or food, or perhaps a specific brand or style of jeans or undergarments. The purveyors of these products use the most powerful tools available to them. They too have read Maslow's theory. What do they use in their marketing strategies? The most common tool used is sex. We've all seen the provocative billboards and advertisements of scantily clad models. Sometimes the product itself has only a subtle reference, but the sexual tone of the message serves as a powerful lure to the targeted audience. Foods are marketed with the intent of scintillating our taste buds, playing on our need to eat, and our desire to eat something that we truly enjoy. We have all, at one time or another, fallen victim to these ads and sought out the product as a direct result of those marketing strategies. The most successful of these campaigns results in brand recognition which sets a standard for that industry. People may ask for a Coke, even though the brand of cola they receive is not that important to them. Many will ask for a Xerox copy of a document with absolutely no preference as to the type of machine that makes the duplication. These successful branding strategies have raised market share and profits in a number of areas. Our goal is to develop an SMS that becomes a culture, an accepted and embraced way of life at our airports. In order to be fully successful, it must be interoperable with the security culture, take into consideration the operational needs of our airport in general, as well as each of the stakeholders. It needs to work during routine operations and enhance our abilities, understanding, and capabilities during emergency operations. The ingredients of the SMS will need to produce a system that meets all of our goals, but a guaranteed success will depend upon satisfying the hierarchy of basic needs for the individuals involved in the processes.

This is a great deal to expect from a SMS, but the stakes are high and the benefits great. Marketing needs to be part of our strategy and, if conducted correctly, may be the key to our success. The tools we use may be different from the companies selling those new fragrances, jeans, or undergarments. We need to find a spokesperson or symbol that will tell our story, carry our message and sell this important product – safety. To do that, we need to find a different kind of sexy. We need to package our goods in a way that it is attractive to our consumers. If we can get them to the table voluntarily, and have the right *pitchman* with a compelling plan that is in all of their best interests, we cannot lose.

ARFF crew do not recognize the value proposition that comes with their career choice. People, in general, respect firefighters and what they represent. Most people, at one time or another, have dreamed of being a firefighter and admire and respect the job that they do. Most people have had a glimpse into the *firehouse culture*, either through movies like *Backdraft*, or through a friend or family member that wears the uniform. As a result, most people have a little envy for the camaraderie, the brother/sisterhood and the culture. They have seen the pumper parked at the supermarket, while the uniformed crew is filling a shopping basket for the crew dinner. They smile at the friendly banter and exchange they hear from the firefighters as they peruse the aisles. They grin at the quantity of food that fills the basket. At the airport, they see the powerful ARFF vehicles responding across the ramp or escorting an aircraft to the gate. They scramble to take a photo of the water salute being performed to honor a retiring pilot. There is an undeniable attraction that makes this occupation sexy. Public perception of firefighters is further enhanced by the fact that interaction with the public is usually a positive experience. The fire department is there to help, and those being helped or witnessing the event are often struck by their expertise, kindness and compassion.

Airports with live fire training facilities are an area of attraction for airport employees. When the airport stakeholders see the smoke rising, they know the firefighters are training again. The ramp vehicles line up at the vantage point with the best view to watch. Who can deny the exhilaration of watching a huge ball of fire being tamed by those massive trucks, or by firefighters walking directly into the silhouette of fire with hand lines pushing the fire back with a blanket of foam (see Figure 12.6).

There is a similar attraction to those in law enforcement, but it is not as universal as that with the fire department. Those that have been on the wrong side of the law at one time or another do not have positive memories of their experiences with police departments. However, for a large portion of the population, the uniform, the authoritative presence and the power and excitement of the job holds a certain fascination. LEOs are important members of an emergency planning group, as well as a SMS planning committee. They bring not only their insight and expertise, but also become part of the glue that brings this diverse group together.

Figure 12.6   Students training on ARFF vehicle with High Reach Extendable Turret (HRET) (Jack Kreckie Photo)

What other marketing tools do we have? The airport is rich with highly respected occupations and individuals, including airline pilots, air traffic controllers, clergy, aircraft mechanics, heavy equipment operators, military personnel, LEOs, wildlife management personnel, emergency medical techniques (EMT's), paramedics and a multitude of others. Each of these groups has an interest in safety, security and effective emergency management. We need to lure each of these groups to our safety table and then draw upon the skills, resources and expertise of each to build a system that best serves all of our needs.

SMS Leadership/Safety Policy

The FAA has provided guidance related to SMS for airport operators. Management's commitment to safety should be formally expressed in a statement of the organization's safety policy. This policy should reflect the organization's safety philosophy and become the establishment of the SMS. The safety policy outlines the methods and processes that the organization will use to achieve desired safety outcomes, will be signed by top management, and will typically contain the following attributes:

- the commitment of senior management to implement SMS
- a commitment to continual safety improvement
- the encouragement of employees to report safety issues without fear of reprisal
- a commitment to provide the necessary safety resources
- a commitment to make safety the highest priority.

This commitment by senior management is an important requirement. This is not just a commitment by airport senior management, but should be made by the management of every company and organization at the airport. It serves as authorization for the appointed safety manager, and an endorsement of the plans and programs developed.

The airport would be well served by forming an airport safety alliance. Members of the alliance would be required to sign a charter which indicates their organization's commitment to safety. The language of the charter should be developed by the stakeholders and attempt to embrace a safety mission statement that satisfies the expressed needs of all the members. The airport safety alliance needs to be managed by a core group that has the commitment of their upper management to commit time and resources to its work. The alliance should have regular meetings and serve as a clearing house for safety information, safety education and distribution of safety-related information. There needs to be a direct line of communications with other groups at the airport, such as any formalized security alliance, so that the important transfer of information can be streamlined.

Under the ICS structure the safety officer has, by design, the 'commitment of senior management'. He or she has the implied approval to cease any operation without getting prior approval of the incident commander when that operation is deemed to be unsafe. The safety officer in an ICS structure is the only one, other than the commander, who wields this type of power. It is a commitment by the senior managers that safety is the top priority. The ICS works quite well and there are certain components that would transfer nicely into an SMS environment. This serves as perhaps another factor that should be considered when attempting to develop a safety culture at your airports. Established groups already well-versed in risk analysis and command discipline, such as your ARFF department, would easily embrace a leadership role in the airport's safety system.

SMS Airport Headquarters

Members of the fire service are problem solvers. On a daily basis, they respond to strange odors, crackling sounds, lockouts, unknown products, accidents, entrapments and any situation people are not sure how to handle. Each problem is approached with prudent judgment, logic, caution and experience. Command of the problem is accepted and established. Hazards are then identified, isolated or mitigated and, when the scene is safe, command is terminated and the teams return to quarters prepared for the next problem. This process is repeated in virtually every fire department in the world multiple times a day. Each is uniquely familiar with the environments of the jurisdiction that they protect and is aware of all of the nuances of the neighborhoods, the industries and the infrastructure of those communities.

All of us who are stakeholders in the aviation industry can see what is happening in the industry in general, or at the airports with which we are most familiar. There is a disturbing trend in the increasing accident rate on our ramps reflected by injuries, damage and even deaths. The costs of the injuries, the damage and delays contribute to the financial demise of profitability in this struggling industry. By assuming a role of leadership, ARFF can contribute to the slowing of this trend, and energize the airport community to develop a safety culture. By reaching out to the airline community and developing partnerships in safety, firefighters have an appeal that can bring these fierce competitors to the firehouse table wherein the common goals and problems can be identified. What will become immediately apparent is that these stakeholders embrace the idea of working with the fire department. They enjoy coming to the fire station and meeting in their environment. They take pride in becoming a part of a proactive safety service and identifying with the fire service. Never underestimate the power of a spaghetti dinner served in the firehouse kitchen, or a fire department T-shirt!

Community Resource Management

Wikipedia definition: 'Crew (or Cockpit) Resource Management (CRM) training originated from a NASA workshop in 1979 that focused on improving air safety.' The NASA research presented at this meeting found that the primary cause of the majority of aviation accidents was human error, and that the main problems were failures of interpersonal communication, leadership and decision making in the cockpit. A variety of CRM models have been successfully adapted to different types of industries and organizations, all based on the same basic concepts and principles. It has recently been adopted by the fire service to help improve situational awareness on the fire ground.

Although first used in aviation, the acronym CRM is used by a number of industries. Customer relations management, cultural resource management, composite risk management, client resource management, and continuous resource management are just a few of the 85 definitions for CRM found in a Google search.

Aviation fire protection crews should be encouraged to expand this definition one more time. ARFF Crews stand in the unique position of having an understanding of cockpit resource management, as well as the fire service model of crew resource management. A hybrid of these cultures that suits the needs of the airport response jurisdiction is community resource management (CRM). By drawing our community together, working together to develop a safety culture, sharing our experiences – our successes and our failures – we build relationships that live long beyond that first firehouse dinner. These relationships will bear fruit every day, but never in such abundance as during the management of an emergency event. The daily benefits will be obvious as one reviews accidents and incident statistics, or simply by driving across the ramp and seeing the products of the airport community-based safety program in action.

Historical Aviation Accident Review and Lessons Learned

Fortunately, aviation is in general very safe. Considering the number of flight operations conducted every year, the number of incidents is extremely low. In light of the number of actual accidents or fires, we see an even better representation of the safety of commercial aviation.

For the ARFF, this means that proficiency will never be developed from all of the plane crashes or actual major aviation-related incidents in which they participate. In spite of this fact, absolute proficiency is exactly what is required in that 'worst case scenario once in a lifetime' incident in order to maximize survivability and minimize loss. Their skill sets will be honed through training, and their tactical experience developed through those incidents in which they are involved and by studying the accidents and fires that occur elsewhere. Common sense, as well as our experience in other activities, tells us that any type of activity or operation is enhanced through familiarity with procedures, players and operations. We could gather a team of the best athletes in any particular sport in the world, and have them led by the very best in coaching staffs. You have the ingredients for a team that is capable of meeting any challenge. Give that team a chance to get to know each other, develop strategies, practice and make adjustments based on their combined strengths and skill sets, and you will have a team that is unstoppable.

Our airports have such ingredients. Each of us in aviation brings certain skills, experience and resources to the table. At an emergency, each of us holds a piece of the puzzle necessary to bring this incident to its best possible resolution. If we have already worked together, understand the strengths, needs and resources of each person in the command post, we can anticipate the needs of the incident, and provide coordinated responses and harmonized solutions. The emergency is not the place to be handing out business cards.

Incident review is a routine event in aviation and in the fire service. We strive to learn from past experiences to follow the best practices and avoid the actions that had less than favorable results. The following incident review provides consideration for areas of emergency operation that may have been made better if a previous relationship had been forged by the stakeholders.

Reviews and commentary in the following report are not meant to be critical in any fashion. In fact, any lessons learned and shared through a review of this incident are gleaned as a result of the honest report, as well as hindsight testimonials as to what worked well and what might have worked better. This report is simply a continuation of that information exchange to perhaps contribute to tactics that may prove effective at a future incident. In this case, the hope is to highlight the potential benefits of relationship building through SMS and CRM, and how they may have been beneficial during emergency management at this incident. Editorial is provided after each section, offering perspective.

This particular incident is an excellent case study for of the following reasons:

- There was no loss of life.
- The incident has been used by the aviation industry to highlight the need for ARFF training on cargo aircraft.
- The Philadelphia Fire Department has taken the lead in effecting change. The PHL chief has lectured extensively at ARFF venues to raise awareness to the lessons learned. PHL has developed a unique database to serve as a tactical tool for incident commanders, providing data that they needed, but could not obtain on the day of the incident.

Accident Review: UPS Airlines Flt. 1307, PHL, Philadelphia, PA

The following information was taken from the NTSB Survival Factors/Airport and Emergency Response Group Chairman's Factual Report, Docket NO. SA-228 EXHIBIT NO. 16A.

Summary

On February 7, 2006, at 2359 (EST), a Douglas DC-8-71F, N748UP, operated by United Parcel Service Company (UPS) as Flight 1307, landed at Philadelphia International Airport (PHL), Philadelphia, Pennsylvania, after the crew reported a cargo smoke indication. The three flight crewmembers were able to evacuate the airplane using the L1 slide. Fire subsequently caused substantial damage to the airplane and numerous cargo containers on board. The three crewmembers received minor injuries.

Notification
The air traffic control tower notified ARFF of an Alert 1 (a reported aircraft emergency or problem) via the 'crash phone' at 23:57:15 EST. The tower reported that UPS was 5 miles southeast of the airport with a smoke warning light in a cargo hold.

Fire Control Time
According to the City of Philadelphia Fire Department dispatch logs, ARFF arrived at the accident site at 2359 local time. A period of 4 hours and 8 minutes elapsed from the initial arrival on scene to the time the incident commander radioed to dispatch for fire control (fire under control) (at 0407 local time).

(Editorial Comment) A control period of 4 hours is the first printed indication that the selected tactics were not as effective as was hoped.

Fire Conditions On-scene
When an ARFF vehicle arrived on-scene, no fire was visible, but smoke could be seen coming from the open L1 door and the outflow vent in the tail. The first indication of visible flame came when firefighters opened the right over wing emergency hatch. Flames were observed rolling on the fuselage ceiling over the tops of the cargo containers. Smoke began emanating from all open exits. All fire was located aft of the over wing exits toward the aft bulkhead. Burn-through of the fuselage roof occurred at several locations between the trailing edge of the wing, aft toward the tail.

Editorial Comment

Hindsight is nearly always 20/20. We have the ability to see the outcome of the selected tactics, pass judgment and then consider alternative methods. In this case, we have an apparently intact fuselage with no signs of significant fire. There is no blistering of paint, no visible deformity to the skin and, based on the smoke report, a containable volume of fire. The first action was to open an over-wing hatch which introduced oxygen. Let us consider the effects of the opposite action. A similar level of effort may have been able to secure the open L-1 door, cutting off the source of oxygen.

Firefighters, in general, will initiate actions to gain access to the fire. From a very basic standpoint, you need to get to it in order to put it out. Later entries in this report will identify the fact that the firefighters lacked the ability to gain rapid access to the area of the cargo involved in fire. Drawing from the concept of CRM developed through the teambuilding accomplished through our SMS, input from UPS personnel may have contributed to a better outcome on this incident. Closing up that aircraft may have slowed the fire by reducing the oxygen available for long enough to develop a strategy drawing from the expertise of the airline, FBOs, and other stakeholders populating the unified command post.

Firefighting Strategy

The ARFF units surrounded the airplane and a water attack was ordered. Access to the main cargo area was obtained via the right over wing doors, and an exterior hand line attack was initiated from this location. Turret streams were applied into the R4 doorway while a Snozzle piercing operation was conducted on the left side. The piercing operation began behind the left aft over-wing exit, in line with the windows, and continued aft toward the tail. The entire operation switched to a foam attack. Eventually hand lines were advanced to the interior of the airplane through the R4 and left side over wing doors until total extinguishment was completed.

Editorial Comment

Remember that we identified the reported '4-hour control time' as an indication that the selected firefighting tactics were not effective. An alternative method may have been to direct some water to the top of the fuselage. In our alternative method of dealing with this fire, we have already secured the doors, limiting the amount of oxygen upon which the fire can draw upon. Now, by directing water to the roof of the fuselage we will reduce the temperature of the metal. Cool metal does not melt. The reaction to the water as it flows over the metal is an excellent method of evaluating the temperatures inside the aircraft. If the water immediately turns to steam in an area of the fuselage roof, that is the area directly over the fire. As time goes on and the fire uses the available oxygen, the amount of heat on the fuselage roof should diminish and the size of the heated area of the fuselage may diminish (a good indicator of effectiveness).

As outside resources arrive, an aerial platform could be positioned to monitor this activity and report conditions to the command post. A thermal imaging camera (TIC), or forward-looking infra red (FLIR) camera (as mounted on many ARFF vehicles) would be an excellent tool to be used in this monitoring position.

If the method is effective we have saved a great deal of effort, reduced our risks and minimized damage. If it is not effective, it will hopefully buy us enough time to gather resources and expertise together and develop an IAP based on the combined knowledge of the stakeholders.

Also indicated in the report, 'The piercing operation began behind the left aft over wing exit, in line with the windows, and continued aft toward the tail.' UPS guidance indicates that the best piercing location is at 10 o'clock and 2 o'clock which is higher than the window line. It is impossible to determine if in the conditions present that day and at this location would have proven more effective, however, it would have been helpful guidance if presented in the unified command post. Better yet, a partnership forged through SMS may have resulted in sharing the information with ARFF prior to the incident.

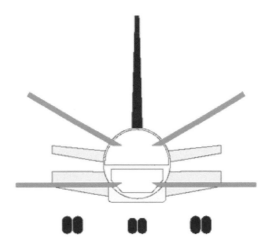

Figure 12.7   Recommended piercing locations as shown in the UPS Aircraft Rescue Fire Fighting Manual

Figure 12.8   Hazmat (hazardous materials) information exchange

The following excerpts are taken from the *Hazmat Information Exchange* section of the PHL report. Actual names of individuals have been changed.

> When Captain (Smith) arrived on scene, he asked the airplane's captain if there was hazmat onboard. The captain notified him that hazmat was onboard; however, he did not have the manifest with him. The captain requested that airport operations obtain the information. According to the Airport Operations Duty Officer, he arrived on the UPS ramp at 0006 and requested the hazmat documents from UPS personnel.

> A UPS representative brought a faxed copy of hazmat documentation to an airport operations agent, who was staged on the UPS ramp. The airport operations agent immediately brought the UPS representative to the scene, and handed the hazmat documentation to the airport duty officer at 0107.

> The UPS representative stated that a UPS ramp supervisor gave him the 'prediction' of what was onboard the airplane. At approximately 0014, airport operations met him at Taxiway Uniform and drove him out to the airplane. When he arrived on scene, a firefighter with a white helmet asked him what hazmat was onboard. He told the firefighter that the hazmat was located in position 14 and 3. He told the firefighter that he could only provide positions of the hazmat, because only the Notice to Captain (NOTOC) discloses the onboard hazmat. The UPS representative also saw a firefighter throw the NOTOC out of the airplane along with some other items. He said he personally picked up the NOTOC and opened it to find four pouch placards, two for hazmat and two for dry ice. He explained what the hazmat onboard was to the firefighter, handed him the two pouch placards for the hazmat, and kept the NOTOC and two pouch placards for dry ice. Later, he handed the rest of the paperwork to a fire lieutenant when he asked for it.

Notification to Captain Chain of Custody

According to the UPS Flight Operation Manual, the NOTOC information is obtained by the captain from the UPS load supervisor prior to departure and placed in a *dangerous goods* pouch, which, in this case, was located on the outside of the lavatory door of the accident airplane. According to interviews of the flight crew (refer to the Operations Group Chairman's Factual Report), the NOTOC was on floor of the flight deck during the flight and the flight engineer picked it up and 'wedged it in the crash axe sheath'.

> According to the ARFF crew, when ARFF arrived on scene, the captain instructed Lieutenant Blue to attempt to enter the cockpit to find the manifest. Lieutenant

Blue entered the cockpit to do a quick search for the manifest and crewmembers, but was not able to find the manifest.

According to a statement from Firefighter White, at a later point, he entered the flight deck to search for the NOTOC again, and found it. He handed it to Battalion Chief Green and he was instructed by Deputy Chief Yellow to handle hazmat operations. According to Battalion Chief Green, he relayed the information on the NOTOC to Lieutenant Black from the hazmat unit for research. According to Lieutenant Black, he handed the NOTOC to a firefighter on his crew and requested information on the particular chemicals from the Materials Safety Data Sheet (MSDS). After receiving information on the chemicals, Lieutenant Black radioed the information to Battalion Chief Green. There was no further handling of the NOTOC until Lieutenant Jones from the hazmat unit requested to see any information regarding the chemicals on board. At this time, he gave the NOTOC to Lieutenant Jones.

The Survival Factors Group received the NOTOC envelope on-scene from Lieutenant Jones on February 10, 2006.

Editorial Comment

In this particular event, hazmat onboard was not a significant hazard to responders or the airport community. There was a greater danger created by the burning of aircraft components than by what might have been onboard. The report would seem to indicate that a significant level of effort and time was committed to tracking down the NOTOC. The conflicting reports would seem to suggest that all information was not being routed through the command post.

Previously in this chapter, the importance of the development of a safety culture and CRM at our airports, and the positive effect it has on emergency management was discussed. The miscommunications regarding the NOTOC during this incident had no obvious effect on the outcome of the event. It does, however, stand as an excellent lesson learned. ARFF departments that immerse themselves in relationships with the airport community will reap the benefits at the incident command post. The community will have learned about emergency management, chain of command, incident site discipline and, more importantly, better understand how important their role is to help achieve the best possible outcome. By coordinating with UPS at the unified command post, a call to the UPS Worldport Global Operations Center could have gleaned all of the dangerous goods information for the flight.

Interviews (Airport Operations)

The incident commander contacted Airport 10 and requested that he contact UPS to track down a manifest for the hazmat on board the airplane. Airport 10 tried to radio an operations agent to go to the UPS hangar, but was not able to get in touch with anyone. He heard an airport operations agent confirm over the radio that he was at Gate 11. Airport 10 then proceeded to UPS via Echo, Sierra and Uniform to stay clear of the airplane.

Airport 10 arrived on the UPS apron at 0006, and flagged down people on the ramp and told them that they had a DC-8 incident. He said they needed 'managers' and the cargo hazmat manifest. He also asked for a mechanic that knew cargo doors, doors of airplane and other general information about the airplane. He advised that the airplane appeared to be on fire and stressed the importance of getting the necessary people in a hurry. While sitting on the ramp, he upgraded the emergency to Alert 2 (at 0014) as directed by his supervisor. At 0020 Airport 10 left the UPS ramp with a 'chief pilot' and a mechanic. He noted that he was 'unable to get a hold of anybody'.

Editorial Comment

The benefits of ARFF involvement in airport SMS planning and CRM teambuilding would have been beneficial, as is evidenced in these comments. The optimal time to learn how to contact the key people needed for an emergency is not during emergency operations. 'Flagging down people on the ramp' can be replaced by the execution of an updated call list and notification system. Through emergency planning and CRM, the people on those contact lists would already know exactly where and to whom to report, what is expected of them upon arrival, what to bring, and would already have corporate authorization to represent the airline in the capacity outlined in the emergency plan.

These interviews provide additional examples of the lack of relationships with the community and how that impacts emergency management. We all need to use this example to remind us that our contact lists need to be inclusive of all operations and be kept up to date.

At 0110, the incident commander asked Airport 10 to identify closest fire hydrant in the area. Airport 10 contacted airport maintenance that confirmed hydrant locations at gates D10, D15, and D8. This information was relayed to the incident commander. The police supervisor (77A) dispatched an officer to check Hog Island Road for a hydrant location as a backup. Fire engines were dispatched to both gates D10 and D15 to run a hose line to support the operation on the runway. Airport 10 requested vehicles on each side of hose line to stop vehicles from running over the hose line.

Editorial Comment

Water supply is a component of the airport's critical infrastructure. Hydrant and water line maintenance is typically a shared responsibility. Fire departments and the water department either operated by the municipality or the airport authority facilities' department oversee the systems and should each have information which includes hydrant locations and flow rates. Tenants on the airport have hydrants in their leased areas and have responsibilities in keeping the hydrants clear. A very basic program in a SMS environment would include an evaluation of hydrant accessibility issues. Nobody will argue that fire hydrants are necessary and should remain unobstructed for use during an emergency, however, tenants with little ramp space and hundreds of baggage carts to store tend to lose sight of that fact during day to day operations. A review of those problem areas through an SMS review can identify parking configurations that work, hydrant visibility programs to help spot hydrants and even relocations during planned construction activities. All of this raises awareness for hydrant locations, and perhaps makes use of airport facilities' personnel during emergencies to escort mutual aid fire apparatus to hydrants to establish water supply.

In addition to hydrant locations, there is a need to be familiar with the water main sizes, capacities and conditions, hydrant flow rates, control valves and airport pumping station capabilities. We need to have a water supply plan to ensure that we can provide constant water supply to the rate and capacity of fire apparatus or building fire protection systems on any area of the field, including the fuel farms. This may include the need for mutual aid agreements (MAAs) or Memorandums of Understanding (MOUs) with mutual aid departments. Depending upon local infrastructure, it may include tankers, large diameter hose wagons, relay pumping, etc. This same information will be extremely helpful in the swift management of a water line break.

The application of the CRM model has benefits in increasing safety, preparedness levels, and collaborative decision making in any application to which it is applied.

Interviews, Aircraft Rescue and Firefighting Personnel

PHL Chief Officer (Two-and-a-half Years ARFF Experience)

This chief said another group of firefighters was standing on a 'pallet truck' working on the cargo door itself. He said that ARFF was 'not familiar with cargo,' which was 'a problem in hindsight'.

Editorial Comment

PHL has taken this valuable lesson and developed plans and programs using technology and training to correct this problem. We should be studying every event and making the same corrections in our operations before we are forced to do so, based on hindsight. PHL is now utilizing a CRM model that employs the cargo carriers, and the airport information technology department, developing systems that have been shared with the ARFF industry.

PHL Captain (Initial IC-Foxtrot 21) (Two Years ARFF Experience)

Foxtrot 21 ordered all ARFF vehicles to take up 'standard position' around the airplane. Foxtrot 21 notified the Fire Command Center (FCC) to 'strike out Box 6355,' which was a code to send predetermined off-airport responders consisting of 4 engines, 2 ladders, 2 chiefs, 2 squads and a deputy chief, to Gate 11. Foxtrot 21 ordered self-contained breathing apparatus (SCBAs) to be used when he found out hazmat was onboard. Foxtrot 2 placed a ladder at the L1 door, over the emergency slide, in order to enter the cockpit to look for manifest and attempt entry into front cargo area. A dual agent (water/powder) line was placed in service through the L1 door.

Editorial Comment

The fire service has predetermined plans and standard operating procedures, which summon additional resources, specialized equipment and manpower to satisfy the needs of the incident. These models create a plan that can increase those resources by striking additional alarms to bring in additional predetermined levels of resources. Modern fire service planning at airports needs to be expanded to automatically draw in non-firefighting expertise and assets available within the airport community. The safety culture created through an effective SMS is the basis for identifying the assets and resources readily available at the airport.

Foxtrot 21 scanned the fuselage with a handheld TIC and found no hotspots. Foxtrot 2 told him that they were unable to locate the manifest because the smoke in the cockpit was too heavy. He called Airport 10 to get a UPS representative and a copy of the manifest. Foxtrot 21 said he was concerned about hazmat volatility and water reactivity, as well as the exact location of Section 15. He also requested the fuel capacity. He was informed the lower cargo hold was clear of smoke and fire, and firefighters told him they could not make entry into the airplane from that area. The crash charts available for ARFF crews were for the passenger version of DC8, not cargo.

Editorial Comment

Cargo position locations such as position 15 or L-15 are standardized among aircraft types and cargo carriers. This information is available through your cargo carriers along with the crash charts for the cargo positions on each type aircraft. This information can be carried in paper form or loaded into mobile data terminals (MDTs) in airport command posts or incident commander vehicles. Cargo airline personnel should be part of the unified command post. Assistance related to cargo positions and interpretation of cargo manifests, dangerous goods manifests, door operations, aircraft load and balance information, etc., are all readily available through the expertise of the airline staff. Planning for events on the ramp, at the freight facilities, or onboard aircraft are projects that should be evaluated through planning in a SMS environment. This exercise not only raises awareness on all of these subjects, but develops the relationships so important in emergency management.

Foxtrot 21 indicated that he did not believe that the responding structural companies or the heavy rescue unit participate in training with the ARFF crews, with the exception of the triennial drill. He indicated that he was not familiar with the manual main deck cargo door opening procedures or cargo airplane, but he does train on other commercial passenger aircraft.

Editorial Comment

These comments in the interview further support the need for CRM as developed through an effective SMS. The relationships developed through SMS provide opportunities for planning, training and problem solving. ARFF is a highly specialized area of firefighting. ARFF departments are not staffed or equipped to the level required to successfully handle a major incident without bringing in outside resources. If we are to include outside response agencies in our airport emergency plan (AEP), we need to include them in our safety systems and planning efforts. This incident illustrates the critical need to include responders necessary for the success of the mission in training events.

Conclusions

This study discusses a number of areas of ARFF that we tend to look at individually rather than collectively. We have applied some of the salient points of those discussions and applied them to a single case study, UPS Flight 1307 at Philadelphia International Airport. Many of the discussions offered provided new ways to think about the tactical challenges the ARFF crew at PHL faced on February 7 and 8, 2006. We do not make determinations as to right or wrong, but rather what lessons with which we can we walk away to better prepare us for our mission.

Lessons Learned

Emergency Preparedness

It is very clear that the planning, management and mitigation of a major aircraft accident or fire requires a huge commitment from numerous agencies, organizations, businesses and individuals from multiple jurisdictions. The need for outreach, planning and practice involving all of the players required to successfully manage these incidents is a mandatory component of our emergency planning model. The ability to operate with this diverse group of stakeholders under the NIMS will better prepare us all to work together under a universally recognized command structure.

Outreach specifically within our airport community (community resource management) throughout the year as part of the development of SMS, ramp safety and awareness will develop a culture that is familiar with the other stakeholders, accustomed to working together and able to think outside their own areas of responsibility.

Teambuilding: Airport Safety Alliance

An ARFF-based Airport Safety Alliance serves as a bridge to bring together airport stakeholders and combine their energy and resources to enhance safety while working on neutral ground. The airport fire station serves as safety headquarters, and has the ability to attract participation from the spectrum of airport tenants. Once gathered, these groups tend to generate their own projects, goals and ambitions. Each has safety concerns that need resolution. In many cases, an authority is required to provide regulatory support within the organization.

Benefits of ARFF-based Airport Safety Alliance

- Proactive approach to risk reduction.
- Development of relationships critical to emergency management.
- Team approach providing 'options' for our stakeholders.
- Experience as 'problem solvers' helpful in development of safety solutions.

Common Trends Affecting Airport Safety

- Increase in vehicle accidents on the ramp.
- Ramp workers with poor driving records being rehired by competition.
- Stakeholders reporting an increase in phantom 'ramp damage'.

Objectives of the Airport Safety Alliance

- Development of a system that increases awareness and safety on the ramp.
- Reduction of the number of accidents, injuries and incidents on the ramp.

Benefits

- Reduction in personal injury
- Reduction in costs associated with damage from accidents
- Reduction in costs associated with down time of damaged equipment, delays caused by accidents
- Making the airport a safer place to work – developing good habits within our community
  - *Approach* Training component
  - Standards
  - Enforcement
  - Measures
- Development and delivery of a safety awareness training module
  - Systems thinking – each of us are part of a system with a role that is critical to the success of the airport mission
  - Safety awareness – review of rules and regulations as well as industry standards, and prudent safety and fire safety practices
  - Examples of recommendations from the safety alliance for establishment of new programs, standards, rules, or requirements
- CDL licenses or equivalencies
- Mandatory safety vests or belts
- Vehicle lighting standards
- Incentive programs
- Evacuation plans
- FOD prevention programs
- Accident investigations/case reviews for the alliance
- Safety hotline
- Information sharing/existing programs
- Team hazard evaluation/recommendations
- Airport-wide safety fair
- Enforcement
  - Internal to your organizations – safety coordinators and supervisors
  - Ramp-wide enforcement
  - Airport law enforcement
  - Airport operations
  - ARFF departments/fueling compliance

Figure 12.9   Demonstration of FOD cleanup tool at safety alliance meeting (Jack Kreckie Photo)

Figure 12.10 Team evaluation of ramp traffic conflict remedies

Figure 12.11 Airport Alliance Safety Fair and Symposium (Jack Kreckie Photo)

- Measures
  - Analysis of statistics for accidents and injuries – past
  - Tracking of accidents and injuries – future
  - Looking for trends, patterns for program modification
  - Looking for measurement of effectiveness of safety program

The ARFF-based Airport Safety Alliance serves as a conduit connecting airport safety stakeholders. Monthly newsletters, safety flyers, accident case reviews and a safety hotline provide the airport community with an opportunity to work together and resolve potential threats to safety. In so doing, relationships are built that pay dividends realized in emergency management. The safety of our airport community should be given the same level of emphasis as is security and quality services to our customers. There is no competition in safety, and during an emergency, we need the cooperation of all the stakeholders to achieve the best possible outcome. Together we *can* make a difference.

References

ARFF Professional Services LLC (2010). *Airport Ramp Safety*. http://www. apssafety.net/id65.html. Milford, MA: ARFF Professional Services LLC.

Vandel, Bob (2004). *Equipment Damage and Human Injury on the Apron*. Alexandria, VA: Flight Safety Foundation.

Chapter 13
Safety Promotion

Kent Lewis

> Safety Promotion requires creating an environment where safety objectives can be achieved.
>
> (FAA, 2010)

Promotion of safety systems is one of the most important discussions that will occur in a high reliability organization. There is much written on policy and risk management, but philosophies and tools will be most effective if the right message is communicated effectively at the right time, among the people who will use these tools and put philosophy into practice. This chapter could be titled Safety Communication or Safety Information, but *Safety Promotion* adds a positive, collaborative character to the process of sharing information within an organization in order to increase the safety knowledge base.

In order to effectively promote operational safety within an organization, it is essential to create an open environment of knowledge sharing and learning. Assessments of human performance and behavior will be the best indicators of safety health, and provide opportunities to develop continuous improvements to a safety management system. Many times a safety program is promoted in a one-dimensional fashion; information is transmitted from organizations or regulators and there is little or no opportunity for feedback from those tasked with interpreting safety philosophy and turning procedures into practice. This approach to *checking the blocks* is most likely found in a pathological or bureaucratic organization. Performance is measured in an audit fashion and when vagaries in the system appear, a *blame and train* mentality is taken to address past behavior, in hopes of preventing future recurrence. This promotion of safety can most often be identified by safety bulletins that exhort increased attention to vigilance, situational awareness and warnings to combat complacency. If any attempt was made to examine the multiple situational factors that lead to divergence from expected system norms, behaviors and expectations, it is usually not shared in communications but rather buried deep in investigation reports. By taking a proactive approach, essential knowledge is shared with the people who will benefit most from the information, and system operators become excited about the collegial aspects of information sharing. Some other indicators of safety health are dedication of resources for safety programs, in terms of direct budget control or human resources, and provisions for training in safety management systems for everyone in the organization, enabled by education and experience of key safety personnel. This approach is farsighted; adequate resources, training and experience are necessary to properly manage and operate a system safely. One current threat to the system is a failure to capitalize off the strengths afforded organizations in the information age. In order for systems to adopt a proactive or generative approach, information must flow freely throughout the entire system; barriers to communication must be removed and safety information should only be used for the purposes of enhancing safety, not for criminal and civil litigation or punitive purposes. This lingering reluctance by States to address best practices can be overcome if international safety values are defined and adopted. There has been much discussion on the important role that a positive safety culture plays within an organization, and that same culture must be embraced by system regulators and operators. With shared values, a systems approach will flourish. Leaders for this initiative will be found at all levels of the system, and those providing information to the pool of operational safety knowledge are the most valued promoters. Positive indicators of this type of system are open access to information and active involvement in the event reporting process by all personnel.

Recipe for Success: A Four-step Plan

1. Communicate
2. Work together
3. Educate
4. Evaluate

Simple enough, right? You probably could stop reading right here, but you've invested this much time in the book already, so hang around for a few more pages. Let's discuss the first ingredient, communications. Why are communications important? Communications are important because that is how information is transferred. A message is sent and we hope someone not only receives it, but also understands the content. Some type of feedback is necessary to ensure that the message is understood, and in aviation we even have controlled vocabulary to acknowledge understanding of a message, 'roger' or willingness to comply, 'wilco'. If the message is not understood or contains incorrect information, then a path is needed to ensure that correct information is obtained. A good system offers opportunities for people to talk and listen, and good resource managers gather information from all sources to make the best decisions. This is important for safety promotion because it is important to transmit the right message in a team environment. Shared cultures, shared goals and shared mental models are essential foundations of safety promotion; everyone on the team needs to be pulling in the same direction or it will be difficult to balance system conflicts between production and protection. When these conflicts develop, high-functioning teams identify threats and deal with them in a safe, efficient manner because they have shared understanding of organizational philosophies, policies, procedures and practices. When operations begin to drift away from expected norms, team members can communicate to find the most effective solutions to return a system to its normal state. An operational example of this would be dispatchers sharing time-critical information about en route turbulence to a flight crew so that proactive measures can be taken to change routes or altitudes, and to ensure the safety of cabin crews and passengers.

Another example is conducting safety seminars, where information on operational issues, ongoing research and lessons learned can be discussed and action plans crafted to address hazards to operational safety. Modern day communications include platforms such as Facebook, Twitter, webinars and YouTube. The model with these mediums is to share early and to share often in a social networking environment. This social contact enhances learning, the medium is simple and consistent, and relevant, timely lessons can be shared. It is critical to mention once more that everyone on the team must be involved in this communication and there must be a joint responsibility to promote an open, informed and just culture of knowledge sharing. 'The organization should promote safety as a core value' (FAA, 2010, AC 120-92, App. 1).

High-functioning teams often define core values as part of their mission and vision statements, one example is the United States Marine Corps, whose core values are 'Honor, Courage and Commitment'. The Marine Corps operational doctrine rests solidly on a foundation of a combined arms concept, it is an air–land–sea combat system charged with the mission of seizing and defending advanced naval bases. Operational safety is an important component of this mission, and resources must be protected in order to complete the mission. The number one responsibility for every unit commander was to take care of your marines. If you took care of your marines, they would take care of the mission. Time and time again, there are proven cases in the military where leading causes of loss were not from combat operations, but rather from training or administrative evolutions. After the Persian Gulf War of 1990–91, more marines were lost the first six months in off-duty mishaps than during the seven months of Operations Desert Shield and Desert Storm. There are no missions in peacetime that justify the loss of personnel or property, and even in wartime, the right mix of knowledge, training, communication, resources and skill will guarantee a successful outcome.

Safety is a core value that should be adopted by high-reliability organizations to ensure the same successful outcome.Organizations must work together, both internally and with external agencies, to accomplish organizational goals. Safety must be inherent, embedded with those goals and placed at the appropriate level to ensure production is also optimized. When production goals increase, safety goals must also increase to maintain a balance. Operational risk decision makers must maintain accountability by ensuring safety stays at the same level or moves to a higher level, adjusting risk controls as warranted. Within a SMS, this can be accomplished by training and dissemination of information (FAA, 2007, 8.3):

- Employees must understand the SMS.
- Employees benefit from safety lessons learned.

- Explain WHY particular actions were taken.
- Develop a positive safety culture.
- Promotion is continuous, just like camouflage.

It may be obvious that the author of this section has a military background, and 'Everything I need to know about safety I learned in the Marine Corps.' There was a great deal of thought put into developing aviation safety programs that combined education of Aviation Safety Officers (ASO) on reporting, aero structures, human factors, fixed wing and rotary wing aerodynamics, mishap investigation and command programs with a commander's leadership course in aviation safety. This was a team concept also, and to borrow the term *gung ho*, which the Marines adopted from the Chinese, teams worked together to safely accomplish the mission. This is the second ingredient in the promotion process. The team forms a cooperative with shared mission, vision and goals and leadership is apparent at every level of the organization. Promotion of safety is not the sole responsibility of a safety manager, rather it is inherent in everyone's mission. There is a visible dedication to this value from management, regulators, manufacturers, employees and other team members. Communications are blended with command, control and information systems to ensure that proper resources are allocated and a clear, consistent message is delivered. Conflict can develop if safety management is not blended properly with business management, because the primary commitment of SMS is not to create a product but rather to maintaining processes or systems. These processes are founded on leadership principles such as knowledge, justice and

> ethical standards of behavior as well as intellectual integrity and sound thinking. Resources are always scarce and there is always competition for them, but the battles over resources are fundamentally contests to assure that these basic processes are properly supported. (Swigger, 2010)

This previous statement was taken from a discussion in a Master's course on governance for academic libraries, as an illustration of the universal nature of team challenges to resource management. *Properly supported* is a case that often gets disputed, but if a team starts out with the same fundamental values and goals, the gaps between production and protection can be identified and sealed.

Information transfer is key to the optimization of resource utilization, and most mishaps have roots in failures of an information system. The right information was not identified and provided to the right people at the right time. Certification standards may not have been communicated or properly developed, or runway conditions reported in an ambiguous manner. Crew alerting systems may not have had salient enough cues, or a quick reference handbook may contain outdated information. There are information search and retrieval challenges of great magnitude in high-reliability organizations, and those who address these challenges up front as a team and communicate efficiently will ensure the greatest success. Information must be shared so safety-critical tasks are properly trained and team members are made aware of hazardous areas of operation. The mission must be well-defined and understood, especially with respect to threats to the operation.

As mentioned in the risk management section, threats can be mitigated through avoidance, transfer, or through appropriate risk controls. Threats can also be controlled by sharing information and acting on *lessons learned* to continuously improve the system. *Lessons learned* must be incorporated into learning systems that educate personnel in State safety systems, manufacturers, airports and air traffic personnel, as well as educational and awareness programs for operational personnel.

The third ingredient in safety promotion is education. Safety is not necessarily a product, but rather a system that contains tightly coupled processes. Education is necessary to develop understanding of the nature of the system, as well as roles and responsibilities to maintain the system in the desired state. Education programs may take many forms, perhaps the form of a seminar, newsletter or an informational website. Regardless of the form, user needs must be recognized and the appropriate messages must be communicated. There are many excellent examples of education materials addressing operational safety; one is the website SKYbrary (http://www.skybrary.aero/landingpage/), a collaboration between Eurocontrol, the Flight Safety Foundation, ICAO, the UK Flight Safety Committee, the European Strategic Safety Initiative and the International Federation of Airworthiness. The US Federal Aviation Administration (http://www.faa.gov/) and Transport Canada (http://www.tc.gc.ca) both have excellent information available on their websites. Flight Safety Foundation (http://flightsafety.org/) and the Aircraft Owners and Pilots Association (http://www.aopa.org/) offer free safety information on their websites. Curt Lewis and Associates also offer a free flight safety information newsletter (http://www.fsinfo.org/). Traditional sources of information cannot be emphasized enough, as public and academic libraries hold treasure chests full of information, and web technologies lend themselves to development of open access journals that report on the latest research.

Communication and education are about the transfer of information, and emerging web technologies need to be examined to better understand the future roles they play in safety promotion. Web 2.0 technologies offer collaborative platforms where team members can share information in an efficient manner beyond the applications of static websites that are *transmit only*. These technologies 'facilitate interactive information sharing, interoperability, user-centered design and collaboration' (Wikipedia, 2010). One example is the aerospace safety wikispace Signal Charlie (http://www.signalcharlie.net/). Utilizing this platform, timely information is shared globally. Web 2.0 technologies allow qualitative aspects of promotion to emerge that complement the quantitative nature of audits, aviation safety reporting systems (ASRS), flight operations quality assurance (FOQA) programs, and aviation safety action programs (ASAP). New web architecture allows information to emerge from a static state into a dynamic platform that is more representative of and compatible with the latest generations of safety systems. This information conduit is an excellent partner to traditional methods of safety promotion such as bulletin boards, forums, seminars, magazines and other multimedia materials.

One last comment on safety management systems in the information age; it is imperative that library and information systems be developed to capture and preserve corporate knowledge, capitalize on the lessons learned and minimize the weaknesses inherent in safety promotion and communications. Information systems facilitate sharing of information, stimulate thinking and enable development of industry standards and recommended practices. This helps to keep us from re-learning lessons the hard way. By sharing information, knowledge is enhanced and the best return on investments in operational safety can be realized.

No systems approach would be complete without a mechanism to evaluate system behavior. This fourth ingredient is essential in developing a SMS. In the simplest form, we want to see if we are doing what we said we were going to do, and observe the system to see if we are getting the desired results. If the answer is no to either assessment, then change is needed to ensure that practice matches procedure, and output meets desired goals. One example for this process is found in the US navy, where teams from the Naval Safety Center or sister squadrons would conduct a safety audit on a unit to determine its safety health. These audits would be in addition to unit evaluations, or in the civilian world quality assurance and air transportation oversight systems. There are many ways to assess an organization's performance, and one of those ways must address measuring the effectiveness of the safety management system. Safety promotion allows us to address issues of mutual concern, such as the challenge of managing human performance in high reliability, tightly coupled systems. In the late 1990s, the FAA, NASA and Department of Defense crafted the National Plan for Civil Aviation Human Factors (Signal Charlie, 2010). This plan has since been superseded by the work of the Commercial Air Safety Team (CAST) but two important goals were identified:

1. reducing error in human–system interactions
2. increase efficiency of human–system performance.

A national agenda was developed that focused on research and application of research. It is within the area of research applications where we find comparative attributes of a twenty-first century SMS:

1. create an environment for change
2. develop human factors (HF) education and training programs at all levels
3. equip personnel and facilities with modern tools and techniques of the HF engineering discipline
4. develop an infrastructure to translate and disseminate human factors products.

With only a small amount of imagination, these tasks can be expanded to meet the promotion goals within a SMS. An example of this plan in action can be found by looking at the FAA Safety Team (Federal Aviation Administration Safety Team, 2010), where government program managers team with volunteer representatives and industry members to promote aviation safety. Two core components of this initiative are promotion of awareness and education for SMS and human factors, and as a team we will develop tools to continuously improve and promote aviation safety. National programs such as these increase access to information from mishap investigations, incident reporting, trend analysis and safety databases. The NASA ASRS database online is the largest repository of aviation safety information, and uses both qualitative and quantitative methods to assess data and provide information to system agents who can effect change (National Aeronautics and Space Administration, 2010). Individuals, as well as large organizations, can access this data and develop a scalable, comprehensive approach to safety management.

It is important to communicate, work together, continue learning and measure the success of a SMS. Awareness and education is required at all levels of an organization, and those charged with development of a SMS should possess the necessary experience and education to mentor fellow team members and guide a learning system. 'Management recognizes that all levels of the organization require training in safety management and that needs vary across the organization' (International Civil Aviation Organization, 2010, 15.9.1). It is necessary for managers to have a deep understanding of safety systems, otherwise there is a risk of a shallow *sticker safety* campaign developing that has no substance or support from top leaders. The message and the medium are important, and must be tailored to the audience. Embry Riddle Aeronautical University, the Transportation Institute and the University of Southern California's Viterbi School of Engineering are just a few examples of organizations that provide quality education on SMS to industry professionals, and this type of education is essential to manage a safety system.

Restricting the flow of information for fear that learners will be overwhelmed is to be guarded against; instead, it is rewarding to see when mutual respect is shared among team members and collegial learning develops as a result. And, while the law of least effort applies in many circumstances, access for all too critical safety information should be the goal for a SMS. Promotion of safety is essential communication between team members that must take place in order to ensure safe operation of the system. Education and assessment are also key ingredients. When we couple the power of knowledge with the right attitude, operational safety performance will attain new heights.

References

Embry Riddle Aeronautical University (2010). http://www.erau.edu/.

Federal Aviation Administration (FAA) (2010). *Advanced Notice of Proposed Rulemaking*. Federal Register, Vol. 74, No. 140/Thursday, July 23, 2009/ Proposed Rules: 36414-7. Federal Aviation Administration.

Federal Aviation Administration (FAA) (2010). *Introduction to Safety Management Systems for Air Operators*. Advisory Circular 120-92. Retrieved March 3, 2010 from http://www.airweb.faa.gov/Regulatory_and_Guidance_Library/ rgAdvisoryCircular.nsf/MainFrame?OpenFrameSet.

Federal Aviation Administration (2007). *Flight Standards SMS Standardization Manual*. Federal Aviation Administration.

Federal Aviation Administration Safety Team (2010). *About the FAASTeam*. Retrieved March 3, 2010 from http://www.faasafety.gov/about/mission.aspx.

International Civil Aviation Organization (2009). *Safety Management Manual. Doc 9858 AN/474,* 2nd edn. International Civil Aviation Organization

National Aeronautics and Space Administration (2010). *Aviation Safety Reporting System Database*. Retrieved March 3, 2010 at http://akama.arc.nasa.gov/ ASRSDBOnline/QueryWizard_Filter.aspx.

Reason, J. (1997). *Managing the Risks of Organizational Accidents*. Aldershot, UK. Ashgate Publishing.

Signal Charlie (2010). *Human Factors*. Retrieved March 3, 2010 from http:// www.signalcharlie.net/Human+Factors.

Swigger, K. (2010). *College and University Libraries. Blackboard Discussion*. Texas Woman's University, School of Library and Information Studies.

Wikipedia (2010). *Web 2.0*. Retrieved March 3, 2010 from http://en.wikipedia.org/ wiki/Web_2.0.

Chapter 14

Safety Management Systems in Aircraft Maintenance

Richard Komarniski

In recent years, SMS programs have been implemented in aircraft maintenance organizations. When one witnesses the results of a successful SMS program in an aircraft maintenance facility, it begs the question: Why wait for ICAO, EASA, FAA and Transport Canada standards and regulations requiring the SMS? The benefits are real and will positively impact many areas of operation, including the bottom line. Aircraft maintenance organizations that have proactively (that is, not waiting until a regulatory authority mandates implementation of SMS programs) implemented SMS programs are years ahead of their competitors in the areas of safety, quality, productivity and profitability due to their implementation of an SMS program. For example, an engine shop in central Canada states engine test cell turn backs caused by human error have been dramatically reduced, and a repair station in Ohio has noted a 15 percent increase in productivity, all contributions of a SMS program. Through process and quality systems such as ISO, six sigma, total quality systems, and Deming (TQM), some repair stations were successful in improving quality in their aircraft maintenance operations. These process and quality system improvement initiatives were generally the responsibility of the Quality Assurance Department, and the success of these quality initiatives was a direct reflection of both management support and management involvement. Now, safety management systems can be thought of as the next generation initiative to improve safety in an integrated fashion across an organization. SMS programs focus and integrate (as appropriate) human performance, human factors and organizational factors with quality management techniques and processes to contribute to achieving sustainable levels of safety satisfaction. A SMS program is an explicit, comprehensive and proactive process for managing risks that integrates operations and technical systems with financial and human resource management.

Implementation of SMS programs results in the design and implementation of organizational processes and procedures to identify air safety hazards and their consequences, and bring the associated safety risks in aviation operations under the control of the organization. Practically speaking, a SMS program is a business-like approach to safety. In keeping with all management systems, a SMS provides for goal-setting, planning and measuring performance. It concerns itself with organizational safety rather than the conventional health and safety approach that many safety programs take. An organization's SMS defines how it intends the management of air safety to be conducted as an integral part of their entire business management activities, with a principal focus on the hazards of the business and their effects upon those activities critical to flight safety. A successfully implemented SMS program is woven into the fabric of an organization. It becomes part of the culture; the way people do their jobs.

The SMS program requires a well-documented approach to safety, along with an effective organization for delivery of the program. This is followed up with a robust system for assuring safety. The organizational structures and activities that become a part of the SMS program are found throughout an organization. Every employee in every department contributes to the safety health of the organization. In some departments, safety management activity will be more visible than in others, but the system must be integrated into 'the way things are done' throughout the establishment. Active monitoring and audit processes are established to validate that the necessary controls identified through the hazard management process are in place and to ensure continuing active and measurable commitment to safety. This will be validated by the use of quality assurance principles to ensure that objectives and goals are being achieved and that proper due diligence has been exercised on hazards identified in the organization. All of the available information on SMS references the need for top management involvement and culture change as being critical to the success of the program. These two factors are co-dependent; one will not happen without the other. When an organization embraces SMS wholeheartedly, it is blatantly obvious! Even an untrained eye would notice that the maintenance shop is bright, clean and organized. Technicians are working with the technical manual open *and* on the right page or have the page displayed on a computer screen and within view. Upon further investigation, aircraft maintenance organizations that have instituted SMS report that warranty claims are down, customer satisfaction is up, deliveries are on time or even early, morale is high and the company is making a profit. An organization has to become vigilant in identifying risks and threats to product quality and proactive in eliminating them. This allows the company to minimize the risk of future occurrences and provides peace of mind to their customers. A SMS provides the framework for the organization to continually and proactively seek ways to identify and defend the organization against hazards and unwanted maintenance safety consequences. Inherent in the SMS program is the recognition that all employees, from senior leadership to front-line staff, are responsible for both organizational and product safety. In order for SMS to be effective, top-level management need to be on board from the start. The accountable executive needs to ensure that the person that is put in charge of the SMS program has the energy, passion, understands what SMS is and what the results of SMS should look like in their organization. The lack of continuity of personnel in some company's SMS programs has caused the program to fail. I have encountered two companies who have replaced SMS managers during the implementation process, creating a poor foundation for a robust program. What does it say about management when the head of the SMS program is relieved of his duties? These SMS programs have been handicapped in their initial stages and may have a difficult time achieving the success we have seen in other companies. For an SMS to be effective there has to be a champion; someone with the authority to commit the resources required to implement, maintain and take responsibility for the SMS. This person must be carefully selected, have the backing of senior management, and good relations with all levels of employees. I have just witnessed what will happen when an airline's accountable executive and his vice presidents do not take the time to educate themselves on the SMS process and understand how to implement the program. SMS is to be orchestrated by these higher levels of authority. As a result, they hired a Director of SMS to develop and implement the SMS program for them. Three years later, they still did not have an effective SMS program. Reports were coming in but a large percentage were complaining sessions, typical health and safety issues, and regular aircraft squawks (snags); all of which have nothing to do with SMS. Some companies are including health and safety, environmental and facility reports that water down the importance of the SMS program. Clearly, the company did not have proper objectives and goals, which are one of the major and foundational components of the SMS. Now the employees are giving feedback such as:

- the SMS program is a 'ratline' to get back at other employees
- we did not get feedback from reports submitted
- it is too much work
- the SMS is nothing but a pilot's protection agency
- we need more positive outcomes.

After three years of misdirection, this company's SMS program requires a major overhaul. It is so much easier to become informed as an accountable executive and vice president before embarking on such a major initiative. There is no excuse for not becoming informed directly and personally. There are excellent seminars and training sessions available to all, even senior management. But again, you have to ensure the speaker or trainer knows what they are talking about and that you know how to apply what you learn to your organization. I was at a conference recently where there was a high-ranking FAA speaker from their SMS Flight Standards program office giving a presentation. His presentation was very vague and when confronted with specific questions relating to SMS, he was dumbfounded and gave the deer in the headlight empty stare in response. He had a lot of theory, but no practical solutions. Take the time to educate yourself. ICAO has released very good guidance material, IS-BAO latest information is excellent and Transport Canada has also released good information since 2007. This medium-sized airline that I have been talking about has invested over a million dollars with no return on investment. In other words, it was a total failure of their SMS program. Not only is this a waste of money, it makes implementing a successful SMS program much more difficult. Maintenance shop owners who approach SMS with comments like 'it is a waste of time', 'regulators will never be able to enforce it' and 'I'll wait until the regulations come out' have been totally misinformed. Still in its infancy, SMS is undergoing scrutiny similar to that which was placed on human factors awareness sixteen years ago. Although some repair stations are planning their eighth recurrent sessions for human factors training, there are still those who are waiting for the final human factors regulation implementation before they will comply. Implementing a new program, whether it is human factors training or SMS, purely because new regulations force compliance may result in a half-hearted effort leading to frustration, mismanagement and possible failure. For example, it also leaves an organization wide open to human factors-caused errors and resulting consequences. Human factors training and SMS implementation are more than just complimentary initiatives. Implementing SMS without a sound human factors foundation would be like building a skyscraper on a sand base. Even if you can build it, it will collapse of its own weight. Management needs to decide what kind of safety management program is best for their organization. Some decisions (those concerning high-level policy and commitment) should be made directly by the accountable executive, while others (procedures and operational documents and systems) are best derived through staff involvement. It is important to build a safety management system to suit each specific organization. One cannot simply copy another organization's SMS and hope for the best. An organization's SMS program must reflect their standards and expectations; it must be managed in accordance with existing procedures, systems and equipment. The accountable manager's involvement is critical to the success of the SMS. The accountable manager needs to inform his team as to the level of risk that he is willing to carry and keep them accountable. For example, if a tool is left behind in an aircraft and the flight controls become stiff because of the misplaced tool, what should the corrective action be?

1. Spend US$1,000 and buy black foam and give everyone 2 hours shop time to shadow their toolbox, along with a policy change.
2. Spend US$150,000 and embed RF chips in all tools in the hangar and have an RF scanner at the door.
3. Spend US$900,000 and send all the technicians' toolboxes home, contract a tool company to set up a tool crib and hire tool crib personnel.

After completing a risk assessment, we find that probability and severity is very high if no action is taken. Before we determine what the corrective action needs to be, we need to determine what level of risk is appropriate. Only the accountable executive can determine what the appropriate action should be. The complexity of the SMS should be tailored to the size and nature of the organization. A large aircraft maintenance organization with many different shops that perform functions ranging from avionics repair and overhaul to engine repair and everything in between requires a detailed SMS that can be implemented in every shop. A small maintenance organization or one that only performs few functions can tailor their SMS so that it is less complex and more easily implemented. A formalized safety program is appropriate for any size of operation. Although the basic principles remain the same, the depth and scope of the program will vary greatly depending on the size and complexity of an organization. Here are a few practical implementation approaches to consider. A small organization may not need to establish a safety committee. Common sense must prevail, but the function intended for that element must still be satisfied. The function of a safety committee is to analyze safety issues, hazards and incidents and recommend corrective action. It may be performed by a single person in a small operation or by a committee in a large operation. A formal program assigns responsibilities and accountabilities and ensures that SMS decision-making process will be adhered to. Implementing a safety management system does not have to be traumatic, but it does require formality, specific personnel assignments and documentation. A safety management program can easily be formalized within existing documentation, such as the maintenance control manual. SMS allows for structured communication systems between maintenance personnel, flight crews, dispatch, cabin staff, fuelers, groomers and management to address concerns and hazards that may affect the aircraft and the organization. It is everyone's responsibility to protect their jobs by protecting the assets that keep them employed. Ultimately, this protects the customers who depend on the airlines to transport them safely every time they fly.

Before one can begin writing a policy manual for an SMS, one will need to identify the existing policies, procedures and processes that can be drawn into the umbrella of an SMS program and compare them to what the SMS requires. This is referred to as a gap analysis. If an ISO or a TQM process has already been implemented, then the implementation of an SMS is probably 65–80 percent already completed. The most common missing link is top management involvement and a very firm commitment to the new business model represented by the SMS program. The total commitment of upper management is necessary for a SMS to succeed. The CEO, Accountable Manager and General Manager must understand the SMS requirements to implement risk assessment and management of hazards as part of their normal responsibilities, and incorporate human factors in their daily meetings and decision making. For some managers this paradigm shift will be difficult, however once embraced these changes will provide phenomenal results, as we have seen already in some organizations. I was pleased to see an operator I was visiting discuss an issue that was affecting the organization. One of the Vice President's spoke up and said it is time for a hazard identification meeting with our SMS manager. The hazards were then identified, the risk assessment employed and the corrective action was defined to address the identified hazard. Decisions were made and documented. Due diligence was shown. Senior management was directly involved and was seen implementing the tenets of the SMS program. An effective implementation strategy for SMS will involve changes in processes and procedures and will almost certainly involve a shift in the corporate culture. The safety culture of an organization is defined as 'the product of individual and group values, attitudes, perceptions, competencies and patterns of behaviour, that determine the commitment to, and the style and proficiency of an organization's health and safety management' (Transport Canada, 2008). Simply put, it is quite literally the way things are done. Every organization has a culture, good or bad, safe or unsafe. The corporate culture is reflected in the mode of operation throughout the organization. Typically, the tone of the culture is established from the top down. If the accountable executive is committed to managing safety risks, then the way that organization operates will reflect this philosophy. In an SMS environment, the accountable executive and all senior managers are accountable for safety. The dedication and involvement of top management towards safety and safety practices should be clearly visible. It is important that senior management is seen to provide a strong and active leadership role in the SMS. Managing safety risks, however, involves more than a personal commitment to make safety one's primary obligation. It includes a commitment to provide the resources necessary to attain the strategic safety objectives established by the organization. The following is a list of activities that demonstrate top management's active commitment to SMS:

- Putting safety matters on the agenda of meetings, from the executive boardroom to the shop floor;
- Being actively involved in safety activities and reviews at both local and remote sites;
- Allocating the necessary resources, such as time and money to safety matters;
- Receiving and acting on safety reports submitted by employees;
- Promoting safety topics in publications, and;
- Probably most important of all, setting personal examples in day-to-day work to demonstrate unmistakably that the organization's commitment to safety is real and not merely lip-service, and by clearly and firmly discouraging any actions that could send a contrary message.

The ideal safety culture embodies a spirit of openness and demonstrates support for staff and the systems of work. Senior management should be accessible and dedicated to making the changes necessary to enhance safety. They should be available to discuss emerging trends and safety issues identified through the SMS. A positive safety culture reinforces the entire safety achievement of the organization and is critical to its success. For an SMS to be successful, it must never be static. Just because the basic components and elements of the SMS are in place, it cannot be considered 'complete'. Your organization is not static: personnel, equipment, aircraft projects, modifications, updates and the operating environment change all the time. As the organization changes, so must the SMS. It must continually evolve using the system outputs and lessons learned. To achieve this state of continuous improvement, it is important to understand that all work done in an organization is the result of process. It has been said that '*The emphasis with assuring quality must focus first on process because a stable, repeatable process is one in which quality can be built on*' (Transport Canada, 2008). In other words, to validate and ensure the effectiveness of a process, the process must exist and be understood, and be followed repeatedly by all personnel. Once it is confirmed that a process exists and is in use, the output or product of that process can be reviewed to ensure that the desired outcome is in fact being realized. Where the result of a process falls short of expectations, that process can then be adjusted to achieve the desired result.

In order to begin implementation of an SMS, there must be a clear understanding of the components that need to be included in the SMS including:

- A sound SMS policy that addresses each of the following items:
  - A corporate commitment to implement an SMS;
  - A commitment to continual improvement in the level of safety;
  - A commitment to the management of safety risk;
  - A commitment to comply with applicable regulatory requirements;
  - A commitment to encourage employees to report organizational and product safety issues without reprisal;
  - Establishment of clear standards for acceptable behavior throughout the organization;
  - Management guidance for setting and reviewing safety objectives and goals;
  - A means to communicate to all employees and responsible parties;
  - Identification of responsibility of management and employees with respect to safety performance.
- A safety reporting policy indicating that there will be no repercussions for coming forward with reactive or proactive reports – an essential element of any SMS is the safety reporting policy. To the extent possible, it should be non-punitive and developed and implemented with all affected parties, from the shop floor to the corporate headquarters. This builds confidence in the system but also provides a clear understanding to all employees of what the safety reporting policy actually is.
  - From a practical perspective, employees are more likely to report events and cooperate in an investigation when some level of immunity from disciplinary action is offered. When considering the application of a safety reporting policy, the organization should consider whether the event was willful, deliberate or negligent.
  - A non-punitive approach to safety reporting does not preclude the use of a general approach to discipline in cases where an employee is involved in similar, recurrent events.
  - The safety reporting policy should also include features to guard against the deliberate abuse of the system, such as using self-disclosure as a means of obtaining indemnity for deliberate violations of both the letter and spirit of the system.
- Identification of roles and responsibilities of the staff specifically addressing what they are required to do to support the program and culture when it comes to:
  - Providing resources essential to implement and maintain the SMS;

- – Ensuring that process needed for the SMS are established, implemented and maintained;
- – Ensuring the promotion of awareness of safety requirements throughout the organization.
- Establishing a set of safety objectives and goals is key to establishing a successful SMS. Safety objectives define what the organization hopes to accomplish with its SMS. Safety objectives are the broader targets the organization hopes to achieve. They should be published and distributed so that all employees understand what the organization is seeking to accomplish with its SMS. Goal-setting is vital to an organization's performance and helps to define a coherent set of targets for accomplishing the organization's overall safety objectives. All organizations have their own ways of setting goals formally and documenting the process. Goals have to be set that empower the employees and get the employees involved in the process of achieving these goals. It is a never-ending struggle to identify and eliminate or control hazards, especially in a complex work environment. We will never run out of things to do to make the system safer for the whole department or organization. To establish objectives and goals, we need to identify high hazards and associated risks and selectively prioritize what we attack. Sound management requires that we identify them, decide how to achieve them and hold ourselves accountable for achieving them. Risk-management procedures can help managers decide where the greatest risks are and help set priorities. Sound safety objectives and goal-setting concentrates on identifying systemic weaknesses and accident precursors, and either eliminating or mitigating them. The safety performance of the operation needs to be monitored, proactively and reactively, to ensure that the key safety goals continue to be achieved. Monitoring by audit forms a key element of this activity and should include both a quantitative and qualitative assessment. This means that a numeric score or grade as well as an effectivity assessment should be applied. The results of all safety performance monitoring should be documented and used as feedback to improve the system. It is widely acknowledged that accident rates are not an effective measurement of safety. A more effective way to measure safety might be to address the individual areas of concern. For example, an assessment of the improvements made to work procedures might be far more effective than measuring accident rates. Performance measurement should be integrally linked to the company's stated overall objectives. This requires the development and implementation of a sound set of safety performance measures and a clear linkage between the safety performance measures. Be realistic during this activity, and follow the basic principles of goal-setting such as writing goals down, stating them positively, prioritizing and being precise (for example, reduce rework by x percent in the first six months or increasing awareness of the program across the organization by a specific date). Performance measures can then

be determined by asking how you will know if you have met your goals. An added benefit of following this process is that the safety goals and performance measures established during this activity will form, or link, to the safety goals and performance measures required under the safety management plan. In addition, the quality assurance department will then be able to use the performance measures to determine effectiveness of current or newly established processes. It is important to realize though, that goals and performance measures will change as the program evolves; they may even change as planned activity (theoretical) moves into practical implementation.

- To ensure that the SMS is working effectively the accountable executive should conduct a periodic review of the SMS processes and procedures. To the extent possible, the review should be conducted by individuals not performing tasks directly related to the SMS. The safety manager for example should not be reviewing the SMS, as he or she is an integral part of the system. The review should also include an assessment of how well the organization is achieving its specific safety goals, the success of the corrective action plans and the risk-reduction strategies implemented. The review is intended to provide a quality review and a continuous improvement function within the SMS. It may be conducted by doing a traditional checklist audit or it may take the form of an effectivity assessment. Whatever the method, the accountable executive should be informed directly of the results. Essentially, this is the accountable executive's report card on how well the system is performing.
- Safety oversight is fundamental to the safety management process. Safety oversight provides the information required to make an informed judgment on the management of risk in your organization. Additionally, it provides a mechanism for an organization to critically review its existing operations, proposed operational changes and additions or replacements, for their safety significance. Safety oversight is achieved through two processes:
  a) Reactive processes for managing safety occurrences. The reactive process responds to events that have already occurred.
  b) Proactive processes for managing hazards, including procedures for hazard identification, active monitoring techniques and safety risk profiling. The proactive method actively seeks to identify potential hazards through an analysis of the everyday activities of the organization. The exception to this rule occurs when a potential hazard has been reported through the organization's safety reporting program. Hazard identification is the act of identifying any condition with the potential of causing damage to equipment or structures, loss of material, or reduction of the ability to perform a prescribed function. In particular, this includes any conditions that could contribute to the release of an un-airworthy aircraft, to the operation of aircraft in an unsafe manner. Hazard identification is an ongoing activity. Hazards emerge and evolve

as a result of changes in the operating environment which occurs frequently. Understanding the hazards and inherent risks associated with everyday activities allows the organization to minimize unsafe acts and respond proactively, by improving the processes, conditions and other systemic issues that lead to unsafe acts. These include training, budgeting, procedures, planning, marketing and other organizational factors that are known to play a role in many systems-based accidents. In this way, safety management becomes a core-business function and is not just an adjunct management task. It is a vital step in the transition from a reactive culture, one in which the organization reacts to an event, to a proactive culture, in which the organization actively seeks to address systemic safety issues before they result in an active failure. Once an event has been reported, or a hazard identified, the procedures for dealing with these issues follow a similar process. See Figure 14.1 for a diagram of the process.

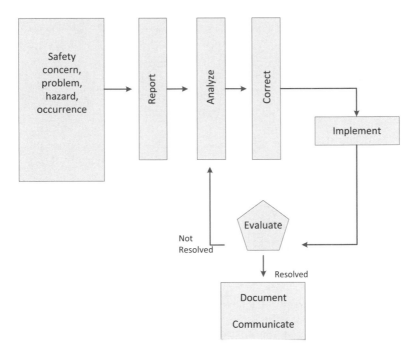

Figure 14.1   The SMS process

Every event is an opportunity to learn valuable safety lessons. The lessons will only be understood, however, if the occurrence is analyzed so that all employees, including management, understand not only what happened, but also why it happened. This involves looking beyond the event and investigating the contributing factors, the organizational and human factors within the organization that played a role in the event. The investigative process should be comprehensive and should attempt to address the factors that contributed to the event, rather than simply focusing on the event itself, the active failure. Active failures are the actions that took place immediately prior to the event and have a direct impact on the safety of the system because of the immediacy of their adverse effects. They are not, however, the root cause of the event; as such, applying corrective actions to these issues may not address the real cause of the problem. A more detailed analysis is required to establish the organizational factors that contributed to the error. We need to ask ourselves, did we set the employee up for failure through the company policies for norms? The reporting system should be simple, confidential and convenient to use and should be complemented with a safety reporting policy. These attributes, accompanied by efficient follow-up mechanisms acknowledging to the report individual that a report has been received, investigated and acted upon, will encourage the development of a reporting culture. The results should be distributed to the individual involved and the population at large where appropriate. Establish a clear definition with examples of what valid safety management system reportables are. Knowing what to report plays a key role in an active reporting program. As a general rule, any event or hazard with the potential to cause damage should be reported. Some examples of these issues are:

- excessive duty times
- rushing through checks
- inadequate tool or equipment control
- incorrect or inadequate procedures and a failure to adhere to standard procedures
- poor communication between operational areas
- lack of up-to-date technical manuals (hard copy or electronic)
- poor shift changeovers
- lack of adequate training and recurrent training
- lack of planning
- lack of basic consumable parts. Corrective action from a SMS finding should result in examples such as:
  - Trolleys with ledges or foam inserts to prevent parts from falling off;
  - Fluorescent orange or green parts bags to prevent fasteners from being ingested;
  - Flags to be placed on flap circuit breaker/levers to prevent flaps from being lowered onto stands or power carts;
  - Aircraft jacking training to create awareness of center of gravity (C of G) limits with engines on or off and also to ensure removal of tail stand

before lowering aircraft off jacks. As in the chart below, we can see that we have a process in place already to deal with a number of reportables. The SMS reportables (see Figure 14.2) focus on organizational hazards because of change and human error events or potential hazards to the aircraft or equipment.

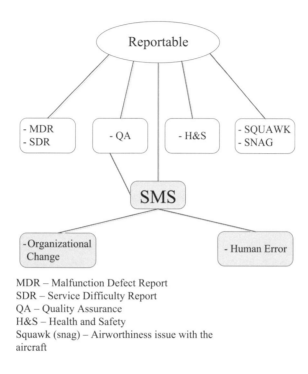

MDR – Malfunction Defect Report
SDR – Service Difficulty Report
QA – Quality Assurance
H&S – Health and Safety
Squawk (snag) – Airworthiness issue with the aircraft

Figure 14.2   SMS reportables

For a SMS to transition from a reactive to a proactive system, it must actively seek out potential safety hazards and evaluate the associated risks. A safety assessment allows for the identification of potential hazards and then applies risk management techniques to effectively manage the hazard. Safety assessments are a core process in the safety management construct and provide a vital function in evaluating and maintaining the system's safety health. A safety assessment activity should be undertaken at a minimum when major operational and organizational changes are planned. If the organization is undergoing rapid change, such as growth and expansion, offering new services, cutting back on existing services, or introducing new equipment or procedures; and when key personnel change. History has shown that all of these create an environment ripe for major safety events to occur. A safety risk profile is a prioritized list of the known risk in your organization. In order to develop a safety risk profile, you must develop a hazard register relating to your organization. Once potential risks have been identified, it is useful to fully understand the impact that they might have if they remain unchecked. In order to determine this, a full risk assessment should be conducted. It should be applied to both the reactive investigations and proactive safety assessments an organization conducts. Your Safety Risk Profile should identify your top 10–12 risks to organizational and air safety as it is impossible to address all risks identified through your system. This methodology allows management to effectively allocate resources where they are required the most. The safety risk profile should be linked to the objectives and goals of your organization. Some organizations that have implemented SMS are claiming to have over a thousand SMS reactive and proactive reports. If the reports are truly SMS reports, the companies would have had a fatal accident by now. I would venture to guess that maybe 10 to 15 percent of the reports were actual SMS reports. For safety management systems to work, it is not the amount of reports issued but about educating the workforce to generate quality proactive reports. We can respond to these reports appropriately and gain the confidence of the employees in accomplishing our goal.

- Training. In order for employees to comply with all safety requirements, they need the appropriate information, skills and training. All employees will require some level of SMS training; the extent to which they are trained will depend on their function in the SMS. For example, a line employee will need to be trained on how to report into the SMS reporting system. This would include how, where and what to report. Additionally, employees should be given basic human factors training to develop an awareness of the individual factors that can impact human performance and lead to errors. This might include coverage of issues such as fatigue, communication, stress, human performance models and lack of awareness. Senior executives and the accountable executive should receive general awareness training related to all aspects of the SMS. The accountable executive is responsible for the establishment and maintenance of the SMS. A general awareness of the SMS is therefore a requirement.
- Quality assurance. In an SMS, the quality assurance program (QAP) elements can be applied to an understanding of the human and organizational issues that can impact safety. In the same way that a QAP measures quality and monitors compliance, the same methods are used to measure safety within the organization. In the SMS context, this means quality assurance of the SMS, as well as quality assurance to ensure compliance to the regulations, standards and procedures utilized by the organization. Quality assurance exists to ensure that all reports have been actioned and that the corrective action has been effective; i.e. no further incidents have occurred. Quality assurance assessment will be conducted to ensure that SMS objectives and goals have been met and are still current.

Implementing and facilitating a safety management system program in an aviation organization is a journey, not a destination. As a company grows in their safety culture the momentum will perpetuate forever.

References

Transport Canada (2008) *Guidance on Safety Management Systems Development AC 107-001.* Ottawa: Transport Canada.

Index